Michel Bilodeau
Fernand Meyer
Michel Schmitt (Editors)

Space, Structure and Randomness

**Contributions in Honor of Georges Matheron
in the Field of Geostatistics, Random Sets
and Mathematical Morphology**

With 121 Figures

Springer

Michel Bilodeau
Ecole des Mines de Paris
Centre de Morphologie
  Mathématique
35 rue Saint Honoré
77305 Fontainebleau
France

michel.bilodeau@ensmp.fr

Fernand Meyer
Ecole des Mines de Paris
Centre de Morphologie
  Mathématique
35 rue Saint Honoré
77305 Fontainebleau
France

fernand.meyer@ensmp.fr

Michel Schmitt
Ecole des Mines de Paris
Direction des recherches
60 Boulevard Saint Michel
75272 Paris Cedex 06
France

michel.schmitt@ensmp.fr

Library of Congress Control Number: 2005927580

ISBN 10: 0-387-20331-1   Printed on acid-free paper
ISBN 13: 978-0387-20331-7

Typesetting: Camera ready by the editors
Printed in the United States of America   (SB)

9 8 7 6 5 4 3 2 1

springeronline.com

Lecture Notes in Statistics   **183**

Edited by P. Bickel, P. Diggle, S. Fienberg, U. Gather,
I. Olkin, S. Zeger

Personal Reminiscences of Georges Matheron

Dietrich Stoyan

Institut für Stochastik, TU Bergakademie Freiberg

I am glad that I had the chance to meet Georges Matheron personally once in my life, in October 1996. Like many other statisticians, I had learned a lot in the years before from his books, papers, and research reports produced in Fontainebleau. From the very beginning, I had felt a particular solidarity with him because he worked at the Ecole des Mines de Paris, together with Bergakademie Freiberg one of the oldest European mining schools.

The first contact to his work occurred in talks with Hans Bandemer, who in the late 1960s had some correspondence with Georges Matheron. Proudly he showed me letters and reprints from Georges Matheron. He also had a copy of Georges Matheron's thesis of 1965, 300 pages narrowly printed. Its first part used Schwartz' theory of distributions (Laurent Schwartz was the supervisor) in the theory of regionalized variables, while the second part described the theory of kriging. Hans Bandemer was able to read the French text, translated it into German language and tried, without success, to find a German publisher. Unfortunately, the political conditions in East Germany prevented a meeting between Hans Bandemer and Georges Matheron.

A bit later, in 1969, I saw George Matheron's book "Traité de Géostatistique Appliquée". It is typical of the situation in East Germany at the time that this was a Russian translation published in 1968, based on the French edition of 1967. (We could not buy Western books and most of us could not read French. At the time the Soviet Union ignored the copyright laws.) I was very impressed by the cover illustration showing a gold-digger washing gold.

Georges Matheron's Traité is a fascinating book, and for me and many others it was the key reference in geostatistics at the time. The progress achieved in this work is perhaps best expressed by the following quote from the postscript (written by A.M. Margolin) of the Russian translation:

> *"We find in the monograph the fundamental ideas and notions of a mathematical theory of exploration, called by Matheron 'geostatistics'. Geostatistics is characterized by a concept which corresponds to the*

*mathematical problems of exploration, by fundamental basic notions
and by a large class of solvable problems.*

*In comparison to classical methods of variation statistics [the sim-
ple methods using mean and variance and Gaussian distribution],
which were until the 1950s nearly the only way of application of mathematics in geological exploration, geostatistics is directed to the true
mathematical nature of the objects and tools of exploration...*

*According to geostatistics, the uncertainty which is characteristic
for results of exploration is a consequence of incompleteness of exploration of the object but not of its randomness. In this probably the
increasing possibilities of the geostatistical theory consist and the principal difference to variation statistics, which is based on the analysis
of geological variables irrespectively to the locations of observation... "*

Not at the time, but later on when I had learnt more about public relations
in science, I admired Matheron's cleverness in coining terms: It was very
smart to call statistics for random fields 'geostatistics' (to use a very general
word, which suggests 'statistics for the geosciences', for a more limited class of
problems) and least squares linear interpolation 'kriging' (originally 'krigeage';
in Freiberg it was for a long time not clear whether it should be pronounced
'krig-ing' or 'kraig-ing' – as suggested by the Russian translation – or 'kreedge-
ing'.)

Georges Matheron's "Random Sets and Integral Geometry" of 1975 was a
landmark in my own scientific development. The mid 1970s were a great time
for stochastic geometry and spatial statistics, which then became more than
just geostatistics. In that time mathematical morphology was also shaped,
and Jean Serra's work in image analysis became known outside Fontainebleau,
both in theory and in applications. The first image analyser, the famous Leitz
TAS, was produced, based on ideas of Georges Matheron and Jean Serra.

We, the scientists in East Germany, were unable to order the book in a
book store, but I got a copy from Dieter König in exchange for a book on
queueing theory. Up to this day, for me (and many others, I believe) Georges
Matheron's book is the key reference in random set theory; the cover of my
copy is now in pieces, and many pages are marked with notes. Unfortunately,
the book has not been reprinted since.

This book gives an excellent exposition of the topics named in the title, it is
very clearly written, and of just the right theoretical level. Georges Matheron's
work also led me to Hadwiger's, whose monograph on set geometry is now
one of my most beloved mathematical books in German language. So Georges
Matheron had an excellent base for the integral-geometric part of his work. It
is a great combination of many mathematical fields, such as integral geometry,
set geometry, Choquet's theory of capacities (in fact Matheron developed
it independently), and ideas of Poisson process based stochastic geometry
(created in particular by Roger Miles) to obtain a new, rich and fruitful theory.
Georges Matheron's book contains many gems. I name here only the theory

of the Boolean model (he writes 'boolean model', with small 'b'), the Poisson polyhedron, and the granulometries.

It was Georges Matheron who made notions of measurability properties rigorous in the context of random sets. This can be explained in terms of Robbins' formula for the mean volume of a compact random set:

$$E\nu(X) = \int_{\mathbb{R}^d} p(x)\mathrm{d}x$$

where $p(x) = P(x \in X)$. Its proof is based on Fubini's theorem and was already given – in part – in Kolmogorov's famous book of 1933. It was Georges Matheron who showed that the mapping $set \rightarrow volume$ is measurable. The σ-algebra with respect to which measurability is considered (based on Fell's topology) is today called Matheron's σ-algebra. Some of his results for models related to Poisson processes were the starting point for further scientific work, when it became clear that methods from the theory of marked point processes can be used to generalize them. Through his book, Georges Matheron has been a teacher and inspirer for a large number of mathematicians in the last three decades.

In 1989 Georges Matheron published the book "Estimating and Choosing". Grown older, I was offered the honour to serve as one of its reviewers. This is a philosophical book, discussing the fundamental question of spatial statistics: "Why does it make sense to perform statistical inference for spatial data, when only a sample of size $n = 1$ is given?" Indeed, very often a statistician has only data from one mineral deposit, or from one forest stand; a second sample taken close to it can typically not be considered as a sample from the same population, because of different geological or ecological conditions. He developed the idea of 'local ergodicity', which is plausible and justifies the statistical approach. Each spatial statistician should read the book. I also enjoyed its sarcastic humor.

In June 1983 Dominique Jeulin came to Freiberg as an invited speaker at a conference. For me and colleagues like Joachim Ohser and Karl-Heinz Hanisch this was the first chance to meet a scientist from Georges Matheron's school. Dominique Jeulin spoke about multi-phase materials, rough surfaces, and image transformations. He is likely the first French person I ever met, and I learned that French English differs greatly from German English. Dominique had to repeat three times his 'Stojáng' until I understood that he asked for me. The contact to him, which is still lively, finally led to the meeting with Georges Matheron.

After the big changes of 1989/90 we had many West German PhD students at Freiberg. One of them, one of the best, was Martin Schlather. Martin had studied one year at Fontainebleau, had written a diploma thesis there, and had been given oral exams by Georges Matheron personally. Thus, he could tell me first hand about Georges Matheron's personality and about the situation at Fontainebleau. Like anybody who ever had personal contact with Georges Matheron Martin admired him, both as a mathematician and as a person. Martin told me that Georges Matheron's work is much more comprehensive than his books and journal publications suggest. So we are all extremely grateful to Jean Serra for publishing a CD with the reports and papers by Georges Matheron. I agree with Martin and Jean that one will find wonderful ideas and solutions to hard problems as one goes through this material.

It was Martin Schlather who encouraged me to travel to the Fontainebleau conference in 1996. This conference was organized by Dominique Jeulin in honour of Georges Matheron, and the lectures are published in the volume Jeulin (1997). There we met Georges Matheron, who did not give a lecture, but was sitting in the front row in the lecture room, obviously listening with attention. One afternoon there was a reception in the municipal hall of Fontainebleau. The maire decorated Georges Matheron (and others) with a medal, for reasons that I did not completely understand, either scientific or political. Georges Matheron was polite enough to bear the ceremony, but quite obviously he did not take it very seriously. He told me anecdotes, and his body language was clear enough. I had no difficulty in communicating with him, he accepted my poor English, and I understood him very well.

In the evening I was honoured by an invitation to the family of Jean Serra, where I also met Mesdames Matheron and Serra. It was an enjoyable, not very long evening, with friendly non-mathematical talk, and a bit of French wine.

In the end, I never orally discussed mathematical problems with Georges Matheron, but I appreciated the contact over many years, through studying his work, and through a few comments that he made on my work. I believe that there is only a small number of international mathematical conferences which Georges Matheron attended. His example shows that a mathematician who does not visit conferences can be nevertheless be influential and widely known. Perhaps, Georges Matheron could have had even more influence. However, did he really want this?

Back to Freiberg, I had the idea of honouring him there. At the time I was the president of the little Technische Universität Bergakademie Freiberg and saw the chance of honouring him with a Dr.h.c. degree. I asked Dominique Jeulin. He liked the idea but warned me that Georges Matheron would probably never come to Freiberg, and if he did, I could not expect his collaboration in a public relations event to the benefit of my university, as I was hoping for. To my regret, I did not pursue the idea any further.

In October 2000, the sad news of Georges Matheron's death spread. I regretted that I had not seen him again after 1996. As a keepsake to Georges Matheron, I asked Jean Serra for Georges Matheron's personal copy of Hadwiger's book – expecting a book with many pencil notes. To my surprise I learned that Georges Matheron had used a library copy.

I am very happy that this volume is now ready, which will honour Georges Matheron, one of the great mathematicians of the 20th century.

A few words about Georges Matheron (1930-2000)

Jean Serra

Centre de Morphologie Mathématique, Ecole des Mines de Paris

That day, Paris was steaming hot, as it may happen in July when there is not the slightest breeze of wind to cool the air. Despite the vigour of his nineteen years, Georges Matheron, while waiting for his turn, was suffocating and finally lost his composure before the board of examiners of the Ecole Normale Supérieure. Fortunately, he came second at the Ecole Polytechnique, which he entered a few weeks later. That was in 1949. He probably had missed the opportunity to work in the best possible environment for him, and at the time he did not know that the path he was heading for would involve him in earth sciences for the rest of his professional life.

Those who have practised geostatistics all know how fascinating it is to discover the mineral world underground, to figure the structures from drillhole variograms, like a blind person fingering an object to guess its shape. However G. Matheron was the first one to know that excitement, all the more as he was creating the mathematical tool while using it to comprehend the earth substratum. Moreover he designed it so that the description of the mineral space and the estimation of mining resources be indissolubly linked, like the two sides of the same coin.

After two years spent at the Ecole Polytechnique, two more years at the Ecole des Mines and another one in the military service, it was in Algeria that the Corps des Mines sent him for his first appointment. He had married one year before, and landed in Algeria with wife and child. He quickly took over the scientific management (1956), then the general management (1958) of the Algerian Mining Survey.

It takes imagination to realize how important it was for a young French engineer. The huge Algerian territory stretches as far as the Saharan South and abounds in orebodies of all kinds. It is one of the reasons why the International Geological Congress has been held in Algiers in 1952, just two years before he arrived. It was also in the early 50's that papers written by three South-African authors, Krige, Sichel and de Wijs, laid the statistical foundations from which G. Matheron would base his theory of *geostatistics*, that was revolutionary at the time.

What does not kill you makes you stronger. In the early 60's, French Algeria collapsed, France recalled its executives to the home country and reorganized its mining research by creating the BRGM (Geological and Mining Survey) in Paris. G. Matheron was assigned a "Geostatistical department", practically reduced to himself. The BRGM did not believe in geostatistics, and the only mining partners came from the CEA ($\sim$ Atomic Energy Commission), namely A. Carlier and Ph. Formery. This solitude was in fact a blessing which enabled him to devote himself to the final development of what will be called later *linear geostatistics*.

The genesis of the following works is instructive, and tells a lot about the personality of G. Matheron. In Algeria, he invented the random functions with stationary increments, while being sceptical about the probabilistic framework. Every deposit is a unique phenomenon, that occurred once only in geological times; besides, when one estimates its reserves, one does not compare its drillings with those of more or less similar deposits. This unique phenomenon, studied in itself, does not offer any more hold to probabilities, than if one wanted to know the proportion of hearts in a deck of cards by drawing only one card once.

It was within that "semi-random" framework that G. Matheron wrote the first volume of his "Traité de Géostatistique Appliquée" (treatise of applied geostatistics) in 1962, which was based upon the Algerian experience, then the second volume (1963) on kriging, which solved the problems of local estimation raised by the uranium deposits of the CEA. Even today, the reader is amazed at such a skilful mastery of first the mathematics, then of the physics of the topic, with the right simplifying approximations. The rule of one-to-one correspondence for instance, in volume I, is still a masterpiece, where each term of the limited expansion of the theoretical variogram (whose estimation is empirically accessible) corresponds, with a known invariable weight, to a term of the limited expansion of the deposit estimation variance (which is sought for).

However this rule could be formulated in a deterministic framework as well as in probabilistic terms. Hence a third book, entitled "La théorie des variables régionalisées et leur estimation", which became his PhD thesis in 1963. The deterministic and random parts were developed successively, with much rigour, and all theoretical conclusions drawn. But G. Matheron waited twenty years before expressing himself, in "estimating and chosing", upon the choice of either approach according to the context – the mathematician was ahead of the physicist.

As the BRGM continued to ignore the practical interest of his sextuple integrals, G. Matheron looked elsewhere to collect followers, through teaching. A "geostatistical option" was created at the Ecole des Mines de Nancy, which provided him with his first PhD student. The latter, Jean Serra, rapidly branched off and oriented methods and applications towards the new field of *mathematical morphology* (random event or deterministic fate?). The quality of the iron ore of Lorraine was defined as much by its grade as by its suitability to grinding, which could only be quantified from the mosaic of the petrographic phases, as they appear under the microscope. Hence the idea of measuring petrographic variograms, and extending the concept of variogram to that of the hit-or-miss transformation, then to that of opening, etc. For more than one year, the master and the disciple, though separated by three hundred kilometers, met each month, sharing their enthusiasm, their notes, and the advancement state of the "texture analyser", which the student was developing with Jean-Claude Klein. Many years later, in 1998, when describing this intense period at the research committee of the Ecole des Mines de Paris, G. Matheron would say : "these were the most beautiful years of my life".

In april 1968, the Ecole des Mines de Paris gave G. Matheron the opportunity to create the "Centre de Morphologie Mathématique" in Fontainebleau, with J. Serra as the other permanent researcher. The events of may 1968 were favourable to them, thanks to all the public founds they released, so that in two years the team grew from two to twelve persons. It spread on both fronts of geostatistics and mathematical morphology. From that time, the first gained international recognition, and proposals for mining estimations coming from the five continents arrived at the Centre. Moreover, from the early 70's, the CMM had been asked to map the sea bed, atmospheric pressures, etc. The application fields broadened and with them the variety of the problems to be solved ; for example, that of submarine hydrography led G. Matheron to invent the universal kriging (1969), then the random functions with generalized covariances (FAI-k, 1973), which both released the constraint of the stationarity hypothesis. Another example : the integration of local mining estimations into operating management programs led G. Matheron to conditional simulations, less accurate than kriging, but which did not smooth the data. Finally, during the 70's, G. Matheron formalised and proposed a definitive answer to the major issue of mining estimation, namely the change of support. In this case, the problem is no longer to estimate the variance of panels with respect to their size, as in linear statistics, but to be able to predict their whole distribution function, in order to fit mining exploitations to the economic conditions (1976).

In parallel with these developments, the mathematical morphology group became independent and evolved to an autonomous centre in the early 80's. Indeed, from the beginning, its applications covered the whole field of optical microscopy, and while metallography and porous media still pertained to earth sciences, medical histology and cytology addressed quite another audience.

G. Matheron did not show more than a polite interest in such applications of mathematical morphology. He did not penetrate them like he had done with mining technology. Morphological applications were too varied, and what excited him was to extract from them some general approaches, which could be conceptualised. This is why were produced the theories of granulometries, of increasing operators, of Poisson hyperplanes and of Boolean sets, which he gathered into the book *Random sets and Integral Geometry*, in 1975. The topic of porous media was the sole exception. This application already had a whole specific physical framework, varying according to the considered scale (Navier-Stokes equation at the microscopic level, Darcy equation at larger scales). How can such changes be linked ? To what extent can random sets provide tractable models ? Throughout his career at the CMM, G. Matheron kept involved with these questions and indicated fruitful directions.

At the beginning of the 80's, his professional life took a more "morphological" turn. On the theoretical level, the geostatistical vein seemed to be exhausted, whereas morphologists had just designed new operators, products of openings and closings, which had the property of being both increasing and idempotent (here the word "morphologists" refers to the members of the CMM team as well as to the American S.R. Sternberg, the German D. Stoyan or the Australian G.S. Watson, among others). As these operators used to apply to both frameworks of sets and numerical functions, G. Matheron situated his approach at the broader level of the complete lattices and constructed a general theory of the increasing and idempotent operators that he called *morphological filtering* (1982-1988). In spite of appearances, these new levels can easily be integrated into the structure of his overall work. Morphological filtering gave a simplified and denoised vision of the numerical functions, as did kriging for mapping. Increasingness and idempotence had replaced linearity and the master observed the consequences of that genetic change.

Since the 90's, multimedia have been the dominant theme in Mathematical Morphology, bringing into focus the three topics of motion, segmentation and colour. Today, these topics still concentrate most of the CMM activities. However, G. Matheron did not take interest in them. Since he inserted mathematical morphology into the lattice framework, he pursued the idea to extend also his random set theory. In order to do so, complete lattices must be first equipped with adequate topologies. G. Matheron's efforts were devoted to this task until his retirement, which was punctuated by an unpublished and last book on compact lattices (1996).

Jean Serra
September 2004

Contents

Part III Mathematical Morphology

List of Contributors

Gerald J. F. Banon
Divisão de Processamento de
Imagens
Instituto Nacional de Pesquisas
Espaciais
São José dos Campos, SP, Brazil
banon@dpi.inpe.br

Junior Barrera
Departamento de Ciência da
Computação
Instituto de Matemática e Estatística
Universidade de São Paulo. São
Paulo, SP
Brazil
jb@ime.usp.br

Michel Bilodeau
Ecole Nationale Supérieure des
Mines de Paris
Centre de Morphologie
Mathématique
35 rue Saint Honoré
77300 Fontainebleau
France
Michel.Bilodeau@ensmp.fr

Prof. Ing. Roberto Bruno
Università di Bologna
Dipartimento di Ingegneria Chimic
Mineraria e delle Tecnologie
Ambientali

Viale Risorgimento,
2 - 40136 Bologna
Italy
roberto.bruno@unibo.it

Jean-Paul Chilès
Ecole Nationale Supérieure des
Mines de Paris
Centre de Géostatistique
35 rue Saint Honoré
77300 Fontainebleau
France
Jean-Paul.Chiles@ensmp.fr

Chantal de Fouquet
Ecole Nationale Supérieure des
Mines de Paris
Centre de Géostatistique
35 rue Saint Honoré
77300 Fontainebleau
France
Chantal.De_Fouquet@ensmp.fr

Pierre Delfiner
TOTAL
2 Place de la Coupole
La Défense 6
92078 Parix La Défense Cedex
Pierre.DELFINER@total.com

G. de Marsily
Université Paris VI et Académie des
Sciences
GDemarsily@aol.com

Edward R. Dougherty
Department of Electrical Engeneer-
ing
Texas A & M University
College Station
USA
edward@ee.tamu.edu

Dominique Jeulin
Ecole Nationale Supérieure des
Mines de Paris
Centre de Morphologie
Mathématique
35 rue Saint Honoré
77300 Fontainebleau
France
Dominique.Jeulin@ensmp.fr

John Goutsias
Center for Imaging Science
The Johns Hopkins University
Baltimore, MD 21218
USA
goutsias@jhu.edu

Frédéric Guichard
DO Labs, 3 rue Nationale
92100 Boulogne
France.
fguichard@dolabs.com

André HAAS
16 rue Paul Mirat
64000 Pau
haas.pau@wanadoo.fr

Henk Heijmans
CWI, P.O. Box 94079 GB
Amsterdam, The Netherlands
henkh@cwi.nl

Wynand Kleingeld
De Beers MRM;
Mendip Court – Bath Road
Wells, Somerset, BA53DG
United Kingdom
Wynand.Kleingeld@dtc.com

Danie Krige
PO box 121 – Florida Hil
1716 South Africa
omricon@iafrica.com

Christian Lantu'ejoul
Ecole Nationale Supérieure des
Mines de Paris
Centre de Géostatistique
35 rue Saint Honoré
77300 Fontainebleau
France
Christian.Lantuejoul@ensmp.fr

Petros Maragos
School of ECE
National Technical University of
Athens
15773 Athens
Greece.
maragos@cs.ntua.gr

Klaus Mecke
Institut für Theoretische Physik
Universität Erlangen-Nürnberg
Staudtstrasse 7, D-91058 Erlangen
Germany
mecke@fluids.mpi-stuttgart.mpg.de

Fernand Meyer
Ecole Nationale Supérieure des
Mines de Paris
Centre de Morphologie
Mathématique
35 rue Saint Honoré
77300 Fontainebleau
France
Fernand.Meyer@ensmp.fr

Ilya Molchanov
Department of Mathematical
Statistics and Actuarial Science,
University of Bern
Sidlerstrasse 5, CH 3012 Bern
Switzerland
`ilya.molchanov@stat.unibe.ch`

Jean-Michel Morel
CMLA, Ecole Normale Superieure
de Cachan
94235 Cachan
France.
`morel@cmla.ens-cachan.fr`

Jacques Rivoirard
Ecole Nationale Supérieure des
Mines de Paris
Centre de Géostatistique
35 rue Saint Honoré
77300 Fontainebleau
France
`Jacques.Rivoirard@ensmp.fr`

Michel Schmitt
Ecole Nationale Supérieure des
Mines de Paris
60 Boulevard Saint-Michel
75272 Paris cedex 06
`Michel.Schmitt@ensmp.fr`

Jean Serra
Ecole Nationale Supérieure des
Mines de Paris
Centre de Morphologie
Mathématique
35 rue Saint Honoré
77300 Fontainebleau
France
`Jean.Serra@ensmp.fr`

Dietrich Stoyan
Institut für Stochastik
TU Bergakademie Freiberg
09596 Freiberg
Germany
`stoyan@orion.hrz.tu-freiberg.de`

Georges Matheron

Part I

Geostatistics

The genesis of geostatistics in gold and diamond industries

Danie Krige[1] and Wynand Kleingeld[2]

[1] Florida Hill, South Africa
[2] De Beers MRM, UK

1 Introduction

Geostatistics has had a phenomenal half a century of development and achievements, and to which George Matheron has made invaluable contributions throughout. The genesis of geostatistics is clearly linked to South Africa and more specifically to gold mining in the Witwatersrand basin, which started towards the end of the 19th century. In the later stages of the development of geostatistics the problems inherent in the valuation of diamond deposits presented a new field for geostatistical contributions.

2 The Influence of Gold and the Origin of Geostatistics

For economic reasons this gold mining was and still is conducted on a selective block basis and calls for intensive and regular sampling of underground exposures of the ore bodies. This resulted in the accumulation over many years of massive data sets conducive to statistical analysis and the study of frequency distribution models. In the pre-geostatistics period ore reserve blocks were valued on the arithmetic averages of samples from the block peripheries; as these blocks were being mined the advancing stope faces inside the blocks were also sampled regularly to yield extensive follow-up block values. Comparisons of these follow-up grades with the original block estimates provide an obvious opportunity for statistical analyses such as frequency distribution studies and classical correlations. However, this opportunity remained dormant until the 1940's.

At that time, extensive exploration of virgin properties in the new South African gold fields by deep drilling was also taking place. Grade estimates for these new mines had to be based on limited sets of drill hole grades with no proper basis for estimating the effects of selective mining to economic cut-off grades. Fortunately, the ideal venue for access to all this data from numerous existing mines and from the new gold fields was provided by the records in the Government Mining Engineer's Department where the first author was privileged to be employed in the late 1940's. He was introduced to the statistical approach for the processing of this data by the earlier initial work by Sichel [27], Ross [26] and De Wijs [3, 4]. This led to a first set of publications ([8, 9]) which in turn introduced, inter alia, Allais [1] and Matheron [15] to the subject.

This paper is, thus confined to ore valuations in the mining field.

2.1 Frequency Distribution Models

The skew nature of the gold grade frequency distributions was first observed by Watermeyer [33] and later studied by Truscott [32]. But these studies were done without the knowledge of the lognormal model and were unsuccessful.

Real progress was absent until the 1940's when Sichel [27] suggested the use of **the lognormal model**. He was a classical statistician and in the mining field he concentrated his efforts on frequency distribution models. He developed the T-estimator with its appropriate confidence limits [28]. This estimator is more efficient than the arithmetic mean, but is strictly valid only for a random set of data which follows the lognormal model exactly. Departures from the 2-parameter lognormal model, as observed in practice, were largely overcome with the introduction in 1960 of the 3-parameter lognormal model [10] which requires an additive constant before taking logarithms. However, there were still cases which could not be covered properly by the 3-parameter lognormal, and led to the introduction of the more flexible Compound Lognormal Distribution [30], originally developed by Sichel for diamond distributions. This model is very flexible and caters specifically for a tail of high values which is much longer than that for earlier models. This development is covered in the second part of this paper.

2.2 The genesis and early development of geostatistics

Geostatistical concepts originated in the late 1940's when Ross applied the lognormal model to a variety of actual gold grade data [26], de Wijs showed how the differences between individual grades depended on their distances apart [3, 4] and particularly when the basic concept of gold ore grades as a variable with a spatial structure was introduced in 1951/2 [8, 9]. The objective of this latter work was to develop more efficient grade estimates for new mines and for ore blocks on existing mines.

The first paper [8] was aimed at finding an explanation for the experience on all the gold mines for many decades, of ore reserve estimates during subsequent mining consistently showing a significant under-valuation in the lower grade categories and the reverse for estimates in the higher grade categories. Classical statistical correlation and regression analyses proved this to be an unavoidable result of block estimates subject to error and to conditional biases [8]. In the proper perspective it was essential to no longer view the peripheral data used for individual block estimates and the ore blocks themselves in isolation. It was essential to see **the peripheral data as part of an extensive spread of data (the data population)** in stopes and development ends in the relevant mine section; also to accept the grade of the ore block concerned as part of a collection of block grades (both intact and already mined out), i.e. as a member of **a population of oreblock grades.**

In this way, **the spatial concept was introduced as well as the concept of support** in moving from individual sample grades (point supports), to block grades. A mathematical model was first set up of the lognormal distribution of actual block values in a mine section. The errors in assigning the limited peripheral grades to the blocks were super imposed on the actual grades to yield the corresponding distribution of block estimates. On correlating these two sets of block values on a classical statistical basis, the averages of the actual block values relative to the corresponding block estimates in grade categories could be observed, i.e. the curvilinear regression of actuals on estimates. This was a theoretical follow-up exercise to simulate the results actually observed in practice. It provided the statistical explanation of the natural phenomena of the unavoidable under- and over-valuation features as mentioned above, i.e. the **inherent conditional biases**. The use of the lognormal model also covered **the curvilinear nature of the regression trend** as observed in practice.

The very fact that a correlation exists between the block estimates and the internal actual grades emphasises the presence of a spatial structure. With the explanation of these conditional biases, the initial application in practice was to apply the trend, or regression, of actuals (or follow-ups) versus estimates –as observed from the mine records or modelled geostatistically– to the orthodox block estimates so as to eliminate these biases. As the regressed estimates were, in effect, weighted averages of the peripheral estimates and the global mean grade of the mine section, it was **the first application of what became known as kriging. It can be labelled Simple Elementary Kriging**, being based on the spatial correlation between the peripheral values and the actual grades of the ore inside the ore blocks, and giving proper weight to the data outside the block periphery via the mean.

During the 1950's several large gold mines introduced regression techniques for their ore reserve estimates on a routine basis. It is instructive to observe that on the gold mines the improvement in the standard of block valuations due to the elimination of conditional biases accounts for some 70% of the total level of improvement achievable today with the most sophisticated geostatistical techniques. It is for this reason, that so much stress is placed on the elimination of conditional bias (so called "conditional unbiasedness").

In the second paper [9], **the spatial structure of the grade data** from 91 drill holes in the main sector of the new Orange Free State gold field was defined by the log-variance/log-area relationship (see Fig. 1). This demonstrated the so-called **Krige formula** (point variance within a large area minus the average point variance within ore blocks = the variance of block values within the large area) as well as the so-called **permanence of the lognormal model** for different support sizes. On this basis the lognormal model for the expected distribution of ore block grades was modelled with a global mean grade as estimated from the 91 drill hole grades. It led successfully to meaningful tonnage and grade estimates for a range of cut-off grades, i.e. the **first version of the now well known tonnage-grade curve**. Without a proper block distribution model the orthodox approach would have based the tonnage-grade estimates directly on the individual drill hole grades with seriously misleading results globally and for individual mines.

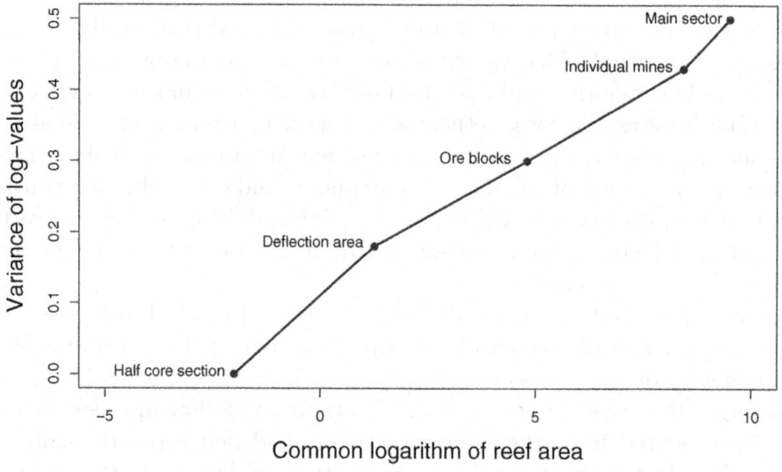

Fig. 1. Dispersion variance versus domain size: example of gold data from the Orange Free State. The horizontal axis represents the area of the domain in logarithmic scale, from 10^{-5} to 10^{10} ft^2

These developments were published in 1951/1952 and aroused world-wide interest in the subject now known as geostatistics, particularly in French circles. Matheron and Duval translated these two papers and re-published them in 1955 together with two personal contributions by them ([15, 5]). This was followed by a paper on exploration prospects in the Sahara by Allais [1]. Matheron [15] in particular covered the more theoretical background underlying the two basic South African papers and the models involved in all this work. He showed, for example, that **the permanence of the lognormal model** can only logically apply where a spatial structure is present and that the positive correlation between the log variances and the mean grades of lognormal distributions – the so-called **proportional effect** – is an inherent feature of the lognormal model. This contribution by Matheron was accompanied and/or followed in the 1950's and 1960's by numerous other notes and publications in French and the introduction of **the theory of regionalised variables**. Matheron's first geostatistical papers in English were in the 1960's ([16, 17]), followed by an English monograph in 1971 ([18]) which covered the theory in detail.

Modelling of the spatial structure is basic to any geostatistical approach. The original approach in South Africa [9] was followed by extensive correlations of pairs of grades for different lags and the results were modelled by **correlograms and covariograms** [11, 12] and used in multiple regression techniques to arrive at the relative weights to be applied to the data available for a block valuation. This was already introduced on a routine basis on some of the large gold mines for ore reserve estimates in the early 1960's and was called weighted moving average estimates until at Matheron's insistence the term **kriging** prevailed. In the mean time Matheron covered a continuous series of further developments of geostatistical models based on the now generally applied **variogram** for defining spatial structures.

The critical need for identifying likely changes in the characteristics of the spatial structure between sections within the ore body, such as grade continuity levels, anisotropy directions, etc. was also already stressed during the 1960's. This basic tenet of geostatistics has been and is still widely met in practice via the linkage of these characteristics with changes in **geological and/or mineralogical parameters** which can more readily be modelled.

2.3 The main basic tenets of geostatistics

Virtually all the fundamental concepts and tenets of geostatistics were established in these early years and are still applicable today.

1. The use of **appropriate parametric distribution models** when practical for confidence limits of estimates of the mean grade. Various nonparametric approaches have been developed, but face the common problem that the pattern of the observed point distribution accepted as the model can be misleading for the upper tail of the distribution unless a very large data base is available.

2. **Spatial structures** generally present in ore bodies and with characteristics associated with geological and mineralogical features.

3. The concepts of **support sizes and types, the proportionality effect** and models for estimating the **SMU (Selected Mining Unit) block distribution** parameters directly or indirectly from the point value distribution.

4. **Kriging** for block estimates. Many types of kriging have been developed but all allow to eliminate - or at least reduce - the conditional bias.

5. If block valuations are done before the actual selective mining stage - when the final data becomes available - the estimates will be **smoothed** and have to be post-processed. Meaningful post-processing techniques were set up after the early stages of the development of geostatistics, as well

as the so-called non-linear geostatistics (see the paper by Rivoirard in the present volume, page 17).

2.4 Conditional unbiasedness

Early developments in South Africa generally retained strong links with classical statistics, particularly through the preference on the gold mines for simple kriging. This is essentially a classical multiple regression technique based on the mean grade for the ore body or local part thereof and on the corresponding spatial structure to provide the necessary covariance's for the solution of the matrix equations involved. However, where block valuations are based on exploratory data to be supplemented at the final mining stage by additional data more closely spaced, the earlier block estimates will be smoothed compared to the final estimates and have to be post-processed before declaring ore reserves and doing mine planning and feasibility studies.

Arising from this problem and the fact that the mean grade as used in simple kriging itself changes within an ore body, a general preference developped for ordinary kriging, which relies only on the data as accessed for the kriging of each block. There is no objection, in principle, to this approach provided the data accessed is adequate to effectively provide a close grade level for the local area encompassing the block. This will ensure conditional unbiasedness but will not overcome the smoothing problem. Also the effort time and cost involved in kriging each block on a relatively large data base required to meet this objective (say 50 instead of only 5, 10 or even 20 values), led to the widespread use of a limited search routine, and lately also to simulation as an alternative. These practices can reduce or eliminate smoothing, but unfortunately, re-introduce conditional biases as prevailed in the pre-geostatistical period and cannot be post-processed unless the conditional biases are first removed. Although such estimates could provide acceptable global tonnage-grade figures for the whole ore body, they could still be seriously conditionally biased for sections of the ore body as will be mined sequentially over short time periods [13] and thus be unacceptable.

3 The influence of the specificity of the diamond mining industry on the development of geostatistics

3.1 Introduction

Diamonds are a unique commodity in the realms of mining and its associated disciplines. The particulate nature of the diamond affects the processes of exploration, evaluation, mining and metallurgy and especially the way in which diamonds are valued and sold.

The evaluation of alluvial diamond deposits specifically drew the attention of some of the greatest minds in the field of geostatistics and this would eventually give rise to novel ways of dealing with sampling and estimation techniques in the case of discrete particle deposits.

The diagram shown in Fig. 2 illustrates the complex nature of diamond deposits compared with other mineral commodities. The average grade for diamond deposits is generally very low and a high degree of geological discontinuity exists, particularly in the case of marine placer deposits where selective mining is absolutely essential.

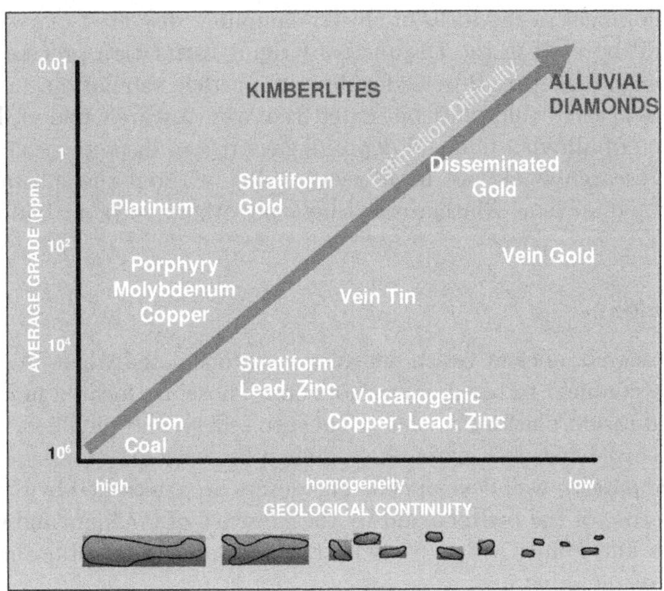

Fig. 2. Plotting the average grade versus the mineralization continuity of various minerals

Groundbreaking work by Sichel during the early seventies [29] gave rise to a statistical approach on how to evaluate these deposits, and models such as the compound Poisson for diamond density distributions [20, 21] and the Compound Lognormal for diamond size distributions were developed. The adaptation of these models had significant impact later on the estimation of other minerals [31].

In the 10 years from 1980 to 1990 a substantial research effort [24] was directed towards understanding the following issues;

1. The complex nature of the geology that gave rise to the discrete particle mineral deposit.
2. The problems associated with the sampling of deposits of this nature and the fact that the sampling could produce non representative results since the sample size is smaller than the trap sites in which the particles occur.

3. The statistical models required to cater for the extremely skewed sampling data, with the emphasis on smoothing the shape of the curve, and to increase sample representavity.
4. Methods to obtain local reserve estimates, including confidence limits, which require a local distribution density function.
5. The need to produce bivariate representation of the probability mass functions for the non-linear kriging procedures used.

The outcome of the research culminated in a geostatistical approach to the development of the ideas of cluster sampling, discrete isofactorial models that could be used in the Disjunctive Kriging estimation process to develop estimates and the Cox Process for discrete particle simulation.

In essence it could thus be stated that the research undertaken on the evaluation of alluvial diamond deposits gave rise to discrete geostatistics.

The research would not have been possible without the substantial input from people such as Matheron, Sichel and others such as Lantuéjoul and Lajaunie.

3.2 Geology

In the research, ancient beach deposits were considered where mineralization is largely confined to basal gravel horizons. Those are located in one or more marine abrasion platforms, usually cut into schists and phyllites.

The shist bedrock is extensively gullied by wave action assuming a characteristic pattern well developped. The gullies are controlled by the slope and the structure of the bedrock and by the presence of boulders and gravel. The pot-holes and gullies act as particle trap sites and can contain high concentrations of the mineral.

Though some trap sites can contain high concentrations they could be surrounded by sites that have low concentration or even be barren. This high degree of variability is related to the complex interaction of geological controls during deposition. The chance of finding a particle in this type of deposit is related to the chance of sampling a trap site and the distribution of particles in the trap site.

The distribution of particles is different for each beach, corresponding to the different marine transgressions and is related to the length of stillstand of the sea which influenced the degree of abrasion of the marine platform and the degree of reworking that took place. It is also influenced by the amount of mineral bearing gravel that was available during the transgression period.

Thus the distribution of particles is directly related to the presence of particles in the gravel, the quantity and quality of the trap sites and to the degree of reworking of the gravel.

In a certain area a characteristic gully pattern is normally formed with parallel gullies at relatively constant distances apart. The trap sites in these gullies also show characteristic patterns with the typical size of a trap site 5m along and 3m across a gully, but variable from area to area.

Fig. 3. Mining an alluvial deposit. The diamonds are trapped in the pot-holes and the gullies of the bedrock

3.3 Sampling

From the outset the sampling of these types of deposits posed significant problems, even at elevated sample support sizes (5 square meters) up to 90% of the samples did not contain any particle. In contrast, there are rare occurrences where several hundred particles were recovered in one sample unit located over a natural trap site such as a deep pot-hole.

Based on the geological model it became obvious that normal sampling theory as applied to homogeneous mineralization was not applicable in the case of marine placers where mineralization was concentrated in trap sites [29].

Other examples occur in vein and alluvial deposits of diamonds and gold. In such deposits, two different factors account for the grade variability at two different scales, firstly the spatial distribution of the trap sites, and secondly the dispersion of the mineralization within the trap sites.

The existence of several scales of variability makes sampling a very complex operation. As a matter of fact, a set of samples of a given size may not account for all the scales of variability. Using many small samples, traps are well detected, but their mineralization contents are poorly assessed. Using a limited number of larger samples, the quality of the traps is better known, but it becomes more difficult to assess the distribution of the traps sites.

The methodology involved in sampling such highly dispersed type ore bodies was addressed in [7]. The paper presents several results on the sampling of highly dispersed type ore bodies and highlights the two major problems encountered when sampling under such conditions.

It also addresses the problem of defining a representative sample support size and making it operatory by resorting to a cluster sampling approach.

During the research it became evident that limited experience exists in the field of in situ sampling, especially when stratification is present.

3.4 Estimation

In his book "Estimer et choisir" [19], Matheron discusses certain fundamental differences between the approaches adopted in statistics and geostatistics. Fundamentally statistics is involved in the estimation of parameters for a chosen probability model, whereas geostatistics is involved with the estimation of a spatial average for a natural phenomenon.

However, in the case of discrete particle deposits where the bases for estimation (variogram and histogram) are not well defined due to the non representative nature of sampling, the necessity of using statistical models in estimation was evident [6]. Matheron highlights that the model is not the deposit and notes that substantial research is needed to explain the variation of model parameters with the geology of a deposit. Such work was carried out by Oosterveld *et al.* [25].

The need for introducing a statistical model for local reserve estimation was clearly indicated and research was done to provide a method to estimate the number of particles expected in an in situ reserve block of specific support size.

The introduction of geostatistics contributed to the understanding and quantification of the risk associated with grade estimation. Uncertainty was defined in terms of confidence limits derived from a modelled probability distribution for the grades of the mining blocks.

The skewness of the sample grades gave rise to skew distributions for the block estimates and their error distributions. This led to research in the field of non-linear kriging, more specifically disjunctive kriging under appropriate discrete isofactorial models [22]. A suitable statistical model which represents a discrete type of particle density distribution had to take into account the distributional characteristics of the trap sites as well as the particles contained within the trap sites. The most important problem in mining geostatistics, i.e. that of change of support, was also addressed. The inference of the parameters is a challenging problem and practical aspects of implementing discrete isofactorial methodology are presented in [14].

The high degree of geological discontinuity also led to research towards a connectivity index in mining. This problem occurs when mining at high cut-off grades where only a fraction of the selective units is above cut-off and where the blocks are split into disjoint patches that cannot be accessed economically during mining [2].

4 Conclusion

The immensity of the South African deposits amongst others that of gold and diamonds has produced the background to the development of geostatistics. The evaluation of deposits drew the attention of the most prominent geostatisticians of our time. Fortunately, we had people such as Matheron and Sichel to assist in the phenomenal development that took place in the last 50 years.

References

1. Allais, M.: Method of appraising economic prospects of mining exploration over large territories; Algerian Sahara Case Study. Man. Sci., **3-4**, 285–347 (1957)
2. Allard, D., Armstrong, M., Kleingeld, W.J.: The need of a connectivity index in mining geostatistics. In: Dimitrakopoulos, R. (ed) Geostatistics for the next century. Kluwer, Dordrecht (1994)
3. De Wijs, H.J.: Statistics of ore distribution. Part 1. J. of the Roy. Netherl. Geol. and Min. Soc., **13-11**, 365–375 (1951)
4. De Wijs, H.J.: Statistics of ore distribution. Part 2. J. of the Roy. Netherl. Geol. and Min. Soc., **15-1**, 12–24 (1953)
5. Duval, R.: A propos de l'échantillonnage des gisements. Annales des Mines, **12**, 76–79 (1955)
6. Kleingeld, W.J.: La géostatistique pour des variables discrètes. PhD Thesis, Ecole des Mines de Paris (1987)
7. Kleingeld, W.J., Lantuéjoul, C.: Sampling of orebodies with a highly dispersed mineralization. In: Soares, A. (ed) Geostatistics Tróia'92. Kluwer, Dordrecht (1993)
8. Krige, D.G.: A statistical approach to some basic mine valuation problems on the Witwatersrand. J. Chem. Metall. Min. Soc.South Afr., **52**, 119–139 (1951)
9. Krige, D.G.: A statistical analysis of some of the borehole values in the Orange Free State goldfield. J. Chem. Metall. Min. Soc.South Afr., **53-3**, 47–64 (1952)
10. Krige, D.G.: On the departure of ore value distributions from the lognormal model in South African gold mines. J. South Afr.Inst. Min. Metal, **61**, 231–244 (1960)
11. Krige, D.G.: Statistical applications in mine valuation. J. Inst. Mine Survey South Afr., **12-2**, 45–84 (1962)
12. Krige, D.G.: Statistical applications in mine valuation. J. Inst. Mine Survey South Afr., **12-3**, 95–136 (1962)

13. Krige, D.G., Assibey-Bonsu, W.: Valuation of recoverable resources by kriging, direct conditioning and simulation. In: Jiang, Y. (ed) Proceedings of the 29th symposium on application of computers in the mineral industries. Balkema, Rotterdam (2001)

14. Lajaunie, C., Lantuéjoul, C.: Setting up a general methodology for discrete isofactorial models. In: Armstrong, M. (ed) Geostatistics. Kluwer, Dordrecht (1989)

15. Matheron, G.: Application des méthodes statistiques à l'évaluation des gisements. Annales des Mines, **12**, 50–75 (1955)

16. Matheron, G.: Principles of geostatistics. Econ. Geol., **58**, 1246-1266 (1963)

17. Matheron, G.: Kriging, or polynomial interpolation procedures? Can. Min. Met. Bull., **11**, 240–244 (1967)

18. Matheron, G.: The theory of regionalized variables and its applications. School of Mines, Paris (1971)

19. Matheron, G.: Estimer et choisir. School of Mines, Paris (1978)

20. Matheron, G.: Quatre familles discrètes. Technical Report N-703. Centre de Géostatistique, Ecole des Mines de Paris (1981)

21. Matheron, G.: Deux autres familles, non moins discrètes mais plus nombreuses. Technical Report N-717. Centre de Géostatistique, Ecole des Mines de Paris (1981)

22. Matheron, G.: Une méthodologie générale pour les modèles isofactoriels discrets. Sciences de la Terre, **21**, 1–64 (1984)

23. Matheron, G.: Two classes of isofactorial models. In Armstrong, M. (ed) Geostatistics. Kluwer, Dordrecht (1989)

24. Matheron, G., Kleingeld, W.J.: The evolution of geostatistics. Proceedings of the 20th APCOM international symposium. SAIMM, Johannesburg (1987)

25. Oosterveld, M.M., Campbell, D.C., Hazel, K.R.: Geology related to statistical evaluation parameters for a diamondiferous beach deposit. South Afr. Inst. Min. Met., 129–136 (1987)

26. Ross, F.W.J.: The development and some practical applications of statistical value distribution theory for Witwatersrand auriferous deposits. MSc thesis, Univ. of the Witwatersrand (1950)

27. Sichel, H.S.: An experimental and theoretical investigation of bias error in mine sampling with special reference to narrow gold reefs. Trans. Inst. Min. Metall. Lond., **56**, 403–473 (1947)

28. Sichel, H.S.: New methods in the statistical evaluation of mine sampling data. Bull. Inst. Min. Metall. Lond., 261–288 (1952)

29. Sichel, H.S.: Statistical valuation of diamondiferous deposits. In: Salamon, M.G.D., Lancaster, F.H. (eds) Proceedings of the 11th APCOM international symposium. J. South Afr. Inst. Min. Met., Johannesburg (1973)

30. Sichel, H.S., Kleingeld, W.J., Assibey-Bonsu, W.: A comparative study of the frequency distribution models for use in ore valuation. J. South Afr. Inst. Min. Met., **92**, 235–243 (1992)

31. Sichel, H.S., Dohm, C.E., Kleingeld, W.J., New generalised model of observed ore value distributions. Trans. Inst. Min. Metall., **104**, 79–124 (1995)

32. Trustcott, S.J.: The computation of the probable value of ore reserves from assay results. Trans. Inst. Min. Metall. Lond., **39**, 482–496 (1929)

33. Watermeyer, G.A.: Application of the theory of probability in the determination of ore reserves. J. Chem. Metall. Min. Soc. South Afr., **19**, 97–107 (1919)

Concepts and Methods of Geostatistics

Jacques Rivoirard

Centre de Géostatistique, Ecole des Mines de Paris

1 Introduction

A mathematician but also an engineer, Professor Georges Matheron has shown an exceptional ability to formalize practical problems, edict relevant concepts and methods in order to find workable solutions. From 1954 to 1963, while working with the French Geological Survey in Algeria and in France, he discovered the pioneering work of the South African school on the gold deposits of the Witwatersrand, and formalized the major concepts of the theory that he named Geostatistics. The "classical" geostatistics (linear geostatistics based on stationary covariance and variogram) was fully developed in his thesis [11]. The regionalized variable under study, as Matheron called it, is then conveniently modelled as a Random Function. However one should not consider the model as the reality, and the analysis of the role to be assigned to the model led Matheron to write "Estimating and Choosing", an essay on probability in practice ([18] for the French version, [20] in English). In the mean time, the bases of both non-linear geostatistics and non-stationary geostatistics were laid out. The reader will find a detailed description of the different geostatistical methods, as well as a valuable bibliography, in the reference book by Chilès and Delfiner [1]. The present article describes the basic concepts of geostatistics, from today's perspective.

2 Linear geostatistics

2.1 The origin

Matheron discovered the work of the South African school during his first professional appointment with the French Geological Survey in Algeria in 1954. One year later, a paper written by R. Duval [4] in collaboration with R. Lévy and G. Matheron, presented the work by Krige [6] in a very concise way in the French journal "Annales des Mines".

The basic ingredients of geostatistics (currently referred to as the support, the dispersion variances, the conditional bias, see further on) were already present in the South-African work (see the paper by Krige and Kleingeld in this volume, page 5). The variances of gold sample values were observed to be higher when these samples were taken within a larger area, e.g. the variance of samples within a mine being higher than the variance of samples within a mining block or panel (here a "sample" is given its physical meaning or its corresponding ore value, which is different from the statistical meaning, i.e. the collection of such sample values). In addition, those variances (more exactly the logarithmic variances, see further on) are linked by Krige's additivity relationship:

variance of samples within mine

= variance of samples within panels + variance of panels within mine.

The fact that the variance of gold values within a given area depends on the "support" on which the variable is defined (sample, panel..., i.e. the 2D generic area in size and orientation, or the generic volume in 3D), has serious consequences when selecting panels from their sampling. As panels are less variant than samples, it follows that the mining panels, selected as being rich according to their samplings, are naturally less rich than these samplings. Similarly panels considered to be poor are richer than their samplings. No bias on samples is necessary to explain this overestimation of rich panels and underestimation of poor ones. It simply corresponds to a particularly dangerous case of conditional bias: conditionally to its samples, the expectation of the value of a panel is not equal to its sampled value, it is lower when the sampled value is high, higher when this value is low.

To avoid the overestimation of selected reserves caused by such a conditional bias, corrected estimators of panels from samples have been developed by the South-African school, based on the lognormal distribution. The lognormal distribution was observed to provide a good fit to gold content values. Let us recall that it is skewed positively, with a tail of large values, and a shape factor given by the logarithmic variance, i.e. the variance of the logarithm of the variable. The higher the logarithmic variance, the more dispersed the values relatively to their mean. Because of the skewness of the distribution, lognormal estimators of the mean of the distribution (computed from the arithmetic mean of log values, i.e. the geometric mean of values) from a limited number of samples, can be significantly better than the arithmetic mean of values. Krige's methodology assumes the permanence of the lognormality of the distributions, e.g. the distributions of sample values within a panel, or within a mine, are lognormal, as well as the distribution of panel values within the mine. The variations of the values of such distributions are then described by their logarithmic variances, these being linked by the additivity relationship mentioned previously.

In addition to the article by Duval, Lévy and Matheron on Krige's work, the same issue of Annales des Mines contained another remarkable article written by Matheron alone [8]. In particular, he derived the permanence of lognormality from a principle of self-similarity when splitting blocks iteratively as considered by de Wijs [3]. This synthesis allowed him to distinguish two terms within the dispersion variance of the support v within the domain V: a term depending on the sole geometry of v and V; and a coefficient of "absolute dispersion" (as he called it) which is an intrinsic characteristic of the mineralization (this parameter will be later replaced by the more general structural tool given by the variogram). During these early years, Matheron produced several statistical reports and studies of deposits. It is worth noticing that the term "geostatistics" appeared explicitly in the title of a study of a lead deposit dated 1955 [9].

Next, Matheron stepped to the linear geostatistics, based on the additivity of variables and linear estimators, without any assumption on the type of statistical distribution. As a matter of fact, the gold content variable initially studied by Krige was the metal accumulation taken vertically across a reef extending in 2D. Supposing that a panel is partitioned into samples with same support v, the mean value over the panel is equal to the arithmetic average of sample values, whatever the statistical distribution of the values, lognormal or not. In other words, the accumulation is an additive variable (this is also the case for the thickness of the reef which gives the ore quantity, but not for the grade, equal to the ratio between the metal and the ore quantities - or equivalently between accumulation and thickness: the grade has to be weighted by the thickness when averaged). In the early South-African school, the estimation of a given block value was brought down to the estimation of the mean value of the statistical distribution made of all possible samples constituting the block, from the samples available at the periphery of the block, and the spatial aspect was not modelled. By contrast, Matheron rather addressed the estimation of the mean value over any block or domain, using located sample data and taking explicitly into account the geometrical configuration of all the elements. This enabled him to formalize the concepts at the basis of geostatistics in a very simple manner.

2.2 Additivity, support and dispersion variances

Let $z(x)$ be a "regionalized variable", as Matheron called it, depending on the location x, traditionally a point in 2D or 3D. In practice such a point often corresponds to the sample support. Assuming that the variable is additive, its "regularized" value over v (e.g. a block) is the arithmetic average of its points:

$$z(v) = \frac{1}{|v|} \int_v z(x)\, dx$$

Similarly we have:

$$z(V) = \frac{1}{N} \sum z(v_i)$$

when considering the regularized value over V, this being partitioned into N blocks v_i with same support v.

The variation of the v_i values within V can be measured by the "dispersion variance" of support v within domain V:

$$s^2(v|V) = \frac{1}{N} \sum_i [z(v_i) - z(V)]^2$$

Similarly the dispersion variance of a point within V is:

$$s^2(O|V) = \frac{1}{|V|} \int_V [z(x) - z(V)]^2 \, dx$$

We also have:

$$s^2(O|v_i) = \frac{1}{|v|} \int_{v_i} [z(x) - z(v_i)]^2 \, dx$$

and the mean of these quantities over the different v_i's within V gives the dispersion variance $s^2(O|v)$ of a point within v. It is easy to demonstrate that such dispersion variances are linked through the additivity relationship:

$$s^2(O|V) = s^2(O|v) + s^2(v|V)$$

This formula is similar to the relationship obtained by Krige on logarithmic variances, but is satisfied for any statistical distribution of the variable, provided that this variable is additive.

In addition, such dispersion variances, which are of crucial importance when the sample support does not coincide with the support of interest (mining block, area to be depolluted...), are related to the "spatial structure" represented by the variogram. But this will be more conveniently developed later within the framework of Random Function.

2.3 Intrinsic approach versus transitive approach

When developing linear geostatistics, Matheron distinguished two approaches, the transitive approach (seldom used), and the intrinsic approach. In the transitive approach, the phenomenon to be studied (orebody, fish stock...) is supposed to be known, for instance, on a regular rectangular grid defined by its orientation, its mesh size, and its origin. The exact boundaries of the phenomenon are considered to be unknown, and the grid is supposed to extend beyond them. The transitive approach allows for instance to estimate the global abundance from such a grid, with an error variance derived from a structural tool called the transitive covariogram (similar to a noncentered covariance, where terms are summed instead of being averaged). The approach can be particularly useful when studying resources with diffuse limits such as some fish or larvae densities in the sea, or when estimating the extension area or the volume of a phenomenon only known at the nodes of a grid. In the case of a regular grid with a given orientation, the only element of randomness is the origin of the grid, which is supposed uniformly distributed over the mesh size.

By contrast, in the traditional intrinsic approach, the domain is considered to be known, and the variable is supposed to have an intrinsic behaviour, independent of the geometry and of the boundaries of the domain. The domain can be considered as a window, allowing to see the variable, which is assumed to extend with the same behaviour beyond the window boundaries. To describe such an intrinsic behaviour, the methodology relies on a form of stationarity, i.e. invariance under translations. Although classical linear geostatistics was essentially completed by the early 60's [10], Matheron did not introduce the convenient framework of Random Function until he completed his thesis [11].

2.4 The intrinsic model and the variogram

The regionalized variable $z(x)$ is then considered as a realization, a possible outcome, of a Random Function (RF in short) $Z(x)$, whose specifications constitute the RF model. This framework allows a simple formulation of stationarity. Thanks to de Wijs's work [3], and to Krige's observation of variance increasing with the area, Matheron privileged the stationarity of the increments over the stationarity of the variable itself. In linear geostatistics the most common model is the intrinsic RF model, defined by the increments $Z(x + h) - Z(x)$ having a zero expectation and a stationary variance (depending on vector h, not on location x):

$$E[Z(x + h) - Z(x)] = 0$$

$$\mathrm{Var}[Z(x + h) - Z(x)] = E\{[Z(x + h) - Z(x)]^2\} = 2\gamma(h)$$

The intrinsic RF model is then completely specified by its structural tool $\gamma(h)$, the variogram, which expresses the mean variability between two points as a function of the vector between them, and which depicts the more or less regular behaviour of $Z(x)$. In practice, a set of distances is chosen, in accordance with the sampling grid. For each distance an experimental variogram value is computed from pairs of data separated by this distance, using implicitly the invariance of increments under translation. This experimental variogram is then interpreted (e.g. some variations being considered as insignificant fluctuations) and fitted by a mathematical function, the variogram model, assuming additional hypotheses on its regularity for instance. As the variogram directly measures the variability between pairs of points, Matheron called it the *intrinsic* dispersion function in early times. The intrinsic RF model, based on increments, allows to express the expectation and variance of linear combinations $\sum_{\alpha} \lambda_\alpha Z(x_\alpha)$, but only when $\sum_{\alpha} \lambda_\alpha = 0$ (as a matter of fact, such linear combinations can be seen as linear combinations of increments; the expectation and the variance of other linear combinations, in particular $Z(x)$, are simply not defined in the model). These admissible linear combinations have a zero expectation, and a variance which depends on the variogram:

$$\mathrm{Var} \sum_\alpha \lambda_\alpha Z(x_\alpha) = - \sum_\alpha \sum_\beta \lambda_\alpha \lambda_\beta \gamma(x_\beta - x_\alpha)$$

To ensure the positivity of such variances, the function $-\gamma(h)$ must be conditionally positive definite. The fitting task is usually simplified by considering a sum of known authorized model components (one of them being the white noise or "nugget effect", a simple discontinuity at the origin).

Such a measurement of variability is not specific to geostatistics. The novelty introduced by Matheron consisted in establishing it within a simple model while accounting for the important concepts of support, regularization, and not the least, the two different following types of variances, and to make all this available to the practitioner: the engineer and the mathematician are in perfect agreement.

If $Z(x)$ is an intrinsic RF, so is the regularized $Z(v)$ over support v, with a "regularized" variogram deduced from the variogram of $Z(x)$. The dispersion variance of the support v within the domain V also depends on the variogram of $Z(x)$:

$$D^2(v|V) = \bar{\gamma}(V, V) - \bar{\gamma}(v, v)$$

and in particular for a point support:

$$D^2(O|V) = \bar{\gamma}(V, V)$$

where $\bar{\gamma}(V_1, V_2)$ stands for $\frac{1}{|V_1||V_2|} \int_{V_1} \int_{V_2} \gamma(x - y)\, dx\, dy$

Such dispersion variances are linked by the additivity relationship, e.g.:

$$D^2(O|V) = D^2(O|v) + D^2(v|V)$$

In practice this explains, or makes it possible to predict the drop of variance when changing the support, from a quasi point sample support to a larger interest support, for instance:

$$D^2(O|V) - D^2(v|V) = D^2(O|v) = \bar{\gamma}(v, v)$$

The other type of variance is the estimation variance, or variance of the error when estimating for instance the value $Z(V)$ over V by the value $Z(v)$ over v (or the contrary). The estimation error $Z(v) - Z(V)$ is an admissible linear combination. Therefore its mean is 0 (estimation is unbiased), and its variance is:

$$\mathrm{Var}[Z(v) - Z(V)] = 2\,\bar{\gamma}(v, V) - \bar{\gamma}(v, v) - \bar{\gamma}(V, V)$$

Using a variation of this formula, the "global estimation variance" can be computed when estimating a whole domain with a regular sampling grid for instance. Conversely, assuming a given variogram, this can help define a sampling grid in order to obtain a given estimation variance.

If we now consider the estimation of $Z(V)$ by a linear combination of samples $\sum_\alpha \lambda_\alpha Z(x_\alpha)$, the estimation error is $Z(V) - \sum_\alpha \lambda_\alpha Z(x_\alpha)$. It is an admissible linear combination iff $\sum_\alpha \lambda_\alpha = 1$ (if not, the error is simply not defined in the intrinsic RF model). Then its expectation is 0 and its variance can be developed:

$$\mathrm{Var}\left[Z(V) - \sum_\alpha \lambda_\alpha Z(x_\alpha)\right]$$
$$= 2 \sum_\alpha \lambda_\alpha \bar{\gamma}(x_\alpha, V) - \sum_\alpha \sum_\beta \lambda_\alpha \lambda_\beta \bar{\gamma}(x_\alpha, x_\beta) - \bar{\gamma}(V, V)$$

By minimizing this estimation variance, one gets an estimator, usually called Ordinary Kriging (OK). Kriging is a generic term for optimal – and generally linear – estimators in geostatistics, and the minimized variance is referred to as the kriging variance (not to be confused with the variance of the kriging estimator, which anyway is not defined in the intrinsic model). One will find an interesting historical study on the origins of kriging compared to other similar methods in a paper by Cressie [2]. The ordinary kriging weights, constrained by the above condition, are solution of a linear system. Such an Ordinary Kriging can be used to estimate the value at a point, the mean value over a block or a domain, or, in the case of mapping, a grid of points or a set of blocks.

2.5 The stationary case

The variogram $\gamma(h)$ may be unbounded and can increase infinitely. When it is bounded (for instance stabilizing on a "sill" level beyond a "range" distance), the IRF can be reduced to a 2nd order stationary RF, characterized by its first two moments being stationary:

$$E[Z(x)] = m$$

$$Cov[Z(x), Z(x + h)] = C(h)$$

In particular the *a priori* variance $\mathrm{Var}\, Z(x)$ is equal to $C(0)$ and we have $\gamma(h) = C(0) - C(h)$. The dispersion variance $D^2(O|V) = \bar{\gamma}(V, V)$ tends towards the *a priori* variance $C(0)$ (absent from a purely intrinsic model) when V increases infinitely in an appropriate way. Then:

$$\mathrm{Var}\, Z(V) = E[(Z(V) - m)^2] = \bar{C}(V, V) = \frac{1}{|v|^2} \int_v \int_v C(x - y)\, dx\, dy$$

tends to zero and $Z(V)$ tends to the mean m (supposed to be ergodic).

The stationary model allows to develop the variance of any linear combination:

$$\text{Var} \sum_\alpha \lambda_\alpha Z(x_\alpha) = \sum_\alpha \sum_\beta \lambda_\alpha \lambda_\beta C(x_\beta - x_\alpha)$$

(the covariance being a positive definite function to ensure the positivity of such variances). The best linear estimator in this model is usually called Simple Kriging (SK). Being unbiased it can be written:

$$\sum_\alpha \lambda_\alpha Z(x_\alpha) + \left[1 - \sum_\alpha \lambda_\alpha\right] m$$

and the kriging weights minimizing the estimation variance are solution of a linear system.

2.6 Questioning the model

For practical applications, an interesting question is whether the mean parameter m is known, and even what meaning it has when we only have a single realization of the RF model, defined over a finite domain (think of the metal grade of an orebody for instance). If the mean is considered to be unknown, it can be removed from the above estimation by setting the condition $\sum \lambda_\alpha = 1$: we are then back to Ordinary Kriging. In practice, stationarity, as usually detected by the variogram stabilizing on a sill, is often not guaranteed for large distances, e.g. throughout the whole domain: stationarity is then only local. In that case Ordinary Kriging is preferred, using datapoints within a moving neighbourhood around the target (point or block of a grid), i.e. a neighbourhood that moves with the target (as opposed to a unique neighbourhood including all datapoints). In practice the choice of the neighbourhood is still a difficult and much debated question on which we will come back later: in theory the larger the neighbourhood, the better the estimation, but also the stronger the stationarity hypothesis. In particular, while Ordinary Kriging is a weighted average of data within the neighbourhood, Simple Kriging also makes use of the mean m, with a complementary weight. This compensates for a possible sparseness of data with respect to the spatial structure, by attracting the local estimate towards the mean in accordance to the stationarity hypothesis.

Now, assuming stationarity over the whole domain does not guarantee that m can be known, nor that it even exists outside the model. It is interesting to follow the evolution of Matheron's thoughts on models through time. Matheron introduced the RF framework for the intrinsic approach in 1965. Although the transitive approach does not use the randomization of the regionalized variable, Matheron ([13], p. 39) described the transitive methods as "being rich in implicit probabilistic contents", as he considered that the operations used for fitting a transitive covariogram "constitute a disguised passage to expectations". Finally Matheron ([18, 20]) went back, stating that, if a methodology is performant, it should mean something on the regionalized variable (the reality !): this questions the objective contents of the RF model, by subordinating it to the regionalized variable. He proposed the operational indexrandom set of methods. The operations of fitting a variogram or a co-variogram are then considered as resulting from the choice of anticipating hypotheses (which may or may not be correct, as could possibly be checked later), that allow to tell more than the data alone. Parameters of the models, that cannot have their counterpart in terms of the regionalized variable, are considered to be purely conventional and must disappear from the final estimation results. This is of course the case of the mean parameter m of the stationary model if the domain V is not large enough to identify the average $Z(V)$, representing $z(V)$, with this mean. Such epistemological developments are worth being considered, if one aims at applying mathematical models.

2.7 Additional remarks

Before going on the other parts of geostatistics, let us go back in more details to some difficult points that concern the choice or the use of models.

Back to conditional unbiasness

Let $Z(V)^\star$ be an estimator of the panel $Z(V)$. Moreover suppose that it is unbiased:

$$E[Z(V) - Z(V)^\star] = 0$$

Conditional unbiasness can be written:

$$E[Z(V) - Z(V)^\star | Z(V)^\star] = 0$$

or equivalently:

$$E[Z(V) | Z(V)^\star] = Z(V)^\star$$

i.e. the regression of $Z(V)$ on $Z(V)^\star$ is linear and coincides with the first bisector. This desirable property ensures that the results obtained when taking a decision on the basis of $Z(V)^\star$ are, on average, as expected (e.g. values selected for being rich, being as rich as expected). This is not the case of the estimator $Z(V)^\star$ taken as the sampled value of $Z(V)$. Having a variance higher than that of $Z(V)$ due to the support effect, it is conditionally biased, with $E[Z(V) | Z(V)^\star] < Z(V)^\star$ when $Z(V)^\star$ is high. This causes overestimation of panels considered as rich, e.g. above a cut-off z:

$$E[Z(V) | Z(V)^\star > z] < E[Z(V)^\star | Z(V)^\star > z]$$

Krige's lognormal estimator aims at correcting this conditional bias. Ideally, if we denote $data = \{Z(x_1), Z(x_2), ...\}$ the values of samples (possibly located outside the panel), the best estimator (unbiased with minimum error variance) is the Conditional Expectation (CE): $E[Z(V)|data]$. If the multivariate distribution $(Z(V), data)$ were normal (any linear combination of these being normally distributed), this regression would be linear, and would coincide with Simple Kriging in a stationary model. However, and particularly with skewed distributions, there is no reason for this regression to be linear. Assuming for instance multilognormality (multivariate distribution of logarithms being normal), the best regression is lognormal, hence based on the geometric mean of the sample values. Although linear kriging was proposed by Matheron as a means to correct in practice the above conditional bias [13], there is no reason why it should do it fully. A practical advantage of linear kriging comes from the reduced hypotheses: only the spatial structure, and possibly the value of the mean parameter are required. As a matter of fact, an expression such as $E[Z(V)|data]$ is simply not defined within the 2nd order stationary RF model. On the other hand, Matheron has shown that the very shape of CE within a model may be used as a heuristic candidate for the estimation, but that it rapidly gets beyond a realistic meaning with a few conditioning datapoints ([18, 20]): CE is guaranteed to be the best estimator only in theory. In addition (Z being additive), if V is partitioned into congruent samples, the best estimator of $Z(V)$ when increasing the number of samples, converges towards the true value, which is the (linear) arithmetic mean. While strict conditional unbiasness seems an unreachable ideal, one can try to approach it. One way is to look for a *linear* regression of $Z(V)$ on $Z(V)^\star$ close to the first bisector (slope close to 1). In Simple Kriging, the estimation error is uncorrelated with data and therefore:

$$\text{Var}\, Z(V) = \text{Var}\, Z(V)^\star + \text{Var}[Z(V) - Z(V)^\star]$$

So Simple Kriging reduces the variability, and the theoretical slope of the linear regression $Cov(Z(V), Z(V)^\star)/\text{Var}\, Z(V)^\star$ equals 1. The mean compensates for the lack of information in the neighbouring data (with a pure nugget effect, Simple Kriging reduces to the mean m: no local estimation is really possible). In Ordinary Kriging, assuming 2nd order stationarity with unknown mean, things are different. The theoretical slope can be notably less than 1, in particular when the spatial structure is poor and the data sparse. In many cases however, this slope takes low values when the neighbourhood is too small (Ordinary Kriging then has too high a variability, with an important conditional bias), but may have a value close to 1 if the neighbourhood is large enough. Hence this slope can be used as a criterion for choosing the neighbourhood.

Back to variances

A question considered by Krige was whether the log variance of samples within a panel could be considered a constant for panels in identical configurations within a given mine. The same question formulated on the dispersion variance of samples within a given panel, in linear geostatistics, would generally receive a negative answer for skewed distributions. Rich panels would generally contain richer and more variant samples: this corresponds to a proportional effect (between the variance and the mean, or a function of the mean), or in statistical language, a heteroscedasticity. This does not imply non-stationarity as is often thought: exponentiating a 1D stationary autoregressive Gaussian process yields to a stationary lognormal process presenting such a proportional effect. When writing $D^2(O|V) = \bar{\gamma}(V, V)$ for instance, linear geostatistics considers an expectation, or average over all possible panels with same support V, then taking the average of different values.

Similarly the kriging variance (the variance of the error which is minimized by kriging) has the meaning of an average of the estimation variance when translating the geometrical configuration (respective locations of data and target to be estimated). It is not conditional on values of the variable observed locally and thus can appear as locally unrealistic: for instance, it is unrealistically small when the estimate is high, and conversely. To have a variance more realistic locally, a correction can be applied, based on the modelling of the proportional effect. On the other hand, non-linear geostatistics can give theoretical access to conditional variances.

Back to lognormal

Lognormality was central in the early development of geostatistics in the 50's. Permanence of lognormality was in particular advocated when working on different supports or within nested areas. In theory however, the average of independent variables with the same lognormal distribution, for instance, is not lognormal. Then, assuming that the variable is lognormal on a given support, it would not remain true for multiple or dividing supports. In other words, assuming that an additive variable is lognormal on some support, there is no chance that this very support coincides with any support available in practice. Since lognormality is not stable, Matheron [12] explored the family of stable distributions that would remain stable with a change of support. However this alternative has not led to practicable results. As one says, simple models are false, but complex models are useless. . . Lognormality, and its permanence, can be expected to hold, although deviations from supposed lognormality can have serious consequences (a third translation parameter in addition to mean and variance, is often considered to enlarge the model). The question of lognormality arises again from time to time. In a note with an eloquent title, Matheron [16] has proposed a variety of enlarged lognormal estimators, obtained notably by exponentiating a linear combination of logarithms with sum of weights equal to 1 such as OK, and assuming or not the permanence of lognormality when changing support. The sensitive estimator referred to as "lognormal kriging" can take many different forms.

3 Multivariate geostatistics

Linear geostatistics can naturally be extended to several variables (such as different concentrations, or a variable and its gradient, etc.) Cokriging allows estimating linearly one variable from a set of variables, known at the same datapoints (isotopic case) or at different datapoints (heterotopic case). Cross structural tools (covariance, variogram) will measure the correlated parts between two variables, or between increments, as a function of the vector separating pairs of points. In the frequently used linear models of coregionalisation, the set of simple and cross covariances (or variograms) is modelled from a linear combination of structural basic components which usually represent different scales. Efficient filtering techniques (kriging analysis) make it possible to extract such a scale component by cokriging from the different variables. One advantage of these filtering techniques, and more generally of kriging and cokriging, is their capability to handle points in irregular geometrical configurations, and not only on regular grids.

Very often, the statistical features of the regionalized variables have a limited scope, since they change with the support on which the variables are considered (this is the case of lognormality, as we have seen before). In particular, the correlation coefficient between two given variables may be small on a small support, and much larger on a larger support – or the contrary. On the opposite, the correlation may be an intrinsic measure of the linear dependency between two variables. This is the case in the model of intrinsic correlation, where simple and cross variograms are proportional to each other, and where the correlation coefficient within a domain does not depend on the support, nor on the domain [11]. Then, in the isotopic case, cokriging is equivalent to kriging.

Similarly, the linear regression $aY(x) + b$ of one variable $Z(x)$ on another variable $Y(x)$ at the same location generally presents little interest, since it changes with the support. However, in the model with residual $Z(x) = aY(x) + b + R(x)$ where $R(x)$ is spatially not correlated with $Y(x)$ – a hierarchical model where $Z(x)$ is subordinated to the master variable $Y(x)$, the linear regression is independent from the support and then has an intrinsic signification. This is the case when the cross-structure between $Y(x)$ and $Z(x)$ is proportional to the structure of $Y(x)$. The decomposition of $Z(x)$ into $Y(x)$ and $R(x)$ has interesting properties in cokriging. When $Y(x)$ is available at every datapoint of $Z(x)$, and in particular when $Y(x)$ and $Z(x)$are available at the same datapoints, the cokrigings of Y and Z can be obtained by kriging Y and R separately.

An instructive example of coregionalisation (that makes the link with non-linear geostatistics) is the case of a concentration $Z(x)$ where the positive values correspond to a geometrical set A with unknown delimitation within the domain under study (ex: metal grade in a vein-type deposit). This set A can be represented by its indicator $1_{x \in A}$, equal to 1 if the point x belongs to A, and to 0 otherwise. If we consider this variable separately, its ideal estimator at point x from datapoints consists in its Conditional Expectation, that is, its probability to belong to A conditionally to the datavalues. The mathematical shape of such a conditional probability is generally unknown. In the context of simulations, it is suggested to introduce such conditional probabilities under given configurations from training images, using multi-point statistics [22]. But most often a pseudo-probability is obtained by kriging the indicator, with a post-processing to eliminate values outside the interval [0, 1]. In addition a relationship may exist between variables, say grade and geometry: for instance the low grades may be preferentially located near the frontiers of the veins. Then the estimation of the indicator can be improved by considering values of Z in addition to values of the indicator. A straightforward exception corresponds to the positive values of $Z(x)$ being distributed within A, independently from its geometry. Then the estimation of the indicator, which has the meaning of a probability, can be performed separately, and complemented by an estimation of $Z(x)$ assuming it is positive (and using only the positive data values). Another exception is cokriging the indicator and Z within a model with residual $Z(x) = m1_{x \in A} + R(x)$ where we assume the absence of spatial correlation between residual and indicator. This corresponds to a reduced hypothesis of independence: internal independence between the variable $Z(x)$ and the set A [11] or absence of border effect, in the sense that the expected value of $Z(x)$ at a point x of A does not depend on whether a neighbouring point $x + h$ belongs to A or not. When data consist in Z values, the cokriging of Z can be obtained by kriging separately the indicator and the residual, and the ratio between the estimate of $Z(x)$ and the estimate of the indicator has the meaning of an estimation of $Z(x)$ in case it would be positive. We will come back to such considerations when dealing with non-linear geostatistics.

4 Non stationary geostatistics

4.1 Kriging with external drift

Consider the model with residual $Z(x) = aY(x) + b + R(x)$. In some circumstances the master variable Y is known everywhere and is used as an auxiliary variable for the estimation of a sparsely sampled target variable Z. Assuming that the coefficients a and b of the linear regression of $Z(x)$ on $Y(x)$ at the same point x are known, Simple Kriging of the residual gives the estimation of Z. If b is unknown, then an Ordinary Kriging of the residual is what is needed. When working with a single realization, the auxiliary variable can be considered deterministic, say $f(x)$ to be consistent with literature, and the model can be written $Z(x) = af(x) + b + R(x)$. This is helpful when a and b are not known (either globally when using a unique neighbourhood, or locally within a moving neighbourhood). A trick when estimating $Z(x)$ as a linear combination $\sum_{\alpha} \lambda_{\alpha} Z(x_{\alpha})$ of data (borrowed from Universal Kriging described further on), consists in imposing the following conditions:

$$\sum_\alpha \lambda_\alpha f(x_\alpha) = f(x)$$

$$\sum_\alpha \lambda_\alpha = 1$$

This ensures that the mean error is 0 whatever the coefficients a and b. In the model, the drift $E(Z(x)) = af(x) + b$ of $Z(x)$ depends upon $f(x)$, hence the name of kriging with external drift (KED). This can be naturally extended to drifts $\sum_l a_l f_l(x)$ where $f_l(x)$ are known shape functions, for instance $(1,$ $f(\mathrm{x}), f^2(\mathrm{x}))$: this was actually the origin of KED when mapping a geological depth with $f(\mathrm{x})$ representing the seismic travel time [1]. This KED method is very popular, probably because of its flexibility: kriging implicitly estimates the coefficients a_l that best fit the drift onto the shape functions. The difficult point may be the estimation of the residual structure to be used in kriging. Indeed, if the coefficients a and b are unknown, the value of the residual is unknown even at points where Y and Z are known. Coefficients and residuals can be estimated, but the structure of the estimated residual is not the true one. The residual structure is often determined indirectly e.g. through cross-validation.

4.2 Universal kriging

By considering increments, the model of intrinsic RF seen previously is far more general than the stationary one, which is constrained to vary around its mean. This freedom is visible when looking at 1D random walks such as a Brownian motion (which is intrinsic but not stationary). However the intrinsic model cannot force the function to show systematic variations (e.g. the sea floor depth increasing from coast line, or the dome-shape top of an oil reservoir). Then it can be helpful to consider a drift of the form $\sum_l a_l f_l(x)$, where the f_l's are known. For instance using the monomials 1, x^2, y^2, xy (in 2D space) corresponds to a quadratic drift centred on the origin. Most often, there is no point to be distinguished as the origin, and the family of functions f_l is taken invariant by translation: the drift can be written $\sum_l a_l f_l(x - x_0)$, whatever the choice of the origin x_0, e.g. 1, x, y, x^2, y^2, xy for a quadratic drift. More generally, monomials of coordinates with degree $\leq k$ allow representing a polynomial drift of degree k (sets of sine and cosine functions with a fixed period, and more generally sets of exponential polynomials, also have this property of invariance under translation). When making the assumption of a polynomial drift of order k, the coefficients a_l's of the drift should be considered unknown *a priori*. Imposing universality conditions $\sum_\alpha \lambda_\alpha f_l(x_\alpha) = f_l(x)$

to the linear estimation $\sum_{\alpha} \lambda_\alpha Z(x_\alpha)$ of $Z(x)$ will ensure the absence of bias, whatever the values of the coefficients: this gives the Universal Kriging.

Yet the remaining problem lies in the estimation of the residual structure when it is not small compared to the variations caused by the drift, since the drift coefficients, and therefore the drift, are unknown. Fitting the drift on the data tends to include some part of the true residual variability, and therefore leads to an underestimation of the residual structure. The variogram of the estimated residuals may in particular display a finite or a small range, whereas this is not the case for the true residual.

4.3 Intrinsic Random Functions of order k

A solution to this problem was offered by the theory of Intrinsic Random Functions of order k ("IRF-k ", [14]). By considering increments, the usual intrinsic RF model studied initially was more general than the stationary one: the constant mean, if any, disappeared, and the family of admissible structures was enlarged from stationary covariances to variograms. Similarly the IRF-k model is defined by considering increments of increments, ..., or generalized increments, that filter out any polynomial drift of order k, and the tools are enlarged to generalized covariances ($-\gamma(h)$ being a generalized covariance of order 0). An IRF-k constitutes an equivalence class of RF that have a given structure and differ by a polynomial of order k. Only those linear combinations that are generalized increments are defined in the model. Their variances can be developed using the generalized covariance and allow linear estimation by kriging, unbiased with minimum variance. In practice the difficult point is determining the structure (the generalized covariance), which is not direct as for the usual variogram, except with regularly spaced data. Indirect fitting of generalized covariances that are admissible linear combinations of odd monomials of $|h|$ is often used. Other generalized covariances are also admissible, in particular some for which kriging coincides with splines.

The interest of the model of Intrinsic Random Functions of order k is not limited to its theoretical clarity, in breaking the vicious circle between drift and residual. As integrating a stationary RF yields an IRF-0, similarly integrating an IRF-k yields an IRF-k+1, and IRF-k's appear as the perfect tool in some Partial Derivative Equations problems where the set of solutions exactly corresponds to the set of RF defining an IRF-k (for instance the set of Random Functions whose laplacian is a given stationary RF exactly constitutes a unique IRF-1).

5 Non-linear geostatistics

5.1 Estimation on point support

While requiring few hypotheses, linear geostatistics has a limited range correlatively. The need for non-linear geostatistics can be felt in different circumstances. Firstly the best estimator of a regionalized variable may be non linear (although any estimation of the regularized variable should be asymptotically linear, as seen above). Secondly, even when using a linear estimation such as kriging, the kriging variance is not conditional on the neighbouring data: with skewly distributed variables, local measures of uncertainties such as estimation variances should depend on the neighbouring data. Finally, and this was the very origin of the development of the non-linear geostatistics in the 70's (after the early lognormal framework of geostatistics), in some cases (selective mining, pollution...), the practical problem is not to estimate the variable itself, but a non-linear function of it: for example its exceeding a given threshold z, which is represented by the indicator $1_{Z(x) \geq z}$. Note that thresholding an estimated value of $Z(x)$ would be an incorrect answer to this problem, as this may very well exceed the threshold while the true unknown value does not.

When its assumptions are acceptable, a very convenient model is the (multi-)gaussian Random Functions one (denoted Y for reasons explained further on), where any linear combination on any finite set of points is normally distributed (which is far more demanding than to require the marginal or even the bivariate distributions to be normal). Firstly, the best estimator of $Y(x)$ from data $Y(x_\alpha)$ coincides with the linear regression, i.e. Simple Kriging (assuming stationarity). Secondly the conditional variance coincides with the unconditional one, i.e. the kriging variance. And finally, the conditional distribution is normal, which makes the best estimation of any function of Y possible through its Conditional Expectation. In particular the CE of the indicator $1_{Y(x) \geq y}$ is equal to the conditional probability of exceeding y, given the data.

Assuming stationarity and a large domain, the marginal distribution corresponds to the histogram of the datavalues. In many cases it is quite different from the famous bell shape curve, and therefore the multi-gaussian model is most frequently used together with a Gaussian anamorphosis: the target variable $Z(x)$ is written as $Z(x) = \phi(Y(x))$, where ϕ is a non decreasing function and $Y(x)$ a stationary standard Gaussian RF, hence the notation. For instance, a lognormal distribution corresponds to an exponential anamorphosis. When the inversion from Z to Y is straightforward (ϕ strictly increasing or nearly so), data and estimation on Z can be expressed in terms of Y. This gives access to the conditional distribution of $Z(x)$ and therefore to the CE of $Z(x)$ or any function, e.g. $1_{Z(x) \geq z}$. The requirements for this theoretically best estimator are severe, as not only the marginal distribution but also the bivariate $(Z(x), Z(x+h))$ or the n-variate ones must be normal (after transformation). Another method for estimating an indicator above a given threshold consists in kriging directly this indicator. The method has been made popular by Journel [5] for its simplicity and the few hypotheses apparently required. Extension to several thresholds allows estimating variables obtained from linear combinations of the different indicators, such as the discretized variable under study or any function of it. On the other hand, kriging an indicator ignores the extra information contained in the Z data or in the other indicators. As a matter of fact, indicators at different thresholds are not independent. The Disjunctive Kriging (DK), proposed early by Matheron [15], is the estimator based on the cokriging of all indicators. This requires the simple and cross structures of indicators, or equivalently the bivariate distributions $[Z(x), Z(x + h)]$. DK is simplified in isofactorial models, where it is obtained by Simple Kriging of factors common to all the bivariate distributions of the model. Through a number of reports, Matheron has undertaken a considerable work developing such models, including the change of support seen in next section. An extensive bibliography can be found in Chilès and Delfiner [1]. Most models correspond to diffusion processes, and refer to a specified statistical distribution (e.g. Gaussian, gamma, binomial negative, with Hermite, Laguerre, and Meixner polynomials as factors). Discrete diffusion models have been developed specifically for diamond mining (see Krige and Kleingeld in this volume). Another model is a hierarchical mosaic model, where the indicators correspond to sets that are nested without border effects and are factorized by indicator residuals [21].

One very particular model consists in a mosaic with independent valuation: each tile of a stationary partition of the space is given an i.i.d. value. The variographic structure of indicators or other functions is entirely based on the probability for two points to belong to the same tile or not. In particular there is no destructuring phenomenon when the cut-off increases or decreases, contrary to what is currently observed on real data. In this particular model, cokriging of indicators equals to their kriging, and DK of any function equals its kriging, with the same kriging weights.

5.2 The change of support

Very often, the question of whether a threshold value is exceeded or not, concerns areas much larger than the quasi point sample support: this may be the whole domain itself, subdomains with specific geometries, or blocks with same support v partitioning the domain. A general solution can be obtained through conditional simulations mentioned further on. However simulations are very demanding in terms of hypotheses.

The case of blocks with same support v partitioning the domain has been studied in details by Matheron. Due to the support effect, the global distribution of values on such blocks is different from the distribution with a point support. They both have the same mean, and their variances are related through the additivity relationship seen above. In addition other relations link the two distributions, which can be derived from the key Cartier's relationship [19, 7]:

$$E\left[Z(x) \mid Z(v)\right] = Z(v)$$

This stipulates that, conditionally to a block value, the expected value of a random point uniformly distributed within this block, is equal to the block value. This relation, or equivalent ones, makes it possible to build *change of support models* under various hypotheses (Gaussian, mosaic, etc.) which aim at predicting the distribution of values on blocks when the point support distribution is known. They can be used in particular to predict global quantities like the proportion of blocks above a cut-off.

The local estimation, i.e. predicting whether a given block $Z(v)$ of the domain, or the N blocks $Z(v_i)$ within a panel, exceeds a cut-off value or not, is a difficult problem. In addition to the simulation approach, different methods have been proposed. A direct approach consists in assuming a given type of distribution (e.g. normal or lognormal) for the block value, only depending upon an estimated value of the block with an appropriate estimation variance. This is in particular justified within a Gaussian model, as the distribution of a block, conditionally to the data, only depends on its simple kriging estimate and variance, and provides the conditional probability of exceeding a cut-off, or more generally the CE estimator of any function of the block. In the usual case where an anamorphosis must be used, things are more complex, but CE can be developed when coupled with a Gaussian change of support model. Similarly, disjunctive kriging can be used under different models equipped with a change of support. EC and DK provide a local estimation consistent with their corresponding global change of support.

The previous non-linear methods are conveniently developed in a stationary framework. In an area with sparse data, the estimation is then strongly attracted by the global characteristics of the domain, just like Simple Kriging is attracted by the mean m. This is in accordance with the stationarity hypothesis, but may be considered too strong. For estimations on point support, the stationary hypothesis made in kriging or cokriging indicators can be reduced if an Ordinary Kriging is used instead of a Simple Kriging for indicators or factors. However, things are more complex when a change of support is used. Alternative methods, more flexible in this respect, have been proposed. One of them consists in firstly estimating the value of the indicator of each block that contains a sample data, and secondly performing the Ordinary Kriging or Cokriging of such a *service* variable. Another method considers the distribution of blocks within a given panel as being uniformly conditioned by the sole estimate of this panel, then using a change of support model up to the panel support. Still another approach is to derive the block distribution within a panel from the more easily estimated distribution of points.

5.3 Information effect and geometrical constraints

Geostatistics is more than the development of mathematical - generally probabilistic - models and methods, and their application. It also includes analyzing the practical problems to be solved, and formalizing them in terms of concepts, which Matheron considered very important. In the mining domain that he studied in details, Matheron proposed an advanced formalization of reserves, in particular for selective mining. As an illustration, consider the case of an orebody divided into equal blocks, each one being possibly selected as ore independently (free selection). It is necessary to account for the support effect in order to predict the number of blocks above a cutoff, from the distribution of point sample values: selectivity on block support is less than it would be on points. Moreover, when a block is to be mined, the decision of selecting it as ore or waste is not made on its true value which is unknown. It is made on an estimate based on the data that are available at this stage (estimation by kriging for instance). The result of such a selection is necessarily worse than it would have been in the ideal case of a selection on the true values: this is the information effect [19]. In practice, additional data will be available at the pre-exploitation stage. When predicting the reserves months ahead, one has to estimate the chances for the future estimated block values to exceed the cutoff, and the correlative true quantity of metal that they contain.

In free selection, blocks are assumed to be possibly selected independently from one another and from their location. In the opposite case, another effect – geometrical constraints - must be considered. An example is given by vertical constraints in a stratiform exploitation, when at a given 2D location, all blocks between the lower and the upper selected blocks must be mined. Another common example is that of the open pit, where mining a given block implies that of all the blocks ahead in the cone defined by the stability slope have been previously mined. Matheron [17] has developed methods for the parameterization of reserves, which allow optimizing such contours of exploitation. They provide a family of technically optimal projects, among which it is possible to choose the best ones under present or hypothetical economical conditions. No probabilistic framework is used here.

6 Simulations

Non-linear methods are more powerful than linear ones. They are however more demanding in terms of hypotheses, reducing the robustness of models with respect to reality. Hence the economy principle of geostatistics, i.e. not using more than is necessary to solve a practical problem [20]. But in many cases, the problem is too complex to be solved directly, even by non-linear methods. Then stronger hypotheses are required, and a solution is to call for simulations, that is, realizations of a complete RF model, considered as plausible versions of the reality - in particular conditional simulations, which honour the values at data points (see the article by Chilès and Lantuéjoul in this volume). The increase in computational power and the possibilities of visualisation have boosted the use of simulations. The proper use of simulations in terms of the practical problem to be addressed, and the choice of the model, i.e. the hypotheses made on reality, are important points to be considered.

7 Perspectives

The need to model complex environments realistically calls for the development of new models, possibly process-based (e.g. diamonds deposited within traps of the sea floor), or defined in a non Euclidean system (stream lines) or in a higher dimensional space. In particular, spatio-temporal models must be considered for variables such as fish or contaminants which vary in space and time. Enlarging to new types of non-stationarity or to new multivariate models is also a promising challenge. A Bayesian approach for geostatistics is convenient to account for uncertainties in the model. It also constitutes a powerful framework for inverse problems in hydrogeology for instance. In this case and in many others, geostatistics offers a consistent approach to analyse several variables jointly and to deal with measurements on different supports. It also provides a flexible approach to combine, merge or filter different variables, measured at the same locations or not, whether on a regular sampling design (e.g. grids, images) or not. Finally geostatistics can also be used for the coupling between observation data and physical models.

References

1. Chilès, J.-P., and Delfiner, P. 1999. Geostatistics: Modeling spatial uncertainty, Wiley, New York, 695 p.
2. Cressie, N., 1990, The origins of kriging, *Math. Geol.*, Vol. 22, 3, p. 239-252.
3. De Wijs, H. J. 1951-53. Statistical ore distribution, Geologie en Mijnbouw, Part I nov. 1951, Part II Jan. 1953.
4. Duval, R., Lévy, R., Matheron, G., 1955. Travaux de D. G. Krige sur l'évaluation des gisements dans les mines d'or sud-africaines. *Annales des Mines*, Paris, vol. 12, p. 3-49.
5. Journel, A. G. 1982. The indicator approach to estimation of spatial distributions. In: Johnson, T. B., and Barnes, R. J., eds, Proceedings of the 17th APCOM International Symposium, Society of Mining Engineers of the AIME, New York, p. 793-806.
6. Krige, D. G., 1951, A statistical approach to some basic mine valuation problems on the Witwatersrand. *Journal of the Chemical, Metallurgical and Mining Society of South Africa*, dec. 1951, p. 119-139.
7. Lantuéjoul, C., 1990. Cours de sélectivité. Lecture Notes, Ecole des Mines de Paris, Centre de Géostatistique, Fontainebleau.
8. Matheron, G. 1955. Applications des méthodes statistiques à l'évaluation des gisements, *in Annales des Mines*, 12, Paris, p. 50-75.
9. Matheron, G. 1955. Etude géostatistique du gisement de plomb de Bou-Kiama, Technical report, Ecole des Mines de Paris, Centre de Géostatistique, Fontainebleau.
10. Matheron, G. 1962-63. Traité de géostatistique appliquée. Mémoires du Bureau de Recherches Géologiques et Minières. Tome I, no 14, Editions Technip, Paris, Tome II: le krigeage, no 24, Editions BRGM, Paris.
11. Matheron, G. 1965. Les variables régionalisées et leur estimation. Masson, Paris. 306 p.

12. Matheron G., 1968, Lois stables et loi lognormale. Technical report, Ecole des Mines de Paris, Centre de Géostatistique, Fontainebleau, 17 p.

13. Matheron, G. 1971. The theory of regionalized variables and its applications, Les Cahiers du Centre de Morphologie Mathématique, no. 5, Ecole des Mines de Paris, 211 p.

14. Matheron G., 1973, The intrinsic random functions and their applications, Adv. Appl. Prob. 5, 439-468.

15. Matheron G., 1973, Le krigeage disjonctif. Technical report, Ecole des Mines de Paris, Centre de Géostatistique, Fontainebleau, 40 p.

16. Matheron G., 1974. Effet proportionnel et lognormalité ou le retour du serpent de mer. Technical report, Ecole des Mines de Paris, Centre de Géostatistique, Fontainebleau, 43 p.

17. Matheron G., 1975, Le paramétrage technique des réserves. Technical report, Ecole des Mines de Paris, Centre de Géostatistique, Fontainebleau, 26 p.

18. Matheron, G. 1978. Estimer et choisir. Les Cahiers du Centre de Morphologie Mathématique 7. Ecole des Mines de Paris, Fontainebleau. 175 p.

19. Matheron G., 1984, The selectivity of the distributions and "the second principle of geostatistics". In Verly, G. et al eds. Geostatistics for natural resources characterization, Reidel, Dordrecht, Part 1, p. 421-433.

20. Matheron G., 1989, Estimating and choosing, Springer-Verlag, Berlin, 141 p.

21. Rivoirard, J., 1989, Models with orthogonal indicator residuals, in M. Armstrong (ed.), Geostatistics, Kluwer, Dordrecht, Vol. 1, p. 91-107.

22. Strebelle, S., 2002. Conditional simulations of complex geological structures using multi-point statistics: *Math. Geology*, v.34, no. 1, p. 1-21.

Prediction by conditional simulation: models and algorithms

Jean-Paul Chilès and Christian Lantuéjoul

Centre de Géostatistique, Ecole des Mines de Paris

1 Introduction

Prediction here refers to the behavior of a regionalized variable: average ozone concentration in April 2004 in Paris, maximum lead concentration in an industrial site, recoverable reserves of an orebody, breakthrough time from a source of pollution to a target, etc. Dedicating a whole chapter of a book in honor to Georges Matheron to prediction by conditional simulation is somewhat paradoxical. Indeed performing simulations requires strong assumptions, whereas Matheron did his utmost to weaken the prerequisites for the prediction methods he developed. Accordingly, he never used them with the aim of predicting and they represented a marginal part of his activity. The turning bands method, for example, is presented very briefly in a technical report on the Radon transform to illustrate the one-to-one mapping between d-dimensional isotropic covariances and unidimensional covariances[1] [44]. As for the technique of conditioning by kriging, it is nowhere to be found in Matheron's entire published works, as he merely regarded it as an immediate consequence of the orthogonality of the kriging estimator and the kriging error.

Although he himself did not use conditional simulations for predicting, he nonetheless supported the researchers of the Centre de Géostatistique who did. As an engineer, Matheron indeed admitted that in real-world situations prediction algorithms can be used heuristically, provided that the conclusions drawn are sifted out with a critical eye. Beyond the presentation of models, methods, and algorithms, this chapter is also an opportunity to examine what conditional simulations really represent.

---

[1] Matheron submitted a paper on the turning bands method to the Advances in Applied Probability, which he later merged with another paper [46]. This project paper, available as [45], mainly focuses on the turning bands operator. It is worth noticing that Matérn [33] already had a similar attitude, presenting the principle of the turning bands in a few lines as an illustration of the relation between covariances in $I\!R^d$ and $I\!R$.

2 Predicting by a Monte Carlo method

2.1 General principle

Reality is considered as a regionalized variable (RV) $z = \big(z(x), x \in V_0\big)$ defined at each point of a domain V_0 of $I\!\!R^d$. This RV may be a permeability, a grade, a geological facies, etc. This is already an idealized representation of the reality but we will not question this simplification. Suppose that this RV is known at some points $x_\alpha, \alpha = 1, ..., N$. We want to know the value of some functional $\mathcal{F}(z, q)$ of that RV as a function of parameters represented by the vector q. For example, $z(x)$ is the permeability at x, q represents a pollutant source, and $\mathcal{F}$ is the breakthrough time to access a protected zone V_1 (lake, catchment area). This breakthrough time largely depends on the heterogeneities of the medium, particularly the continuity of the high-permeability zones – which can act as conduits, or the low-permeability zones – which are flow barriers. The problem is too complex to be addressed analytically. One approach is thus to generate a pseudo-reality that possesses the same type of variability as the real field, and compute the breakthrough time corresponding to that pseudo-reality. That breakthrough time is then considered as a possible value for the true breakthrough time. We call this pseudo-reality an image.

If the pollutant source is of limited extent, its impact is very different depending on whether it is located in a low-permeability pocket – acting as a containment zone, or in a high-permeability zone. It is thus important to take all the information on local permeabilities into account. This can be achieved by using an algorithm producing images that honor the data.

A single image gives a single answer and does not tell us how far it departs from reality. This calls for a Monte Carlo method [64]: several images are built, which give different breakthrough times; the distribution of their values can then be used to build confidence limits for the true value – this is at least what we hope.

2.2 Conditional simulations

Since it is difficult to conceive deterministic algorithms generating realistic and varied images, stochastic algorithms are used instead. The conditional simulations produced are defined within a probabilistic context. The RV z is considered as a realization of some random function (RF) $Z = \big(Z(x), x \in V_0\big)$. In other words, denoting by $\big\{Z_\omega, \omega \in \Omega\big\}$ the set of realizations of Z, the RV z is the realization corresponding to a particular state ω_0 of Ω: $z = Z_{\omega_0}$. A non-conditional simulation is any realization Z_ω. For some states of Ω, the realization honors the data, namely satisfies $Z_\omega(x_\alpha) = z(x_\alpha)$ for each $\alpha = 1, ..., N$. A conditional simulation is any of these realizations.

If a large number of independent conditional simulations are built, and if each of them has its value $\mathcal{F}(Z_\omega, q)$ computed, then the distribution of the obtained values tends to the conditional distribution of $\mathcal{F}(Z, q)$. In particular, their average tends to the conditional expectation. The prediction by Monte Carlo is thus a very general and powerful method.

2.3 Kriging or simulating?

Predicting by a Monte Carlo method is however a heavy approach (a large number of simulations must be generated). Moreover, it is optimal only when the spatial distribution of Z is known, which is not the case for usual applications. Three questions arise:

- Is it possible to avoid the Monte Carlo method for estimating the distribution of $\mathcal{F}(Z, q)$, or at least its mean and variance, even at the price of an approximation?
- Are there model classes for which all models of a given class lead to the same results, at least in a first approximation? If so, how to choose the most appropriate class and its parameters?
- And if not, how to choose the most appropriate model and its parameters?

The answer to the first two questions is generally negative, and we will see that the answer to the third one mainly depends on the ”thematician”'s or geostatistician's skill. This is why Matheron never showed much interest in the latter question, which cannot get a definite statistical answer. He expended much effort to obtain positive answers to the first two questions, for problem classes as broad as possible.

The first problem concerns the prediction of the value taken by z at an unobserved point x_0, or more generally by a linear functional of the RV z, for example its average $z(v)$ in a block v. If only linear predictors are considered, a quadratic optimum minimizing the estimation variance is given by **simple kriging**. This optimum is an approximation to the conditional expectation, and coincides with it if Z is a Gaussian RF (GRF). Only the mean and the covariance of Z are required to derive this optimum. These parameters are reasonably accessible from the data provided some homogeneity assumption (stationarity) that can be weakened by various means. Consequently, kriging does not depend on the other characteristics of the RF and is a robust answer to the problem at hand.

The prediction of a nonlinear function of $z(x_0)$ or $z(v)$, such as the indicator above a threshold, can similarly be obtained by **disjunctive kriging** (DK), provided that predictors of the form $\sum_\alpha f_\alpha\big(z(x_\alpha)\big)$ are considered. Except in special, more favourable cases, DK provides an approximation to the conditional distribution. Its implementation requires only bivariate distributions. When Matheron defined DK, the research about possible bivariate distribution models was in its infancy. He therefore devoted much effort to the development of bivariate models covering the various needs of applications (continuous or discrete variables, diffusive, mosaic, or intermediate models, support effect, etc.). As Matheron showed it, choosing a type of bivariate distribution and its parameters remains reasonably possible. Since DK does not depend on multivariate distributions beyond bivariate ones, it remains a robust solution to the problem at hand.

When problems increase in complexity, it is no longer possible to call for similar simplifications (note in particular that a breakthrough time measured on a kriged model has nothing to do with the actual breakthrough time, due to the smoothing effect of kriging). We then resort to the Monte Carlo approach. In either case (direct calculation of the conditional expectation or Monte Carlo), the result depends on the whole spatial distribution of the RF, a distribution whose choice remains largely arbitrary (beyond bivariate distributions). Conditional simulations thus possess a more or less heuristic character, which will have to be kept in mind when analyzing the results of a prediction by Monte Carlo.

2.4 Applications of simulations

Simulations are used first of all for methodological purposes. Matheron resorted to non-conditional simulations to quantify how conditional expectation is approximated by DK, or to assess the validity of a change of support model [54]. Simulations can also be used to explore the range of fluctuations that can be expected from a model – when no direct calculation is possible, or inversely to determine the type of information about the model that can be retrieved from a single realization, thus leading to efficiency tests of inference procedures or *a posteriori* validation tests of a model.

The other application of simulations, more precisely of conditional simulations, is prediction by numerical means. The objective is usually to predict the result of a complex process applied to the simulated field: flow and transport simulation (hydrogeology, petroleum reservoir), exploitation scenario (mining), etc. It can also consist of assigning a precision to the predicate in terms of variance or confidence limits. More simply the objective can be to visualize what the reality looks like.

2.5 Necessity of a model

Three arguments can be put forward to justify resorting to a stochastic model:

Data representativity

Any trained geostatistician knows that assessing the marginal distribution of a variable is not a trivial task. When data are scattered, the sample histogram may present artefacts which are simply due to the presence of clusters in specific areas. And even though a fine grid of data may be available in some domain V, thus enabling us to know the regional histogram in V, the representativity of that histogram for another domain remains questionable if V

is not much larger than the range of the RV under study: in particular, the tails of the distribution may be under- or over-represented. The situation is of course much more critical for multivariate distributions. Sample distributions, e.g. from a training image, may capture the anecdote of a specific situation rather than the essence of the phenomenon.

Distributional consistency

Multivariate distributions are usually no longer directly accessible when their number of support points becomes large. However, they may be required in the prediction exercise. To overcome this difficulty, the simplest approach is to resort to a stochastic model. Once the model type has been chosen and its few parameters have been fixed, the whole spatial distribution, and hence all its multivariate distributions are specified. If the model depends on a limited number of parameters, we can expect them to be reasonably estimated from the data. For a GRF for example, no other parameters than the mean and the covariance are required. For a Boolean model, rather than the covariance, the object distribution and the Poisson process intensity can be used. The advantage of this approach is that it automatically delivers a consistent set of multivariate distributions.

Predictable quantities

More formally, a stochastic model is completely specified by a probability space $(\Omega, \mathcal{A}, P)$ where Ω is the set of all possible realizations, $\mathcal{A}$ is a family of subsets of Ω (called "events") satisfying all the axioms of a σ-algebra and P is a probability on $\mathcal{A}$. Besides its role in the definition of P, $\mathcal{A}$ determines what quantities are measurable; it other words, it determines what quantities can be predicted by Monte Carlo techniques. Indeed, everything cannot be predicted. For instance, the standard definition of arc-connectivity is not a measurable concept. It must be slightly restricted if we want to predict the probability that two given points belong to the same connected component of a random set [29].

The crucial choice is that of the model type. As it will be shown below, such a choice cannot be easily made using geostatistical tools. In practical applications, the model type is *a priori* chosen. For that choice to be as relevant as possible, the geostatistician must have a large variety of models at his disposal as well as a good knowledge of their properties (including when the model is implicit) and of the algorithms to simulate them.

3 Models

3.1 Design

A general survey of Matheron's list of publications readily shows that he devoted much effort to the design of prototype models but very little to their possible extensions or ramifications. A typical example is the Boolean model that he introduced in [36] as a porous medium prototype while he was investigating the emergence of Darcy's law starting from the Navier-Stokes equation. Such a model can be extended by releasing the independence assumption on the object locations (attraction or repulsion between objects...) and on the objects themselves (correlations between closely located objects). If a numerical marker is assigned to each object, then new models can be designed by replacing the union between objects with another composition rule such as addition (shot noise model [33, 76]), supremum (Boolean random function [25, 67, 77]) or superimposition (valued dead leaves model [24]).

Prototype models are interesting in that they depend on a small number of parameters and possess very good stability properties (algebraic, morphological, stereological, etc.), which make them mathematically tractable. Matheron often resorted to them to test physical assumptions.

A few elaborate models have also been designed by Matheron in the context of specific applications, such as Ambarzumian processes for modeling the evolution of fluvio-deltaic sedimentation as a function of marine subsidence [40], or generalizations of Sichel distribution for modeling the sample distribution of diamonds in alluvial deposits [50, 51].

3.2 Statistical characterization

The most celebrated contribution of Matheron in this domain is undoubtly his introduction of a hitting functional to characterize the statistical properties of a closed random set (see [47] and also the paper by Molchanov [65] in this volume). Closed random sets are put forward rather than open random sets because they include locally finite point, line and flat processes which are very useful for applications. This standpoint is also shared by Kendall [26], even though both random set families can be deduced from each other by complementation. More specific is the characterization of the equivalence classes of random sets that have the same topological opening and the same topological closure. It can be used to describe natural phenomena such as porous media [36].

Matheron did not work much on the statistical characterization of random functions, except on three occasions. Following Doob's ideas [12], he showed that a closed random set can be equivalently characterized by its hitting functional or the spatial distribution of its indicator function when it is separable. He also addressed the problem of predicting the extrema of a random function on a compact subset. A maximum (resp. a minimum) does exist if the random function has upper semicontinuous (resp. lower semicontinuous) realizations, which boils down to saying that its subgraph (resp. supergraph) is a random closed set [38]. The interest that he later developed for a theory of lipchitzian random functions appears in [63].

Matheron also studied the statistical characterization of certain random structures, including random tessellations [35], random measures and capacities [37], random filters and topologies [38]. Structures like random populations of objects were surprisingly not specifically investigated, but they can be seen as closed random sets on the space $\mathcal{K}(\mathbb{R}^d)$ of the compact subsets of $\mathbb{R}^d$ [72]. The characterization obtained differs from that of Carter and Prenter [8].

3.3 Internal consistency

Only in rare circumstances is a model specified by the functional (spatial distribution, hitting functional) that characterizes it. For instance a two-dimensional Boolean model of random disks is preferably specified by its Poisson intensity as well as its radius distribution, rather than by its hitting functional. A drawback of this approach is that the different ingredients used in the specification of the model may not be compatible. This problem of **internal consistency** was investigated by Matheron in several papers and reports [52, 56, 59]. A few examples encountered in geostatistical studies are given below:

Covariance of a random set

Is any function of positive type allowed as a covariance model for a stationary random set? It is well known (e.g. [36]) that the average directional derivative at the origin of a random set covariance is proportional to the specific boundary content of the random set[2]. This automatically excludes the Gaussian covariance $h \longrightarrow \exp\{-|h|^2\}$ as a set covariance because all its directional derivatives vanish at the origin. In contrast to this, the exponential covariance is authorized whatever the workspace dimension (see the first two examples of Fig. 3). The spherical covariance is authorized in one dimension as a convex function on $[0, \infty[$ [58]. Whether or not it is authorized in more than one dimension is still questionable.

Covariances and point distributions

More generally, which covariance models are compatible with a given point distribution? At present, only very partial answers are available. For instance, a lognormal random function cannot admit a spherical covariance if its coefficient of variation is too large[3] [59].

---

[2] The specific boundary content of a stationary random set (specific perimeter in $2D$, specific surface area in $3D$) is the mean $(d-1)$-volume of its boundaries per unit volume.

[3] It should be pointed out however that this example is not critical because such a lognormal random function is usually specified by the parameters of its lognormal distribution and the covariance of its underlying GRF (for which every covariance model is allowed).

Distributions on different supports

The change of support problem can be seen as another internal consistency problem. In many fields of applications (mining, epidemiology...), it frequently happens that samples of different sizes (or supports) are available, or that the support of the quantity to be predicted is different from that of the samples. To fix ideas, let Z be a stationary random function defined on $I\!R^d$. Let also $Z(v)$ and $Z(V)$ denote the average of Z on supports v and V:

$$Z(v) = \frac{1}{|v|} \int_v Z(x)dx \qquad Z(V) = \frac{1}{|V|} \int_V Z(x)dx$$

If $|v| < |V|$, one expects the distribution F_v of $Z(v)$ to be more scattered than that F_V of $Z(V)$. Indeed, there exists a stochastic ordering relating the distributions of Z at various supports [52]. More precisely, if v "divides" V, then

$$\int_{-\infty}^{+\infty} \varphi(z)dF_v(z) \geq \int_{-\infty}^{+\infty} \varphi(z)dF_V(z) \tag{1}$$

for all convex numerical function φ. From this set of inequalities, it is easy to see that $Z(v)$ and $Z(V)$ have the same mean, that the variance of $Z(v)$ is greater than that of $Z(V)$. Of course, many other inequalities relating the moments of Z can be derived.

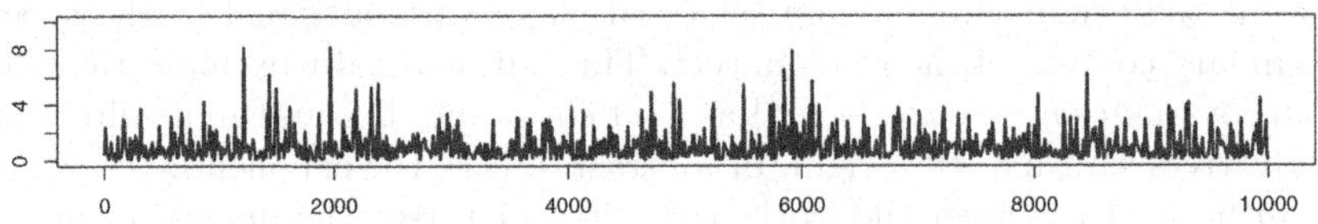

Fig. 1. Realization of an exponential diffusion process

As an illustration, consider an exponential diffusion process, a realization of which is shown on Fig. 1. It turns out that an explicit formula is available for the distribution of Z on all supports [53]. Fig. 2 shows the distributions obtained on support sizes ranging from 0 to 32. A continuous transition from an exponential distribution to a quasi-normal one can be observed.

3.4 Statistical inference

In the book "Estimating and Choosing" dealing with probability in practice, Matheron [59] makes a clear distinction between the prediction of a regional phenomenon ("Estimating") and the estimation of model parameters ("Choosing") that possess not a real but only a conventional significance. In what follows, we examine first how to estimate the parameters of a model, then we turn to the problem of how to choose one model.

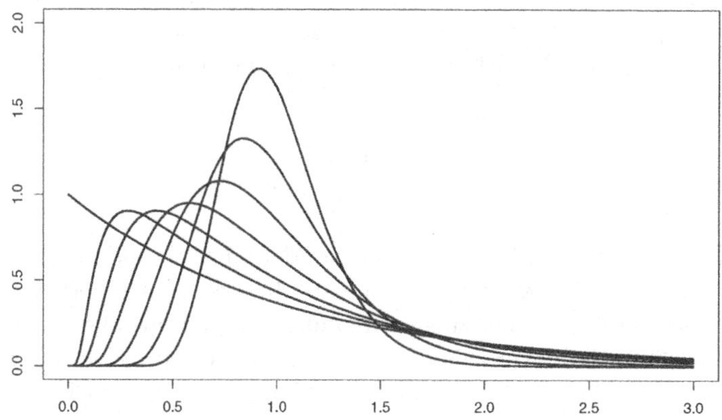

Fig. 2. Distributions at supports of size 0, 1, 2, 4, 8, 16 and 32

Estimation of model parameters

Here the stochastic model type (Gaussian, Boolean, etc.) is supposed to be known. If its parameters are few, then they may have an actual counterpart in the reality and thus can be reasonably estimated provided that data are numerous enough [59]. In the case of a GRF, Matheron shows that the behavior of the variogram at the origin has an objective significance. Similarly, Switzer proposes several methods for testing the objectivity of variogram parameters (sill, shape and nugget effect), which enables him to generate confidence regions for these parameters [79]. In contrast to this, a decomposition of the variogram in nested components has not the same level of objectivity, because different models can fit similarly the sample variogram.

Once the model parameters have been fixed, the results of the Monte Carlo procedure are conditional on that model. They do not take the uncertainty on these parameters into account, and the confidence limits obtained may look too narrow. An alternate approach is to work within a bayesian framework:

(i) postulate a prior distribution for the model parameters;
(ii) determine their posterior distribution from the data;
(iii) generate simulations with model parameters independently selected for each simulation according to that posterior distribution.

The results are then conditional on the chosen prior distribution (and on the data). As an illustration, Handcock and Wallis [18] present a nice analysis of a meteorological field in a Bayesian framework, even though their choice of a prior distribution is questionable from a physical standpoint. In order to cope with this difficulty, a noninformative prior distribution is often chosen in accordance with the physics of the problem (see Mosegaard and Tarantola [66] for a discussion on the concepts of noninformative prior distribution and homogeneous distribution).

The statistical inference problem may be complicated by the fact that closely located data produce redundant information. Consider for instance a standard Gaussian process Y with mean m and exponential covariance (unit scale factor). The average of Y on a segment of length ℓ

$$Y\big([0,\ell]\big) = \frac{1}{\ell} \int_0^\ell Y(x)dx$$

is an unbiased predictor of m. One may wonder what should be the value of ℓ to have a probability above 95% that the absolute difference between $Y\big([0,\ell]\big)$ and m is less than 0.05. The answer is much larger than one would usually expect ($\ell \geq 3000$). However this does not constitute a problem for Matheron insofar as he considers that the quantity of interest is the spatial integral

$$y\big([0,\ell]\big) = \frac{1}{\ell} \int_0^\ell y(x)dx$$

and not the mean m itself.

Choice of RF model type

Regarding the choice of the stochastic model type, the situation is quite different. There is no general tool for discriminating between several candidate model types. In flow studies, for example, the connectivity of low (or high) permeability zones is an important factor. The variogram is not an indicator of connectivity since random sets with very different connectivities can share the same point distribution and the same variogram (an enlightning example with its consequences on flow and transport is given by Zinn and Harvey [83]). Resorting to multivariate distributions is often considered, but the sparsity of data in usual situations makes the differenciation between possible model types somewhat illusory.

To fix ideas, consider the following three random set models, a realization of which is depicted on Fig. 3:

- a standardized GRF with covariance $\sin\left(\frac{\pi}{2}e^{-|h|}\right)$ that is thresholded at 0;
- a tessellation of Poisson polyedra that are randomly and independently valued to 0 or 1;
- a dead leaves model [39, 24] of balls. These balls have random radii with complementary distribution function

$$1 - F(r) = \frac{1}{r}\frac{\sinh r}{\cosh^3 r}$$

and are also randomly and independently valued to 0 or 1.

All three models are stationary with the same point distribution and the same exponential covariance. Consequently, they have exactly the same bivariate

Fig. 3. Realizations of three random sets with the same trivariate distributions. Left, a thresholded GRF; middle, a Poisson tessellation; right, a dead leaves model

distributions. Indeed, being autodual[4], they have also the same trivariate distributions.

Even though all models do not look plausible from the application standpoint, we may nonetheless wonder how they can be discriminated.

Some attempts with quadrivariate distributions (4 vertices of a regular tetrahedron or 4 points regularly arranged on a line) turned out to be unsuccessful.

At the end of the day, 1000 simulations of each model were drawn. On each simulation, the proportions p_0 and p_1 occupied by the largest connected component of each phase are determined. This gives two points (p_0, p_1) and (p_1, p_0) that are reported on a scatterplot. Fig. 4 shows the 3 scatterplots obtained. Clearly the model based on Poisson polyhedra distinguishes clearly from both other models. The dead leaves model and the thresholded GRF can also be discriminated, but the difference is not so important.

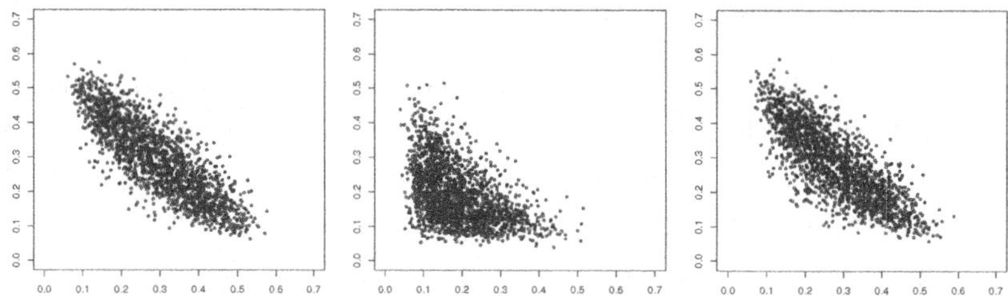

Fig. 4. Scatterplots of the proportions occupied by the largest components for the three models of Fig. 3

---

[4] A separable random set is said to be autodual if it has the same spatial distribution as its complement. The inclusion-exclusion formula shows that the trivariate distributions of an autodual random set are completely specified by its bivariate distributions.

Except situations where good quality training images are available, the choice of the model type is left to the user. This choice may be made easier by additional knowledge, such as the geology of the site under study, or hydraulic test results. The model type contains rich information that it conveys to the simulations. Correlatively, there is a risk incurred in the choice of a model type, especially if it has not been propped up. If in doubt, several contending models can be considered in order to compare the variability of the results produced.

4 Algorithms

4.1 Model or algorithm?

Many algorithms can be found in the "stochastic imaging" literature to produce images with some randomness (e.g. the sequential indicator simulation[5] [11]). According to the authors, those algorithms have the potent advantage of avoiding the design of a model, some attributes of which fail to be experimentally accessible. The model is implicitly and completely specified by the algorithm.

In a sense, this argument is valid. When considering all outputs produced by the algorithm, one obtains a family of multivariate distributions that satisfies all of the consistency relationships of a spatial distribution. Now, one may wonder which model is actually simulated. Suppose for instance that the algorithm is applied to an autodual random set with exponential covariance. Is this one of the three models of Fig. 4, or a mixture of them, or even a totally different model? Answering that question is all the more difficult since Emery established that the multivariate distributions obtained do not have a clear status; they can be considered as deriving from a model, but not necessarily from a spatial model [13]. All things considered, it does not seem reasonable to recommend using such an algorithm without a careful study of its properties.

It may be interesting to mention that Matheron suggested using stochastic imaging algorithms, not for producing spatial simulations but for heuristically

---

[5] The sequential indicator simulation, or SIS, is an algorithm designed for simulating conditionally an indicator random function. It is based on the approximation of the conditional distribution at x by a Bernoulli distribution with a mean equal to the indicator kriging at x. It can be shown that the indicator covariance is well reproduced if simple kriging is used and if the kriged values are in $[0, 1]$. This latter condition is very restrictive (unidimensional completely monotonic covariances). As shown by Emery [13], the indicator kriging approximates the conditional expectation reasonably well for a mosaic model with i.i.d. valuations, but rather crudely for a truncated GRF. This suggests that SIS produces outcomes close to those of a mosaic model. As a matter of fact, this is generally not the case, unless drastic conditions are satisfied (unidimensional exponential covariances; points are orderly simulated).

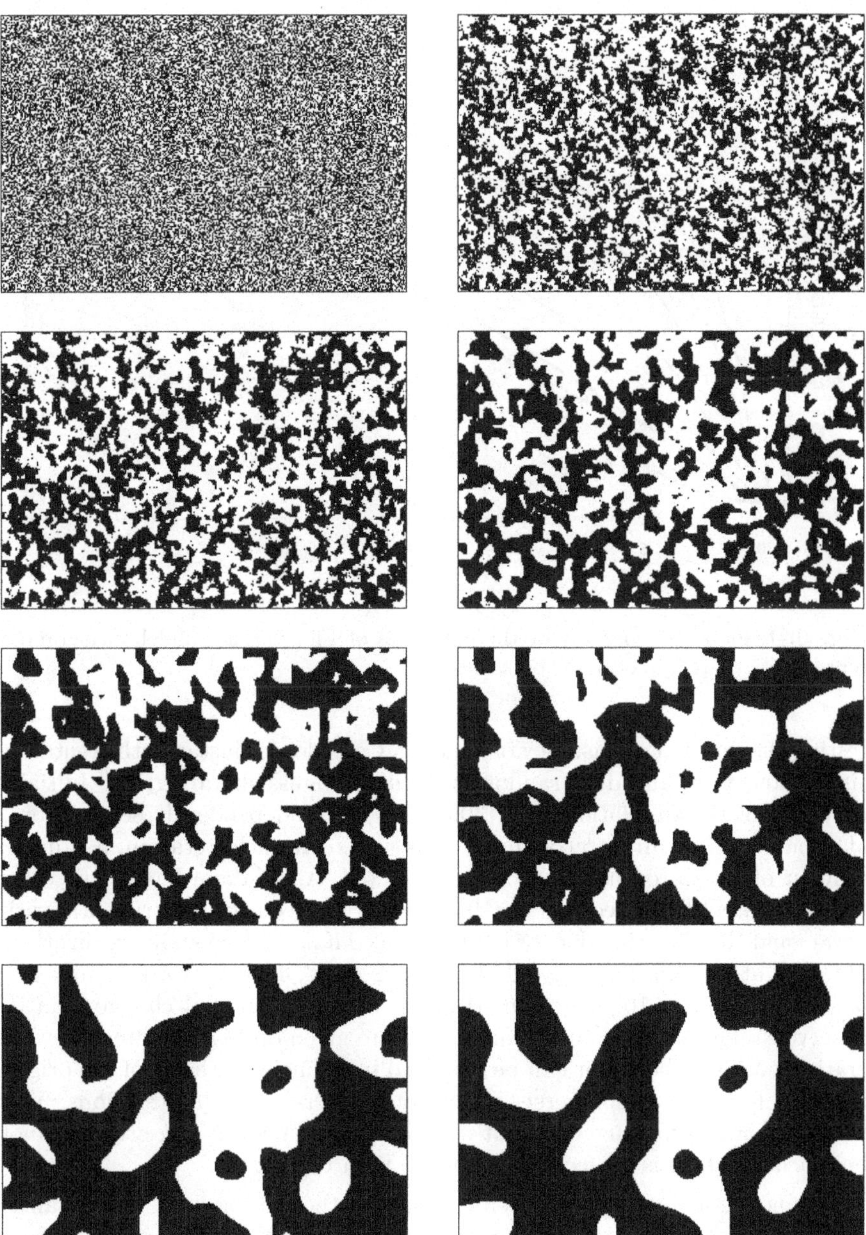

Fig. 5. Tentative simulation of a random set with a Gaussian variogram. This figure shows the outcomes produced at the initial stage and after 10^6, 2×10^6, 4×10^6, 8×10^6, 16×10^6, 32×10^6 and 64×10^6 iterations

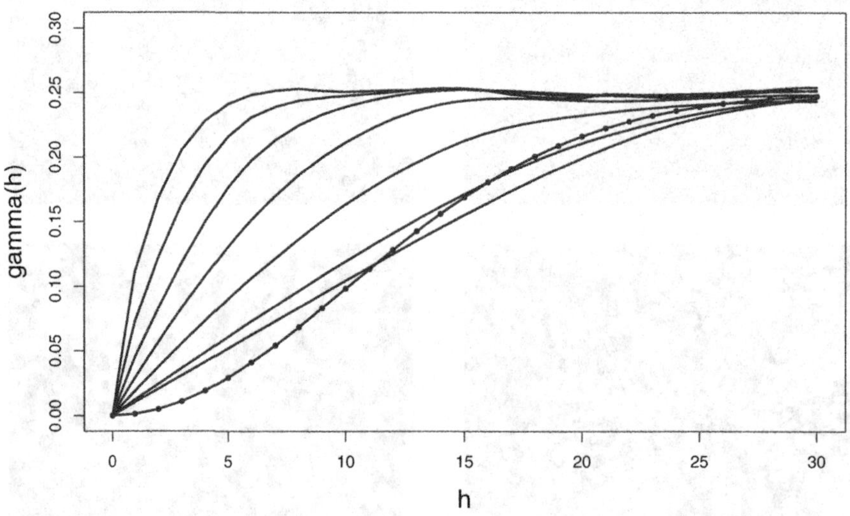

Fig. 6. Regional variograms of the outcomes of Fig. 5. The model variogram is represented by points overlaid by a line

testing the internal consistency of a model. Consider for instance the tentative simulation of an autodual random set with a Gaussian variogram[6] Initially all pixels of the simulation field (300×200) are independently set to 0 or 1 at random. Then an iterative procedure starts. At each iteration, a random number of pixels (on average 1.5) are selected. All of them have their value complemented if this reduces the discrepancy between the experimental variogram and the Gaussian one, or left unchanged if not. Fig. 5 shows realizations obtained at the initial stage and after 10^6, 2×10^6, 4×10^6, 8×10^6, 16×10^6, 32×10^6 and 64×10^6 iterations (beyond this, only limited changes can be observed). The corresponding variograms are given on Fig. 6. After a few iterations, a linear behavior can be observed in the neighborhood of the origin, and this linear behavior persists even after several millions iterations. This tentative simulation suggests that a parabolic behavior at the origin cannot be reached, which is perfectly compatible with theory.

The same algorithm has also been applied to other variograms. For instance, outcomes with an exponential or a spherical regional variogram can be obtained in less than one million iterations (see Fig. 7). This suggests that there actually exist random sets with exponential (which was already known) or

---

[6] Considering autodual random sets only is not a limitation. If X is a stationary random set with proportion p and variogram $p(1 - p)\gamma$, then the random set Y equal to X or X^c with the same probability 0.5 is autodual with variogram 0.25γ.

spherical variograms. Of course, this approach is not constructive; it gives a clue, but cannot be considered as a proof of existence.

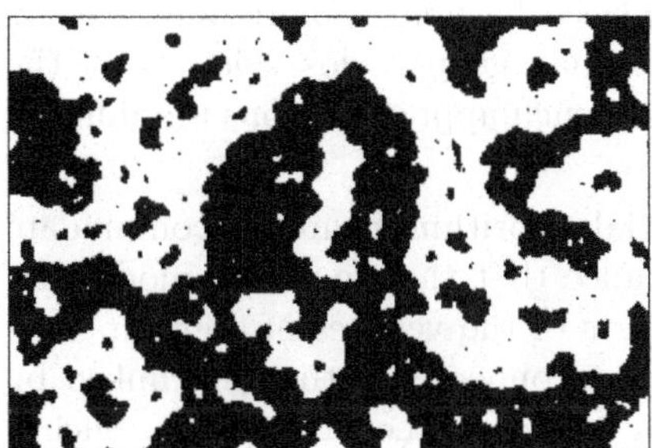 

Fig. 7. Tentative simulation of random sets with an exponential (left) or a spherical (right) variogram. The scale factors of both variograms have been chosen to give them the same integral range (or correlation area). As a result, the exponential outcome has its specific perimeter 2.1 times larger than that of the spherical outcome, which is expressed by its broken aspect

From now onward, all algorithms considered are assumed to derive from spatial models.

4.2 Sequential or parallel?

Several algorithms are sometimes available to simulate a given model. Which one should be chosen? Among all possible criteria (correctness, accuracy, stability, speed, memory requirement,...), an important one is its capability to be implemented on parallel processors. One can distinguish between

1. **sequential** algorithms, for which the distribution used to simulate the nth point of the simulation field depends on the available conditioning data as well as the simulated values at the $n - 1$ points already processed.

2. **parallel** algorithms, for which each point is simulated conditionally to the available data but irrespectively of the values taken by the other points of the simulation field.

3. **distributed** algorithms, which are intermediate between sequential and parallel algorithms.

A sequential algorithm is applicable only when the simulation field is finite. It requires its points to be linearly ordered. This order relation can be chosen arbitrarily; in particular it can be either deterministic or random. However its choice is not indifferent insofar as it determines the conditional distributions to be simulated. One particular choice may lead to notable simplifications.

Usually the expression for a conditional distribution becomes more and more complicated (if not intractable) as the number of already processed points increases. Here lies the main criticism against sequential algorithms. In practice, the conditional distributions are replaced by approximate ones, e.g. by accounting only for the conditioning information in a close vicinity of the point to be simulated. The consequences of such approximations are difficult to assess.

It should also be pointed out that sequential algorithms cannot accommodate all types of constraints. Suppose for instance that the stochastic model is a random set and the simulation field consists of the vertices of a graph. Two points of the simulation field are said to be connected if they are linked by an arc of the graph, all vertices of which belong to the same phase. Such a connectivity constraint cannot be addressed sequentially because it does not say which arc or which population of arcs connects both points. More generally "regional" contraints cannot be taken into account by sequential algorithms.

4.3 Discrete or continuous?

The previous section has already pointed out the importance of the simulation field. In common practice, three different types of field can be encountered, namely a limited number of points with arbitrary locations, a regular grid, and a compact subset with nonempty interior. An algorithm is said to be **continuous** if it is applicable to the third type of simulation field; otherwise, it is said to be **discrete**.

Discrete and continuous algorithms do not work the same way. A discrete algorithm is designed to assign a value directly to each point of the simulation field. In contrast to this, a continuous algorithm produces a set of basic ingredients that summarizes the simulation content. These ingredients can subsequently be used to derive the value at each point of the simulation field.

The main advantage of continuous algorithms is that they make simulations available at any spatial resolution. The impact of spatial resolution on certain measurements can therefore be tested. This is important in practice because the scale that affects measurements may not be well known.

All continuous algorithms can be used to simulate discretely, but the converse is not true. There exist discrete algorithms that cannot be extended to continuous ones. For instance, Cholesky's algorithm can be used to simulate a GRF at a limited number of points. Several thousand points can be considered provided that the covariance matrix is not too badly conditioned. In the case of a regular grid, an algorithm based on FFT techniques can be recommended [10]. Both algorithms express the fact that a GRF is the Fourier transform of an orthogonal Gaussian measure.

All algorithms developed by Matheron are continuous. The most typical example is the so-called **turning bands** algorithm to simulate a GRF [44, 46].

In the stationary case, the covariance C (supposed to be continuous) can be written as

$$C(h) = \int_{S_{d-1}} C_\theta(<h, \theta>) \, d\varpi(\theta)$$

where $d\varpi$ is the orientation distribution of the spectral measure associated to C, and C_θ is a unidimensional covariance for each orientation $\theta \in S_{d-1}$. The turning bands method is in fact a stereological device for reducing a multidimensional simulation into unidimensional ones. If $\theta_1, ..., \theta_n$ have been independently drawn from $d\varpi$, and if $X_1, ..., X_n$ are independent stochastic processes with respective covariances $C_{\theta_1}, ..., C_{\theta_n}$, then the central limit theorem asserts that the random function

$$Y^{(n)}(x) = \frac{1}{\sqrt{n}} \sum_{k=1}^{n} X_k(<x, \theta_k>) \qquad x \in \mathbb{R}^d$$

tends to be Gaussian as n tends to infinity.

The practical implementation of this algorithm was not exactly Matheron's primary concern. Indeed, he never studied its rate of convergence towards gaussianity. Assuming that the covariance C_θ of X_θ is proportional to the autoconvolution[7] of some numerical function f, he proposed to simulate X_θ as a shot-noise process

$$X_\theta(t) \propto \sum_{p \in \mathcal{P}} f(t - p) \qquad t \in \mathbb{R}$$

starting from a homogeneous Poisson point process $\mathcal{P}$. If C_θ has a bounded support, say $[-a, a]$, then replacing $\mathcal{P}$ with a periodic point process with period a simplifies the implementation (see Fig. 8) and reduces the execution time dramatically [30]. If C_θ has no bounded support, then the simulation of X_θ is only approximate because the support of f is also unbounded. In such a case, other techniques for simulating X_θ are required. The spectral method [75]

$$X_\theta(t) \propto \cos(Ut + V) \qquad t \in \mathbb{R}$$

where $U \sim d\mathcal{X}_\theta$ (the spectral measure of C_θ) and $V \sim \mathcal{U}([0, 2\pi[)$ is effective only when C_θ is differentiable at the origin. In three dimensions, Emery [14] remarks that an exponential covariance can be seen as a mixture of spherical covariances[8]. This allows him to simulate the stochastic processes with unidimensional covariances associated to exponential, or mixtures of exponential covariances like those of a spherical covariance with a randomized scale factor. Migration techniques are also available [30].

---

[7] The autoconvolution of f is the convolution of f by its reflection $\check{f}$ around the origin.

[8] This result can also be seen as a direct consequence of the algorithm proposed by Hammersley and Nelder [17] for simulating shot-noise processes with exponential covariance.

Fig. 8. Simulation of a stochastic process to get a three-dimensional GRF with spherical covariance

4.4 Conditional or not?

Even if the predictor presented in this paper is based on conditional simulations, non conditional simulations are nonetheless worthwhile being considered. At least three arguments can be put forward:

1. Many models can often be devised to represent a natural phenomenon or some physical reality. **Selecting** one among them is not always easy on account of their complexity and of the number of parameters they involve. Using non conditional simulations and their display can facilitate the choice of the most appropriate model. Moreover, this gives the possibility to better grasp the role of each parameter, the range of values it can take, its sensitivity... Finally, it provides information about the statistical fluctuations that one should expect from a model.

2. A **posterior validation** of the model can be obtained by replicating non conditional simulations at the data locations. Each data value is then assigned a score using the histogram produced at its location. The score histogram can serve to quantify the adequation of the model to the data set.

3. Conditional simulation algorithms are often **constructed** starting from non-conditional ones. The most classical example is the **kriging technique** (or regression technique) for simulating a GRF Z defined on $I\!\!R^d$ and subject to the conditions $\left(Z(x_\alpha) = z_\alpha,\ \alpha \in A\right)$. It rests on the fact that Z can be written as the sum of two independent GRF, namely the (probabilist versions) of the simple kriging on the data and a residual. Since the kriging is linear, its calculation is straightforward. As for the residual, it can be obtained as the difference between a non-conditional simulation z^S and its kriging estimate on the simulated values $\left(z^S(x_\alpha) = z_\alpha^S,\ \alpha \in A\right)$ at the data location. Finally, the conditional simulation algorithm can be written as

$$z^{CS}(x) = z^S(x) + \sum_{\alpha \in A} \lambda_\alpha(z_\alpha - z_\alpha^S) \qquad x \in I\!\!R^d$$

which involves the kriging coefficients $(\lambda_\alpha,\ \alpha \in A)$.

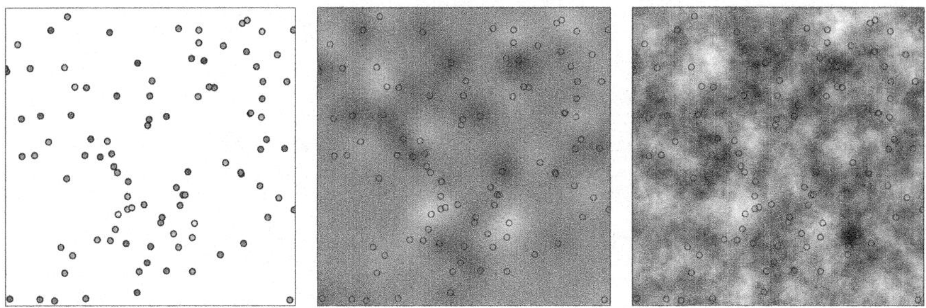

Fig. 9. A conditional simulation of a GRF can be obtained by adding a simulated residual to the simple kriging (middle) on the data (left). The result is depicted on the right

4.5 Iterative or non-iterative?

As seen above, the algorithm for simulating a GRF conditionally is straightforward. This is more the exception than the rule. Most often algorithms are iterative:

1. Iterative algorithms are simpler to design than non-iterative ones. They rest on simple principles. Their construction is often based on pre-existent algorithms such as the Metropolis-Hastings algorithm or the Gibbs sampler.

2. They can accommodate a wide range of conditions and need minor modifications as a new set of conditions or constraints is added. They can also handle compatibility problems between the model and the data.

In order to illustrate both arguments, the Markov chain restriction principle is presented. Let Ω a state space, supposed here to be discrete to simplify notation. Each state $\omega \in \Omega$ is assigned a probability $p(\omega)$. Because of a set of conditions or constraints, only the states within the subset Ω_c of Ω are allowed. The objective is to simulate the conditional distribution p_c that p induces on Ω_c.

Suppose that p can be simulated as the limit distribution of a Markov chain on Ω with transition kernel P. This suggests the following iterative algorithm. The transition kernel P is applied to the current (allowed) state to generate a candidate state. This state is accepted as a new state only if it is allowed (see Fig. 10). Such an algorithm defines a Markov chain on Ω_c with transition kernel

$$P_c(\omega, \omega') = P(\omega, \omega') + 1_{\omega = \omega'} P(\omega, \Omega \setminus \Omega_c) \qquad \omega, \omega' \in \Omega_c$$

The p_c-irreductility of Ω_c and the p-reversibility of P are sufficient conditions to ensure that the limit distribution simulated by this algorithm is precisely p_c [30].

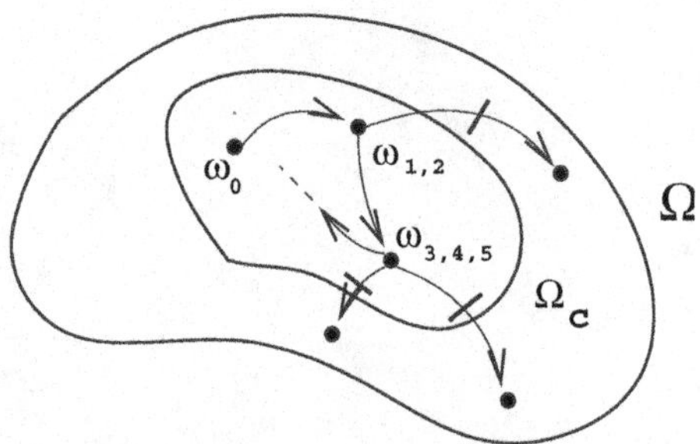

Fig. 10. Conditional simulation by restricting a transition kernel. Starting from the current state ω, a transition kernel is run on Ω to generate a candidate state ω'. The new generated state is ω' if $\omega' \in \Omega_c$ and ω if not

The question that now comes is how to initialize this algorithm. A natural idea is to run the transition kernel P starting from an arbitrary state of Ω until an allowed state has been generated. This idea is all the more interesting because in many practical situations the compatibility between the model and the data cannot be ascertained beforehand. Indeed, the number of iterations required to get an allowed state indicates how little compatible they both are. In the non compatibility case, this algorithm never terminates.

Other case by case initialization procedures can be considered, such as in the following two examples.

Conditional simulation of a Voronoi tessellation

Consider a Poisson point process $\mathcal{P}$ in $\mathbb{R}^d$ with intensity function θ. Associate to each point of $\mathcal{P}$ (or "germ") a subset of $\mathbb{R}^d$ (or "cell") defined as follows. This is the set of all points strictly closer to its germ than to any other germ. The cells along with their boundaries constitute a partition of $\mathbb{R}^d$ called a Voronoi tessellation.

The problem addressed here is the simulation of a Voronoi tessellation, subject to the condition that each pair of points of a finite subset $C \in \mathbb{R}^d$ is known to belong either to the same cell or to two different cells.

A Voronoi tessellation is characterized by its population of germs. To avoid edge effects, the intensity function is assumed to have a finite integral, say ϑ. In that case, it is possible to define a typical germ as a germ with random location (p.d.f. θ/ϑ), and a population of germs is made of a Poisson number (mean ϑ) of independent typical germs. A simple way to simulate iteratively this model is based on Metropolis algorithm. Let ω be the current population, and let $\sharp\omega$ be the number of germs it contains. With respective probabilities

$$\frac{\vartheta}{\vartheta + \sharp\omega + 1} \qquad \frac{\sharp\omega}{\vartheta + \sharp\omega} \qquad \frac{\vartheta}{(\vartheta + \sharp\omega)(\vartheta + \sharp\omega + 1)}$$

a new germ is added to the population, an old germ is randomly selected to be removed, or an old germ is randomly selected to be replaced by a new one.

This algorithm is reversible (inheritance from the Metropolis algorithm). Moreover, the set Ω_c of allowed populations is irreducible (because germs can be replaced). Accordingly, this algorithm can be made conditional by applying the restriction principle (see Fig. 11). To initialize it, it is convenient to introduce the partition $(C_i, i \in I)$ of C defined by the equivalence relation $c \mathcal{R} c'$ iff c and c' share the same cell. For any $i \in I$, let X_i be the set of all points in $\mathbb{R}^d$ strictly closer to each point of C_i than to any point of $C \backslash C_i$. An initial population can be obtained by picking one point in each X_i. This procedure is effective provided that all X_i are non-empty, which holds iff the conditioning data are compatible with the tessellation model.

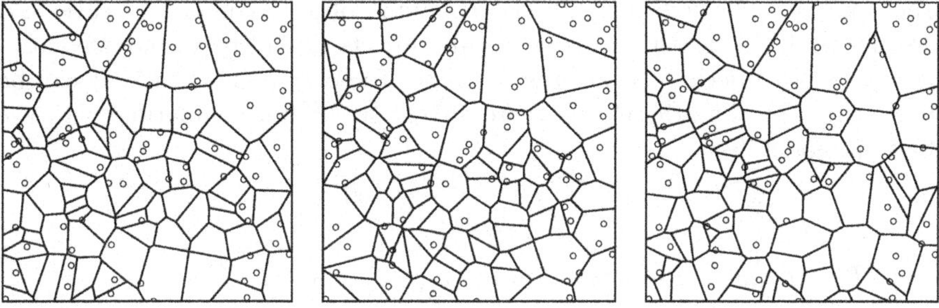

Fig. 11. Three conditional simulations of a Voronoi tessellation. The intensity function of the germs is $\theta(x, y) = 180 - 100x - 60y$ for $0 \le x, y \le 1$

Conditional simulation of a Boolean model

Two ingredients are required for the construction of this model. The first one is a Poisson point process $\mathcal{P}$ in $\mathbb{R}^d$ (intensity function θ). The second one is a family $\big(A(x), x \in \mathbb{R}^d\big)$ of objects, i.e. random non-empty compact subsets. These objects are independent but not necessarily identically distributed (the hitting functional of $A(x)$ is denoted by T_x). By definition, a Boolean model is the union of the objects located at the points of the Poisson process

$$X = \bigcup_{x \in \mathcal{P}} A(x)$$

The problem considered here is the simulation of the Boolean model, subject to the condition that two finite subsets C_0 and C_1 must belong to X^c and X respectively.

The Boolean model X is specified by the population of objects that constitutes it. If θ is assumed to have a finite integral[9], say ϑ, then it is possible to define

---

[9] This situation occurs when θ is locally integrable and the simulation field is bounded.

a typical object as a randomly located object (p.d.f. θ/ϑ), and a population of objects is made of a Poisson number (mean ϑ) of independent typical objects. As a consequence, the same iterative algorithm as in the previous example can be applied to get a non-conditional simulation of the Boolean model, up to the fact that $\sharp\omega$ acts as a number of objects instead of a number of germs.

As previously, this algorithm is reversible. The set Ω_c of allowed populations of objects is also irreducible due to the fact that the concatenation of two allowed populations is also allowed[10]. Accordingly, the restriction principle can be applied to produce conditional simulations (see Fig. 12). Regarding the initialization, a simple procedure consists of simulating typical objects one after the other. Any object that hits C_0 is automatically discarded. The procedure is continued until all the remaining objects completely cover C_1. In order to avoid starting with too many objects, it can be recommended to keep only the objects that are the first ones to cover points of C_1. It can be shown that the conditioning data are almost surely not compatible with the Boolean model iff

$$\sum_{C \subset C_1} \frac{(-1)^{\sharp C}}{T(C_0 \cup C)} = 0$$

where T is the hitting functional of the typical objects, i.e.

$$T(K) = \int_{\mathbb{R}^d} \frac{\theta(x)}{\vartheta} T_x(K)dx \qquad K \in \mathcal{K}$$

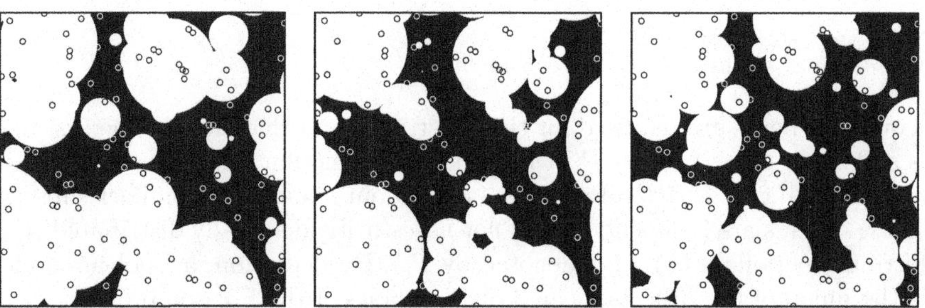

Fig. 12. Three conditional simulations of a Boolean model. The objects are disks with exponential radius (mean 5). On average 100 objects are hitting the simulation field (100×100)

The description of both algorithms is not complete without a proper specification of their rate of convergence towards their limit distribution. It turns out that this rate of convergence is specified by the second largest eigenvalue

---

[10] This stability property makes the object replacement in the algorithm not strictly necessary.

of the transition kernel that rules the evolution of the number of germs or objects during the conditional simulation. This eigenvalue can be estimated empirically by finding the function of the number of germs that possesses the maximal autocorrelation and determining its integral range [7, 30].

5 And now?

5.1 Model enrichment

The models presented above are simple prototypes. In practical applications, the parameters often display some spatial evolution: the marginal distribution (e.g. the proportions of the various facies) varies vertically, and even laterally; the size of the objects of a Boolean model as well as its intensity may also vary, etc. The spatial evolution of these parameters can be represented either by ordinary functions or by random functions. Most simulation algorithms can be adapted to account for such generalizations (e.g. [30] for several generalizations of the Boolean model, and [1] for plurigaussian simulations).

The inference of the model parameters, however, calls for special tools and methods. Spatial variations of the proportions of both facies of a truncated Gaussian simulation can be obtained by means of a regionalized truncation threshold. The inference of the stationary covariance of the underlying GRF is addressed by Matheron et al. [57]. That approach can be generalized to several facies of a plurigaussian simulation, but the inference of the coding parameters remains a difficult task. Boolean models can also accommodate spatially variable facies proportions, for example by regionalizing the Poisson intensity or the object distribution. Benito Garcia-Morales proposes a method for the inference of the regionalized intensity of a Boolean model based on a deconvolution process [2, 3].

5.2 Process-based and stochastic models

The geological formations as well as their properties result from complex processes (crystallization, sedimentation, alteration, etc.) which, unlike physics, are not always ruled by well established laws. Nevertheless, an increasing number of geological and geochemical processes are understood and can be modelled. These models do not deliver reliable deterministic predictions – too many parameters remain poorly known, – but they contain a valuable source of information about the morphology of the formations and the spatial variability of their properties. Since this information is qualitatively much richer than that of the simple stochastic models currently used, the challenge is to mix the process-oriented and stochastic approaches. The first attempts to conduct that type of work are due to Matheron [40], Jacod and Joathon [22, 23] in the simulation of sedimentary sequences. Their potentialities were then rather limited. Supported by a better understanding of sedimentary processes and the fast improvement of data processing capabilities, this approach has recently known a new surge of interest. In particular, Lopez simulates the deposition of meandering channelized sedimentary architectures at the scale of oil reservoirs [32]. A space-time evolution model of the meandering channel centreline is derived from hydraulic equations. This channel centreline is then wrapped using a facies model that combines geometrical models, field observations and stochastic processes, to mimic the development of point-bar deposits, oxbow lake deposits, crevasse splays, overbank deposits, etc. Successive channel migrations, controlled by the erodibility of the terrain, and aggradation contribute to progressively build the whole fluvial architecture. This model is flexible and realistic enough even if it is based on a limited number of parameters. Current work is devoted to the conditioning on borehole and low-resolution seismic data.

A similar approach needs to be developed for other reservoir environments, above all carbonate reservoirs, which represent 60% of oil resources. These reservoirs are very heterogeneous and of a great complexity. The processes generating the calcareous rocks (biological growth and sedimentation) present a large variability and must be complemented by diagenetic alteration processes. These are often guided by conduits allowing the migration of fluids, the latter being determined by the reservoir architecture.

5.3 Data integration

In the Earth sciences as well as in other domains, measurement methods bring increasingly rich and varied information (remote sensing, petroleum seismic, borehole imagery, core sample analysis, flow tests, production data, etc.). Geostatistics provides a framework for the development of solutions to integrate that information. The methods are adapted to the diversity of data. In simple situations a cosimulation in the framework of a standard multivariate model is the solution. MCMC algorithms are useful to honor global constraints such as production data. The principle is to generate a series of conditional simulations according to a Metropolis dynamics allowing the maximization of some criterion of adequacy to that global information. Hu [19] developed the gradual deformation method to fasten the iterative process. In the case of a GRF, the principle is to consider the set of simulations defined by $S_t = S_1 \cos t + S_2 \sin t$, where S_1 and S_2 are two independent Gaussian simulations, and $t \in [0, 2\pi[$ is a parameter. The optimal value of t wrt some objective function is selected. The process is iterated after replacing S_1 by S_t and generating a new independent simulation S_2. This algorithm can be generalized to more than two basic simulations. It can also be used globally or locally, and has been extended to several types of nongaussian RFs [20]. A similar approach has been developped by Sénégas [73] to quantify the errors incurred while reconstructing a relief from pairs of stereoscopic images. It does not rely on any optimisation procedure, and can be seen as a multichoice generalization of the Metropolis-Hastings algorithm.

Physical systems are ruled by physical equations. When several variables are linked by a system of partial differential equations and boundary conditions, their statistical properties are closely linked. Much effort has been devoted to the development of consistent models and simulation methods, especially to solve inverse problems in stochastic hydrology (RamaRao *et al.* [69]; LaVenue *et al.* [31]; Gómez-Hernández *et al.* [16]; Roth *et al.* [71, 70]) and geophysics (Tarantola [80]; Iooss *et al.* [21]).

5.4 Sequential data assimilation

Complex space-time phenomena are characterized by the strongly nonlinear dynamics of processes involving a large number of variables. The system of partial differential equations with space and time boundary conditions does not master the complexity of these processes totally, either because it is based on simplifying assumptions or because the boundary conditions are not exactly known. It is thus wise to introduce statistical techniques that can guide these models to assimilate the flow of measurements emanating from automatic devices. This is the aim of data assimilation techniques. From the point of view of the designer of deterministic numeric models, data assimilation can be seen as an algorithm for updating the model state as new data are available. From the statistician's point of view that numerical model may help us in improving the operational predicting by using knowledge of nonlinear relationships between the various data sources. The simplest data assimilation technique is the Kalman filter, which amounts to a series of cokriging and applies to linear processes. The Ensemble Kalman Filter is based on conditional simulations. Like for autoregressive processes, the evolution of the system is described by error terms or innovations. The sequential assimilation of the data (in time) allows an updating of the model parameters and therefore gives robustness to the method with regard to a misspecification of the parameters. There is of course some limitation to that robustness and it is advisable to use consistent models for the covariances of the various error terms, especially when there are interferences between the structures of different variables, which is a common situation. Geostatistical concepts were shown to contribute to improve sequential data analysis techniques in several applications to oceanography and air quality (Sénégas *et al.* [74]; Wolf *et al.* [82]; Bertino [4]; Bertino *et al.* [5, 6]).

5.5 Exact simulations

Despite the publication of an abundant and important literature over the last twenty years, determining the rate of convergence of an iterative simulation algorithm remains an arduous task. This prompted Propp and Wilson to design an algorithm with backwards iterations and coupling in the past in order to get an "exact" simulation after a finite number of iterations [68]. Their pioneering paper has opened the door to new methodological and applicative developments, and several authors designed more general or different algorithms for simulating various models from stochastic geometry exactly. Among them, Kendall and Thönnes successfully simulated exactly a Boolean model with bounded objects [27], whereas Kendall and Møller did the same for a wide class of point processes [28]. It should be noted that Kendall prefers referring to "perfect" rather than "exact" simulations to express the fact that the number of iterations required may be rather large. This should not be a major problem as it suffices to run an iterative algorithm starting from one exact simulation to get as many as desired. Exact simulation algorithms undoubtly constitute an exciting field of investigation even though it is probably too early to get a precise idea of the real possibilities they can offer regarding conditioning and data integration.

References

1. Armstrong, M., Galli, A., Le Loc'h, G., Geffroy, F., Eschard, R.: Plurigaussian simulations in Geosciences. Springer, Berlin (2003)
2. Benito García-Morales, M.: Non stationnarité dans les modèles de type booléen: application à la simulation d'unités sédimentaires. PhD Thesis, Ecole des Mines de Paris (2003)
3. Benito García-Morales, M., Beucher, H.: Inference of the Boolean model on a non stationary case. In: 8th annual conference of the International Association for Mathematical Geology. Berlin (2002)
4. Bertino, L.: Assimilation de données pour la prédiction de paramètres hydro-dynamiques et écologiques: cas de la lagune de l'Oder. PhD Thesis, Ecole des Mines de Paris (2001)
5. Bertino, L., Evensen, G., Wackernagel, H.: Combining geostatistics and Kalman filtering for data assimilation of an estuarine system. Inverse problems, **18–1**, 1–23 (2002)
6. Bertino, L., Evensen, G., Wackernagel, H.: Sequential data assimilation techniques in oceanography. International Statistical Review, **71**, 223–241 (2003)
7. Besag, J., Green, P.J.: Spatial statistics and bayesian computation. J.R. Statist. Soc., **55-1**, 25–37 (1993)
8. Carter, D.S., Prenter, P.M.: Exponential spaces and counting processes. Zeit. für Wahr. und ver. Gebiete, **21**, 1–19 (1972)
9. Chilès, J.P.: On the contribution of hydrogeology to advances in geostatistics. In: Monestiez, P. *et al* GEOENV III – Geostatistics for environmental applications. Kluwer, Dordrecht (2001)
10. Chilès, J.P., Delfiner, P.: Geostatistics: modeling spatial uncertainty. Wiley, New York (1999)
11. Deutsch C.V., Journel A.G.: GSLIB: Geostatistical software library and user's guide. Oxford University Press, New York (1998) (2nd edition)
12. Doob, J.L.: Stochastic processes. Wiley, New York (1953)
13. Emery, X.: Simulation conditionnelle de modèles isofactoriels. PhD Thesis, Ecole des Mines de Paris (2004)

14. Emery, X.: Private communication (2004)

15. Evensen, G.: Sequential data assimilation with a nonlinear quasi-geostrophic model using Monte Carlo methods to forecast error statistics. J. Geophysical Research, **99–C11**, 17905–17924 (1994)

16. Gómez-Hernández, J.J., Sahuquillo A., Capilla, J.: Stochastic simulation of transmissivity fields conditional to both transmissivity and piezometric data. Journal of Hydrology, **203**, 162–174 (1997)

17. Hammersley, J.M., Nelder, J.A.: Sampling from an isotropic Gaussian process. Proc. Cambridge Phil. Soc., **51**, 652–662 (1955)

18. Handcock, M.S., Wallis, J.R.: An approach to statistical spatial-temporal modeling of meteorological fields. JASA, **89–426**, 368–390 (1994)

19. Hu, L.Y.: Gradual deformation and iterative calibration of gaussian-related stochastic models. Math. Geol., **32–1**, 87–108 (2000)

20. Hu, L.Y.: Gradual deformation of non-gaussian stochastic simulations. In: Kleingeld, W.J., Krige, D.G. (eds) Geostatistics 2000. Cape Town (2000)

21. Iooss, B., Geraets, D., Mukerji, T., Samualides, Y., Touati, M., Galli, A.: Inferring the statistical distribution of velocity heterogeneities by statistical travel-time tomography. Geophysics, **68–5**, 1714–1730 (2003)

22. Jacod, J., Joathon, P.: Use of random-genetic models in the study of sedimentary processes. Math. Geol., **3–3**, 219–233 (1971)

23. Jacod, J., Joathon, P.: Conditional simulation of sedimentary in three dimensions. In: Merriam, D.F. (ed) Proceedings of the International Sedimentary Congress. Plenum Press (1972)

24. Jeulin, D.: Dead leaves model: from space tessellation to random functions. In: Jeulin, D. (ed.) Advances and applications of random sets. World Scientific, Singapore (1997)

25. Jeulin, D., Jeulin, P.: Synthesis of rough surfaces of random morphological functions. Stereo. Iugosl., **3–1**, 239–246 (1981)

26. Kendall, D.G.: Foundations of a theory of random sets. In: Harding, E.F. and Kendall, D.G. (eds) Stochastic geometry. Wiley, London (1974)

27. Kendall, W.S., Thönnes, E.: Perfect simulation in stochastic geometry. Pattern Recognition, **32**, 1569–1586 (1999)

28. Kendall, W.S., Møller, J.: Perfect simulation using dominating processes on ordered spaces, with application to locally stable point processes. Adv. Appl. Prob., **32**, 844–865 (2000)

29. Lantuéjoul, C.: Conditional simulation of object-based models. In: Jeulin, D. (ed.) Advances and applications of random sets. World Scientific, Singapore (1997)

30. Lantuéjoul, C.: Geostatistical simulation: models and algorithms. Springer, Berlin (2002)

31. LaVenue, A.M., RamaRao, B.S., Marsily, G. de, Marietta, M.G.: Pilot point methodology for automated calibration of an ensemble of conditionally simulated transmissivity fields. Water Resources Research, **31–3**, 495-516 (1995)

32. Lopez, S.: Modélisation de réservoirs chenalisés méandriformes; approche génétique et stochastique. PhD Thesis, Ecole des Mines de Paris (2003)

33. Matérn, B.: Spatial variations. Stochastic models and their applications to some problems in forest surveys and other sampling investigations. Meddelanden från Statens Skogsforskningsinstitut, **49-5**, Almaenna Foerlaget, Stockholm (1960). Lecture Notes in Statistics, **36**, Springer, Berlin (1986) (2nd edition)

34. Matheron, G.: Les variables régionalisées et leur estimation. Masson, Paris (1965)

35. Matheron, G.: Ensembles et partitions aléatoires. Technical Report N-73. Centre de Morphologie Mathématique, Ecole des Mines de Paris (1967)

36. Matheron, G.: Eléments pour une théorie des milieux poreux. Masson, Paris (1967)

37. Matheron, G.: Mesures et capacités aléatoires. Technical Report N-75. Centre de Morphologie Mathématique, Ecole des Mines de Paris (1968)

38. Matheron, G.: Filtres et topologies aléatoires. Technical Report N-77. Centre de Morphologie Mathématique, Ecole des Mines de Paris (1968)

39. Matheron, G.: Schéma booléen séquentiel de partition aléatoire. Technical Report N-83. Centre de Morphologie Mathématique, Ecole des Mines de Paris (1968)

40. Matheron, G.: Les processus d'Ambarzoumian et leur application en géologie. Technical Report N-131. Centre de Morphologie Mathématique, Ecole des Mines de Paris (1969)

41. Matheron, G.: Théorie des ensembles aléatoires. Cahier du Centre de Morphologie Mathématique, Fasc. 4. Ecole des Mines de Paris (1969)

42. Matheron, G.: La théorie des variables régionalisées et ses applications. Cahiers du Centre de Morphologie Mathématique, Fascicule 5, Ecole des Mines de Paris (1970)

43. Matheron, G.: Ensembles fermés aléatoires, ensembles semi-markoviens et polyèdres poissonniens. Adv. App. Prob., **4-3**, 508–541 (1972)

44. Matheron, G.: Quelques aspects de la montée. Technical Report N-271. Centre de Morphologie Mathématique, Ecole des Mines de Paris (1972)

45. Matheron, G.: The turning bands: a method for simulating random functions in $I\!R^n$. Technical Report N-303. Centre de Morphologie Mathématique, Ecole des Mines de Paris (1972)

46. Matheron, G.: The intrinsic random functions and their applications. Adv. Appl. Prob., **5**, pp. 439–468 (1973)

47. Matheron, G.: Random sets and integral geometry. Wiley, New York (1975)

48. Matheron, G.: Compléments sur les modèles isofactoriels. Technical Report N-432. Centre de Géostatistique, Ecole des Mines de Paris (1975)

49. Matheron, G.: Remarque sur la fabrication des aléas. Technical Report N-653. Centre de Géostatistique, Ecole des Mines de Paris (1980)

50. Matheron, G.: Quatre familles discrètes. Technical Report N-703. Centre de Géostatistique, Ecole des Mines de Paris (1981)

51. Matheron, G.: Deux autres familles, non moins discrètes mais plus nombreuses. Technical Report N-717. Centre de Géostatistique, Ecole des Mines de Paris (1981)

52. Matheron, G.: The selectivity of the distributions and the second principle of geostatistics. In: Verly, G. et al. (eds) Geostatistics for natural ressources characterization. Reidel, Dordrecht (1984)

53. Matheron, G.: Isofactorial models and change of support. In: Verly, G. et al. (eds) Geostatistics for natural ressources characterization. Reidel, Dordrecht (1984)

54. Matheron, G.: Changement de support en modèle mosaïque. Sciences de la Terre, Série Informatique Géologique, **20**, 435–454 (1984)

55. Matheron, G.: Change of support for diffusion-type random functions. Math. Geol., Série Informatique Géologique, **17-2**, 137–165 (1985)

56. Matheron, G.: Suffit il pour une covariance d'être de type positif? Sciences de la Terre, Série Informatique Géologique, **26**, 51–66 (1987)
57. Matheron, G., Beucher, H., de Fouquet, C., Galli, A., Guérillot, D., Ravenne, C.: Conditional simulation of a fluvio-deltaic reservoir. SPE paper 16753. 62nd Annual Technical Conference and Exhibition of SPE, Dallas, 591–599 (1987)
58. Matheron, G.: (1988) Simulation de fonctions aléatoires admettant un variogramme concave donné. Sciences de la Terre, Série Informatique Géologique, **28**, pp. 195–212
59. Matheron, G.: Estimating and choosing. Springer, Berlin (1989)
60. Matheron, G.: The internal consistency of models in geostatistics. In: Armstrong, M. (ed) Geostatistics. Kluwer (1989)
61. Matheron, G.: Two classes of isofactorial models. In: Armstrong, M. (ed.) Geostatistics. Kluwer, Dordrecht (1989)
62. Matheron, G.: Une conjecture sur la covariance d'un ensemble aléatoire. Cahiers de Géostatistique, Fasc. 3. Ecole des Mines de Paris, 107–113 (1993)
63. Matheron, G.: Treillis compacts et treillis coprimaires. Technical Report N-5/96/G, Centre de Géostatistique, Ecole des Mines de Paris (1996)
64. Metropolis, N., Ulam, S.: The Monte Carlo method. J. Amer. Stat. Assoc. **44**, 335–341 (1949)
65. Molchanov, I.: Early years of random sets. This volume
66. Mosegaard, K., Tarantola, A.: Probabilistic approach to inverse problems. In: Lee, W. (ed) International Handbook of Earthquake and Engineering Seismology, Academic Press, San Diego (2002)
67. Préteux, F., Schmitt, M.: Boolean texture analysis and synthesis. In Serra, J. (ed.) Image Analysis and Mathematical Morphology, Theoretical Advances. Academic Press, London (1988)
68. Propp, J.G., Wilson, D.B.: Exact sampling with coupled Markov chains and applications to statistical mechanics. Rand. Struct. and Alg., **9**, 223–252 (1996)
69. RamaRao, B.S., LaVenue, A.M., Marsily, G. de, Marietta, M.G.: Pilot point methodology for automated calibration of an ensemble of conditionally simulated transmissivity fields. Water Resources Research, **31–3**, 475–493 (1995)
70. Roth, C., Chilès, J.P., Fouquet, C. de: Combining geostatistics and flow simulators to identify transmissivity. Advances in Water Resources, **21–7**, 555–565 (1998)
71. Roth, C., Fouquet, C. de, Chilès, J.P., Matheron, G.: Geostatistics applied to hydrogeology's inverse problem: taking boundary conditions into account. In: Baafi, E.Y., Schofield, N.A., (eds.) Geostatistics Wollongong '96. Kluwer, Dordrecht (1997)
72. Schmitt, M.: Estimation of intensity and shape in a nonstationary Boolean model. In: Jeulin, D. (ed.) Advances and applications of random sets. World Scientific, Singapore (1997)
73. Sénégas, J.: Méthodes de Monte Carlo en vision stéréoscopique. Application à l'étude de modèles numériques de terrain. PhD Thesis, Ecole des Mines de Paris (2002)
74. Sénégas, J., Wackernagel, H., Rosenthal, W., Wolf, T.: Error covariance modeling in sequential data assimilation. Stoch. Env. Res. and Risk Ass., **15**, 65–86 (2001)
75. Shinozuka, M., Jan, C.M.: Digital simulation of random processes and its applications. J. of Sound and Vibr., **52–3**, 591–611 (1972)

76. Serra, J.: Fonctions aléatoires de dilution. Technical Report C-12. Centre de Morphologie Mathématique, Ecole des Mines de Paris (1968)
77. Serra, J.: Boolean random functions. In Serra, J. (ed.) Image Analysis and Mathematical Morphology, Theoretical Advances. Academic Press, London (1988)
78. Stoyan, D., Mecke, K.R.: The Boolean model: Matheron till today. This volume
79. Switzer, P.: Inference for spatial autocorrelation functions. In: Verly, G., et al. (eds) Geostatistics for Natural Resources Characterization. Reidel, Dordrecht (1984)
80. Tarantola, A.: Inverse Problem Theory: Methods for Data Fitting and Model Parameter Estimation. Elsevier Science Publishers B.V., Amsterdam (1987)
81. Verlaan, M., Heeminck, A.W.: Tidal flow forecasting using reduced rank square root filters. Stochastic Hydrology and Hydraulics, **11–5**, 349–368 (1997)
82. Wolf, T., Senegas, J., Bertino, L., Wackernagel, H.: Data assimilation to three-dimensional hydrodynamics: the case of the Odra lagoon. In: Monestiez, P. *et al* GEOENV III – Geostatistics for environmental applications. Kluwer, Dordrecht (2001)
83. Zinn, B., Harvey, C.F.: When good statistical models of aquifer heterogeneity go bad: A comparison of flow, dispersion and mass transfer in multigaussian and connected conductivity fields. Water Resources Research, **39–3**, 1051 (2003)

Flow in porous media: An attempt to outline Georges Matheron's contributions

J.P Delhomme[1] and G. de Marsily[2]

[1] Schlumberger Water Services
[2] Université de Paris et Académie des Sciences

1 A brief assessment

During the last three decades of the 20th century, the application of a stochastic approach to flow in porous media has certainly been a very active research area, all over the world. Most of the eminent scientists working abroad in the domain (e.g., Dagan, Neuman, Gelhar, Gutjahr...) happened to visit the Ecole des Mines de Paris before 1980 and, more precisely, they spent some time in two small research labs that were located close to each other in Fontainebleau: the Laboratory for Mathematical Hydrogeology, and the Center for Geostatistics. All of them therefore had the opportunity to meet Georges Matheron who was, at that time, heading the Center for Geostatistics. Strangely enough, these visitors then perceived Matheron more as the inventor of kriging than as the author of "Elements pour une Théorie des Milieux Poreux", a book he had written as early as 1967[41], i.e. only two years after his first book on random functions (RF) and on the estimation of regionalized variables.

Georges Matheron's interest in porous media was formed quite early in his career. His pioneering work in both geostatistics and mathematical morphology definitely originated in mining industry problems, but in the 60s, the petroleum industry offered Georges Matheron one of those problems with a high potential for both down-to-earth applications and abstract speculation that always triggered his best theoretical work. The question was: was it possible to characterize the geometry of a porous medium (i.e., its texture) through some image analysis technique and derive its permeability from the textural characteristics? Appearing in thin sections as a binary (pore/grain) two-dimensional image, a porous rock could be described by the mathematical morphology techniques that Matheron was developing at that time. There is no doubt that this aspect dominated in Matheron's attraction for this research topic. The connection between pore morphology and fluid flow parameters was never satisfactorily established, but Matheron unraveled the knot of how Darcy's law comes out of Navier-Stokes' law. Carried along by his momentum, he then investigated the scale change (or effective permeability) problem, and created a first link between geostatistics and physics in his 1967 book on flow in porous media. This book was initially published as a series of four articles in 1966-1968 in the "Revue de l'Institut Français du Pétrole".

The initial limited impact of this book on the researchers working on groundwater flow did not only proceed from the fact that it was quite difficult to read – a common feature of most of Matheron's theoretical writings. Another reason was that this work somehow missed the main problems in subsurface flow. Hydrogeologists had little interest in thin-section analysis or even in local-scale issues. In their daily practice, they were dealing with shallow aquifers in unconsolidated sediments, coarse alluvium or chalk. Their problems were those of full-scale aquifers, with large-scale heterogeneities related to the sediment deposition processes and complex boundary conditions, not that of the pore scale. Even the oil industry, where plugs were taken on cores to measure local permeability values and thin-sections were sometimes analyzed, Matheron's contribution was used mostly for the upscaling problem; the passage from pore-scale $2D$ images to the permeability tensor was only obtained in the late 1990s, with scanning images and the development of parallel super-computing tools.

In the final analysis, if Georges Matheron has undoubtedly been instrumental in the development of a stochastic approach to flow in porous media, it is in many other ways than by his 1967 book on porous media. Although he did not personally publish much on fluid flow afterwards, i.e. only two papers on dispersion, in 1980 with Marsily [47] and in [16], an article on Darcy's Law in 1984 [48], one on 1-D macroscopic dispersion in [49], a last note on upscaling in [50], and a late paper on the inverse problem with Roth, Fouquet and Chilès in [59], Matheron's decisive influence is to be found in the theoretical guidance he constantly provided to the Fontainebleau researchers, from Delhomme's initial introduction of kriging and geostatistics into groundwater hydrology [10, 11, 12, 13] to Dong's work on the application of generalized covariances [17], in his constant ability to suggest new ideas and, when needed, to correct misconceptions, for example, when some researchers initially attempted, in their enthusiasm for the new stochastic approach, to erroneously impose a stationary covariance on the head fluctuations around a regional slope. But, in the end, Georges Matheron's main impact on stochastic hydrogeology may well reside in the fact that, without kriging and conditional simulations, this domain may simply never have come into being, at least not in the shape it now has. See also Neuman [55] on this aspect.

In this paper, we will first summarize Matheron's work at the pore scale, about Darcy's law, upscaling and hydrodynamic dispersion, and then his major contributions to the field of macroscopic stochastic hydrogeology.

2 Emergence of Darcy's law

The term "emergence" was, we believe, first used by Matheron to mean that Darcy's law "emerged" from the underlying Navier-Stokes law, which applies at the pore scale. "Emerging" means here that it is the outcome of the averaging which is made by the flow in a large volume of porous medium, just as, for instance, the normal distribution "emerges" from the averaging of random numbers, through the Central Limit Theorem.

Darcy's law is an empirical linear relationship between the macroscopic filtration velocity $\langle \mathbf{u} \rangle$, and the macroscopic pressure gradient $\langle \operatorname{grad} p : \langle \mathbf{u} \rangle = -(k/\mu) \ \langle \operatorname{grad} p \rangle$. Here, $\langle . \rangle$ stands for a spatial or an ensemble average. k is called the permeability (dimension squared length), a parameter specific to each soil or rock, which can be a scalar or a second-order tensor if the medium is anisotropic, and μ is the dynamic viscosity. Physically, Darcy's law is the result of the integration of the Navier-Stokes equations in the very complex geometry of the pore space. Navier-Stokes is the general equation of fluid mechanics for Newtonian fluids. However, since this pore geometry is, in general, unknown, it is impossible to systematically derive Darcy's law and the value of the tensor k from Navier-Stokes, except when a very simple geometry of the pore space is assumed (e.g., cylindrical tubes or fissures of constant aperture, see e.g. [36].

The general linear form of Darcy's law, and some properties of the permeability tensor k, can however be rigorously established. This is done for an incompressible fluid in steady-state flow while assuming that the microscopic velocity $\mathbf{u}$ is small enough to neglect the inertial term in the Navier-Stokes equations (this is quite acceptable in practice since the flow velocity in porous media is, in general, very small). Furthermore, these equations are written without the body forces F, only for the sake of simplicity. With these assumptions, the Navier-Stokes equations are reduced to a linear version, namely the Stokes equations. In the pore space, these equations then reduces to :

$$\mu \nabla^2 \mathbf{u} = \operatorname{grad} p \quad \text{and} \quad \operatorname{div} \mathbf{u} = 0$$

Here, ∇^2 is the Laplace differential operator $\Sigma_i \partial^2/\partial \mathrm{x}_i^2$, p is the microscopic fluid pressure, and $\mathbf{u}$ the microscopic fluid velocity vector. Let the porous medium be considered as a stationary and ergodic random set, and let $\omega(\mathrm{x})$ be the microscopic porosity $-\omega(x) = 1$ in the pores and 0 in the grains. The solution of the microscopic flow problem in the entire domain can be described as finding a stationary random velocity $\mathbf{u}(\mathrm{x})$ that satisfies :

$$\mu \nabla^2 \mathbf{u} = \omega(x) \operatorname{grad} p; \quad \operatorname{div} u = 0$$

$$\mathbf{u}(x) = \omega(x)\mathbf{u}(x); \quad E(\mathbf{u}) = \langle \mathbf{u} \rangle$$

To establish the existence and uniqueness of this solution, Matheron [41] proposed the use of a variational principle to represent the energy dissipation by the viscous forces. The power dissipated per unit volume at the microscopic level is :

$$W = -\mathbf{u} \cdot \operatorname{grad} p \tag{1}$$

It is then possible to show that the random velocity $\mathbf{u}$, which minimizes the mathematical expectation $E(W) = -E(\mathbf{u} \cdot \operatorname{grad} p)$ while satisfying div $\mathbf{u} = 0$ and $\mathbf{u}(x) = \mathbf{u}(x)\omega(x)$, is also the solution of the Stokes equation. If $\mathbf{u}$ is then extended to the grains ($\mathbf{u} = 0$ (in the grains), the relationship $E(W) = -E(\mathbf{u} \cdot \operatorname{grad} p) = -\mu E(\mathbf{u} \cdot \nabla^2 \mathbf{u})$) can be extended over the whole space. Furthermore, for a stationary $\operatorname{grad} p$ and a stationary $\mathbf{u}$ with div $\mathbf{u} = 0$, we have:

$$E(\mathbf{u} \cdot \operatorname{grad} p) = E(\mathbf{u}) \cdot E(\operatorname{grad} p) \ . \tag{2}$$

Then :

$$E(W) = -E(\mathbf{u} \cdot \operatorname{grad} p) = -\langle \mathbf{u} \rangle \cdot \langle \operatorname{grad} p \rangle \tag{3}$$

which means that the averaging conserves the energy : the average of the microscopic energy dissipation is equal to the energy dissipation at the macroscopic level.

It can then be shown that the macroscopic Darcy's law $\langle \mathbf{u} \rangle = -(k/\mu)\langle \operatorname{grad} p \rangle$ derives from the linearity of the Stokes equation $\mu \nabla^2 \mathbf{u} = \operatorname{grad} p$ and from this conservation of energy. Furthermore, it can be shown that the permeability tensor k is symmetric and positive definite.

This contribution by Matheron in his 1967 book is a major theoretical result.

3 Upscaling

At the macroscopic level, the permeability $k(x)$ can also be regarded as a Random Function, for studying the behavior of heterogeneous porous media where the macroscopic parameter $k(x)$ varies in space. It is then of interest to again average the permeability in order to obtain the equivalent homogeneous permeability at a larger scale. Again, using the variational principle, Matheron gives an upper and a lower bound to this average permeability $\langle k \rangle$ at the large scale:

$$[E(k^{-1})]^{-1} < \langle k \rangle < E(k) \tag{4}$$

Expressed in words, the average $\langle k \rangle$ always lies between the harmonic and the arithmetic mean of the local permeability value. The harmonic mean is the obvious average for one-dimensional flow. This result had been known since the early 1940s, but Matheron re-established it in a different way with the variational principle.

Furthermore, it is possible to show [41], for porous media; [32] in electrodynamics) that, in two dimensions, and for macroscopic parallel flow conditions, the average permeability $\langle k \rangle$ is exactly the geometric mean if the multivariate probability distribution functions of k is log-symetric, e.g. log-normal :

$$\ln\langle k \rangle = E(\ln k)$$

The geometric mean, which is an arithmetic averaging in the log space, always lies between the harmonic and the arithmetic means. This result has been extended to three dimensions, using a perturbation approach, initially as a conjecture, e.g. by King [29], Dagan [8], Indelman and Abramovich [27], de Wit [15], and now confirmed by Noetinger [56]. According to these authors, the general expression for the average value of a log-normal permeability distribution is thus, for an isotropic medium in 1, 2 or 3 dimensions:

$$1/\langle k \rangle = \langle 1/k \rangle \qquad \text{in 1-D, the harmonic mean}$$
$$\ln\langle k \rangle = \langle \ln k \rangle \qquad \text{in } 2D, \text{ the geometric mean}$$
$$\langle k \rangle = \langle k^{1/3} \rangle R^3 \qquad \text{in 3-D, a power average with exponent } 1/3$$

These expressions are of course approximations (higher-order developments), but Matheron in 1993 stated that the 1/3 power is authorized in 3-D. Most field studies show that the experimental PDF of the permeability of rocks is indeed log-normal, therefore these expressions are now commonly used. For radial flow systems, or for transient conditions, average permeabilities have not yet been established theoretically but the above results have been shown to be applicable by numerical experiments with flow models. See also [58, 51, 9] for the problem of up-scaling permeabilities.

For the radial flow case, Matheron worked in steady-state following the approach developed by Schwydler [63, 64, 65, 66, 67], as he could read the Russian literature. He concluded that the average permeability can vary anywhere between the harmonic and the arithmetic mean, depending on the local value at the well, and on the distance to the outer (assumed circular) limit of the domain. More generally, for the non-uniform flow, Matheron [41] stated :

"Unfortunately, one has to conclude that in a medium with regionalized permeabilities, there is no macroscopic Darcy law able to generally describe non-uniform flow".

This statement has been questioned many times, since the mathematical development on which it is based is unpublished, most likely lost, and according to Matheron himself, particularly cumbersome.

In 1993, Matheron produced a last paper on this issue, giving bounds for the average permeability form the Schwarz inequalities.

4 The work on dispersion

Matheron's interest for the Theory of Dispersion probably resulted from a series of discussions in the early 1990s with Jean Fried, a member of Emsellem's group in Fontainebleau who also worked with Matheron, when Jean Fried published his book on dispersion [22].

His interest arose because the classical dispersion theory assumes that the heterogeneity of the velocity field generates a dispersive flux with respect to the average velocity, which is Fickian, i.e. can be expressed according to Fick's Law :

$$F = -\mathsf{D} \cdot \operatorname{grad} C \tag{5}$$

where F is the solute dispersive flux, D is the anisotropic dispersion tensor, and C the volumetric concentration. In a certain way, this theory states that Fick's law "emerges" from the distribution of the microscopic velocity, just as Darcy's law "emerges" from Navier-Stokes' law. In Bear's classical book [1], it is stated that this dispersion tensor may need a certain time to reach its "asymptotic" value, but that effect had always been ignored. Matheron questioned the statement that there was necessarily a constant asymptotic value and worked on a stochastic definition of the velocity field.

As hydrodynamic dispersion is the result of the heterogeneity of the velocity field, the stochastic approach seemed particularly well suited to represent this variability. Transport was studied in the ordinary space R^n $(n = 1, 2, 3)$ with the following simplifying assumptions :

1. The velocity variations of the fluid in the medium is the dominant mechanism, molecular diffusion is negligible.
2. The Eulerian microscopic velocity field $\mathbf{u}$, which is assumed unknown, can be regarded as a stationary random process, i.e. $\mathbf{u}$ is a vectorial stationary random function, and $\mathbf{u}$ is conservative, i.e. $\operatorname{div} \mathbf{u} = 0$. This means that the flow is in steady state conditions and that the porosity is constant.
3. A slug of tracer is injected at time $t = 0$ at the origin $\mathbf{X} = 0$ of the system (here, $\mathbf{u}$ and $\mathbf{X}$ are vectors, with components u^i or X^i). A subscript t denotes the time : $\mathbf{X}_t$). The transport can be described by giving, as a function of time, the position $\mathbf{X}_t$ of a particle injected at time $t = 0$ at the origin. Kolmogorov [31] has shown that if the particle is transported by advection and diffusion (Brownian motion), the probability density $\rho(\mathbf{X}_t)$ of the particle is identical to the concentration obtained by solving the classical transport equation for a slug injection of tracer.

Let $\mathbf{V}(t) = \mathbf{u}(\mathbf{X}_t)$ be the Lagrangian velocity, i.e. the velocity of a particle following its trajectory along a flow path. Matheron (unpublished note, 1981; see also [16] has shown that if $\mathbf{u}$ satisfies assumptions 1, 2, 3 given above, then $\mathbf{V}$ is a stationary random function with the same probability distribution function as $\mathbf{u}$. We can now write :

$$\mathbf{X}_t = \int_0^t \mathbf{V}(\tau)d\tau \ .$$

We then have :

$$E(\mathbf{X}_t) = \int_0^t E\left[\mathbf{V}(\tau)\right]d\tau = tE(\mathbf{V}) = tE(\mathbf{u}) = \bar{\mathbf{u}}t$$

where $\bar{\mathbf{u}} = E(\mathbf{u})$. Thus the average position of the particle is just the average velocity multiplied by the time. Let us now determine the variance of this position; this variance is now an $n \times n$ matrix. Superscript T denotes the transposition of a vector :

$$Var(\mathbf{X}_t) = E\left\{[\mathbf{X}_t - E(\mathbf{X}_t)]^T [\mathbf{X}_t - E(\mathbf{X}_t)]\right\}$$

$$Var(\mathbf{X}_t) = \int_0^t \int_0^t \left\{E\left[\mathbf{V}(\tau)^T V(\tau')\right] - \bar{\mathbf{u}}^T \bar{\mathbf{u}}\right\} d\tau d\tau'$$

$$Var(\mathbf{X}_t) = \int_0^t \int_0^t E\left\{[\mathbf{V}(\tau) - \bar{\mathbf{u}}]^T [\mathbf{V}(\tau') - \bar{\mathbf{u}}]\right\} d\tau d\tau'$$

$$Var(\mathbf{X}_t) = 2 \int_0^t (t - \tau)C(\tau)d\tau$$

where $C(t)$ is the $n \times n$ covariance matrix of the components of the Lagrangian velocity $\mathbf{V}$ taken with a time lag t.

The variance of the position of the particle is the equivalent of the "spreading" of the pulse of tracer around its mean position; it is therefore related to the dispersion tensor, as shown by Einstein [18] :

$$\mathsf{D} = \frac{1}{2}\frac{d}{dt}[Var(\mathbf{X}_t)] = \int_0^t C(\tau)d\tau$$

Matheron drew very important conclusions from this simple result :

1. The dispersion tensor D is a function of time, and not a constant. As each component of the tensor varies with time, there is *a priori* no reason why the principal directions of this tensor should remain constant, as was generally thought. The longitudinal direction (the major principal component of the tensor) will in general not coincide with the direction of the average velocity, as was usually assumed.

2. If the covariance matrix $\mathbf{C}$ of the Lagrangian velocity is well behaved, i.e. $\mathsf{C}(t) \to 0$ sufficiently rapidly as $t \to \infty$, one can assume that the integral of $\mathsf{C}(t)$ will become constant as $t \to \infty$. Thus one can, in general, expect that after a certain time, a constant dispersion tensor will emerge, which would be called an asymptotic Fickian behavior.

3. The dispersion tensor is a direct function of the Lagrangian velocity field. It is not a function of the properties of the porous medium only. If a new Eulerian velocity field is created, e.g. going from parallel flow to radial flow, or changing the vertical/horizontal velocity ratio, then the Lagrangian velocity field (and the flow path) will be changed and also the covariance matrix of this velocity. The dispersivity tensor (the components of the dispersion tensor divided by the average velocity) is therefore not an intrinsic property of the medium, independent of the flow field. This was a major contribution to the dispersion theory, not sufficiently recognized. See also [36].

4. Only in the case where the probability distribution function of the Eulerian velocity field **u** is multi-Gaussian is it possible to show that the transport equation equivalent to the particle position is :

$$\sum_j \sum_k \int_0^t C^{jk}(\tau)d\tau \frac{\partial^2 C}{\partial x^j \partial x^k} - \sum_j \bar{u}^j \frac{\partial C}{\partial x^j} = \frac{\partial C}{\partial t}$$

5. This is similar to the dispersion equation where the dispersion tensor is made a function of time. For all other distributions of velocity, there is no equivalent dispersion equation for early times until the asymptotic behavior is reached. There is in fact very little reason why the Eulerian velocity field should have a Gaussian distribution. One can therefore conclude that there is no correct dispersion equation representing transport for early times.

In 1980, Matheron and Marsily published an example of this theory applied to the stratified case [47], where the velocity field was perfectly parallel to the stratification. They therefore considered a $2D$ velocity field, $u(z)$, constant in the x direction, parallel to the bedding, but random in the perpendicular z direction. The formal calculation of the longitudinal dispersion coefficient in the x direction as a function of time is then relatively straightforward. This random velocity field was assigned a covariance function in the z direction. Using the above approach, but with the inclusion of a local diffusion by adding a Brownian motion to the particle displacement, it was easy to show that the existence of a constant asymptotic dispersion was only possible if the integral of the covariance $\int_0^\infty C(z)dz$ of the velocity was zero. Otherwise, the dispersion coefficient would increase constantly. A similar result had been obtained by Gelhar et al [24], but in a different form. The type of covariance with a zero integral is called a "hole effect" covariance in geostatistics (because it is initially positive, then becomes negative and then tends towards zero, in order that the integral be zero), and is relatively rare. Thus, for stratified media, there would not, in general, be an asymptotic constant dispersion coefficient. Although the case of a stratified medium with a velocity not strictly perpendicular to the bedding (i.e. with a component of the velocity orthogonal to the bedding) was shown to give a constant asymptotic dispersion coefficient, for any type of covariance of the velocity field, the "abnormal" case without any asymptotic behavior opened the way for very profound changes in the analysis of dispersion, taken over by physicists, and used for other media, for which Matheron's and Marsily's 1980 paper [47] is considered as the initiator of the work on "abnormal diffusion" (see [3, 4], and has been cited more than 200 times in the literature.

5 The very beginning of macroscopic stochastic hydrogeology

In the early 1970s, researchers in hydrogeology were facing a dual problem in their first attempts to use computers to build numerical models of aquifers: how to estimate, from the scarce data obtained by pumping tests, the transmissivity (T) values to put as a first guess into each mesh of the discretized models, and how to estimate, from more easily available and therefore more numerous piezometric data, the head (H) values to use as calibration data for the models. Two other concerns were: initially, to assess the confidence level to ascribe to the head values used in the model calibration, and eventually to evaluate the uncertainty attached to the head level and flow rate forecasts given by the numerical models.

The closeness of Matheron's research team was a real blessing for Delhomme, who was by then a new PhD student, arriving in the small research group on Mathematical Hydrogeology headed by Emsellem and Marsily. Having benefited from Matheron's teaching of geostatistics, he started to compute variograms and to use kriging, soon observing that it was better to work on transmissivity data in log scale, and that the existence of a drift in piezometry had to be faced with adapted tools. Concurrently, Marsily was starting to tackle the inverse problem: i.e., find a permeability (or rather transmissivity, i.e. permeability times thickness) field that, at the same time, matches the pumping test data and is consistent with the piezometric measurements. Matheron's kriging technique was also to prove, later on, to be fundamental in this work, as will be shown below.

Whereas the range (or integral scale) values initially found from raw transmissivity variograms were desperately small, it soon appeared, through the work on Y=lnT, that, apart from the very small scale variability captured by the nugget effect, quite large ranges existed in aquifers. Delhomme [12] mentions values from one kilometer for unconsolidated alluvial aquifers to more than 10 km for consolidated aquifers, such as limestone or chalk. In an extensive survey of 20 unconsolidated aquifers and 10 consolidated ones, Hoeksema and Kitanidis [26] give similar orders of magnitude.

Matheron in his 1974 note [45] on proportional effect and lognormality gave a clue to this need to work in log-transmissivity. If $Y(x) = \ln T(x)$ is multinormal with average $m(x)$ and variance σ^2, then: the variogram $\Gamma_T(h)$ of T is proportional to $(1 - exp(-\gamma_Y(h)))$, where γ_Y is the variogram of Y. This explained why short ranges could be observed in raw T variograms, even in cases where the $\ln T$ variogram was unbounded, at least within the size of the aquifer: if $\gamma_Y(h)$ is linear at the scale of the study, then $\Gamma_T(h)$ is exponential and the apparent range of T is actually related to the slope of the linear variogram, i.e. to the variability of T.

Another issue was not clear at that point and became so only after more thinking. Rewriting for instance the partial differential equation describing an aquifer in steady-state flow in the absence of recharge or pumping, i.e. :

$$\operatorname{div}(T \operatorname{grad} H) = 0$$

as :

$$\Delta H = -\operatorname{grad}(lnT)\operatorname{grad} H$$

made it clear that the gradient of lnT was intrinsically involved in the flow equation, regardless of whether T is lognormally distributed or not. But such deep insight into the partial differential equation was not current in the early 70s. Regarding piezometry, the focus was still on the impact of a drift on the raw variogram. The variogram constructed from the head data of the Crau aquifer (South of France) shown by Delhomme [10] is apparently anisotropic, with a rapid increase in the NE-SW direction that corresponds to the general groundwater flow. In this particular case, one may consider the NW-SE variogram as free of the drift effect. But is the assumption of an isotropic underlying variogram for H correct? Furthermore, there is not always a direction without a drift. It was thus necessary to reconstruct the underlying variogram by separating the drift from the residuals in order to use universal kriging for the head, and ad hoc methods were imagined. However, the solution was not unique.

An overall conceptual framework that could handle both transmissivity and head problems was slow to emerge. Georges Matheron appears to have been instrumental in developing it, through his work on intrinsic random functions of higher orders (IRF-k). An IRF-k is a random function with stationary increments of order k. It soon became apparent that this was particularly well suited to represent the non-stationary head. Automated identification techniques were developed for the generalized covariance (GC) functions of IRF-k's, which made it possible to routinely use kriging for piezometric surfaces, based on local head measurements [10, 11]. Boundary conditions which are very strong constraints on piezometric surfaces were introduced in kriging, by specifying that the gradient component normal to a no-flow boundary is zero [13, 6] p.322-323).

6 The importance of the conditioning effect

Before the end of the 70s, another technique, besides kriging, also due to Matheron's creative mind, caused a breakthrough in hydrogeology: conditional simulations. The principle is: first, construct a non-conditional simulation of a Gaussian random function, i.e., a realization of an RF that has the same (generalized) covariance function as the studied phenomenon but does not otherwise take the data into account; then, pass to a conditional one by adding kriging errors picked from the non-conditional simulation to the kriging estimator of the studied phenomenon. The latter simulation retains the structural features of the former and is calibrated on the data at the same time. To construct non-conditional simulations of a Gaussian random function, Matheron also invented the turning band method (1973) that consists in adding up a large number of independent simulations defined on lines scanning the plane and with constant values orthogonal to those lines. This method was later made known in the English-language literature by [34] and became quite popular. Journel's sequential simulation algorithm, which directly generates conditional simulations, was only developed in the 1980's [28, 14], see also Delfiner and Haas, this volume, page 89).

In 1975, Freeze had the idea of using Monte-Carlo simulations to study how the uncertainty on transmissivity could translate into an uncertainty on head, through a flow simulator. He had however disregarded two aspects : the spatial correlation of transmissivity and the conditioning on $\ln T$ local measurements. The authors vividly remember Matheron, after having read Freeze's paper which concluded that the uncertainty in flow modeling was enormous due to the uncertainty on transmissivity, urging them to perform conditional transmissivity simulations to estimate the reduction on the uncertainty brought by conditioning on T measurements. He was also aware of Gelhar's approach, initiated in 1976 [23], where he included spatial correlation and used the perturbation method to solve analytically the stochastic flow equations, but Matheron considered that the variance of $\ln K$ or $\ln T$ in real media was much too large for the perturbation method to be used. Only stochastic Mont-Carlo simulations were, to Matheron, appropriate. Delhomme (1979a) used conditional simulations to characterize the effect of conditioning, by studying a head field (H_C) that corresponds to a log-transmissivity field (Y_C) conditioned on $\ln T$ measurements. His paper was a breakthrough in stochastic modeling and, with more than 200 citations to date, is probably one of the most cited references resulting from Matheron's ideas in hydrogeology.

Looking for the reduction of the head variance due to conditioning by Y data, Delhomme [12] considered a rectangular aquifer with no-flow conditions on three boundaries and a prescribed head condition on the 4th one, with a uniform recharge over the rectangle, the variogram of log-transmissivity Y being linear with a nugget effect. From the same 30 data, he generated 50 different conditional simulations Y_C of Y. After generating the values of Y_C over the grid meshes of a discretized aquifer model, he solved the flow equation numerically and determined the head values H_C for each realization. From those 50 grids, the second-order statistical moments of the head were computed, and maps of the expected value and standard deviation of H_C were drawn. Delhomme stated that the conditioning effect was small in his case study. The reason was that most data points were located in the central part of the aquifer. Therefore, little conditioning was exerted on Y near the aquifer outlet boundary that was controlling the head in the entire domain. This demonstrates, if needed, that an aquifer problem is a whole, where head boundary conditions are of paramount importance. In his concluding remarks, Delhomme mentioned the need for conditioning to the measured values of both transmissivity and head, which implicitly leads to solving an inverse problem.

In an internal report [2], p.3-8), Delhomme also described conditional simulations that were directly computed, this time, from head data; the problem there was to prepare different plausible piezometric maps to be used as input for the inverse problem of the Oise aquifer, near Origny-Sainte-Benoite, France. A single generalized covariance was initially used over the whole aquifer. Surprisingly, in low-gradient areas, the head surfaces showed some local troughs and humps that could not be related to any pumping or recharge. The idea emerged to locally adjust the generalized covariance, so as to get rid of the unwarranted fluctuations.

The underlying problem was in fact that, when a single variogram is used for kriging over the whole domain, the kriging standard deviation map reflects merely the spatial pattern of data points, and the same holds true for the amplitude of fluctuations around the kriging estimates in conditional simulations. What was wrong in the initial assumption of a unique variogram was that, in the Oise aquifer, the hydraulic gradient was not as constant as, for instance, in the Crau aquifer and that the variogram, for the piezometric head, must actually be proportional to the square of the hydraulic gradient (the drift magnitude). The smaller this gradient, the smaller the amplitude of the head fluctuations around the drift. But all that had not yet been formally established, by the time of the Origny study.

7 From IRF-k's to stochastic PDE's

As described above, a general conceptual framework was slow to emerge. Matheron [43], in his 1971 article on the IRF-k theory, had noted that if Q is a continuous IRF-k, there exists a unique twice differentiable IRF-$(k+2)$ H satisfying the Poisson equation $\Delta H = Q$. Matheron also stated that there was a relation between the generalized covariances of H and Q and the generalized cross-covariance of Q and H. For instance, if Q is a stationary RF with zero mean and covariance $C_Q(h)$, the generalized covariance of H is solution of: $\Delta^2 C_H(h) = C_Q(h)$, and the generalized cross-covariance of Q and H is $-\Delta C_H(h)$.

The Poisson equation could have represented the fluid flow in a porous medium of constant unit permeability, H being the head and Q the pumping rate. But this was a temporary dead-end in hydrogeology: a uniform permeability and a spatially continuous pumping rate were not very realistic assumptions and Matheron did not address the problem of boundary conditions, which is an essential one in groundwater flow.

Although Matheron did not initially suggest addressing the aquifer problem with his theory, IRF-k's provided the correct conceptual model to represent the non-stationary solutions of stochastic PDE's. A description of the groundwater flow problem using stochastic PDE's had been introduced, by the end of the 1970's, by researchers like Gutjahr and Gelhar [25] or Dagan [7]. In response to their first attempts, Matheron helped them to clarify the approach, showing that, for a globally unidirectional flow in a 2D infinite aquifer, if the log-tranmissivity Y is stationary, the head perturbation H is not: it is an IRF-0. The variogram of H is smooth at the origin and, even if the covariance of Y is isotropic, the variogram of the head fluctuations H is anisotropic and increases faster perpendicularly to the direction of flow than parallel to it.

Some additional interesting properties of stochastic partial differential equations were found in the 1980's, regarding the stochastic relationship between log-transmissivity and head gradient fluctuations. First, they are locally uncorrelated in the sense that the cross-covariance of $Y(x)$ and $H(x+h)-H(x)$ is by definition zero at $h = 0$. Parallel to the flow, it is positive for all non-zero lags in the downstream direction, and negative in the upstream direction, starting by a linear increase (resp. decrease), then reaching a maximum (resp. minimum) and finally decaying to zero at infinity. The grad H disturbance created by a local change in Y is thus localized. Conversely, this cross-covariance is zero for all lags, perpendicular to flow.

This important remark has to be put in relation with what Neuman observed in 1980 [54] about the spatial structure of the errors on log-transmissivity estimates, after numerically solving an inverse problem. He found that, in the case of a marked head gradient in one direction, the correlation between errors on log-transmissivity decreased more slowly parallel to flow than perpendicularly to it. Head fluctuations actually do not bring information on log-transmissivity across streamlines, only along them. It was also found that the regional hydraulic gradient is involved linearly in the cross-covariance, which means that the information brought by the head was inversely proportional to this regional hydraulic gradient. In a large but finite aquifer, the regional hydraulic gradient depends on the location with respect to the boundaries and the nature of those boundaries. The steeper the head gradient, the bigger the amplitude of the head fluctuations around the drift, as stated above, and the more information on log-transmissivity is carried by the head when solving the inverse problem.

Coping with aquifer boundary conditions has always been a difficult problem. The derivation of the head generalized covariance and the cross-covariance by Dong [17], using the perturbation approach, was still based on infinite aquifer conditions, and tricks were used to handle the boundary conditions. For prescribed head boundaries, an array of fictitious head measurement points along the boundaries was used. For no-flow ones, Delhomme's technique [13] could have been used, with an array of pairs of fictitious data points, perpendicular to the boundaries. Only in 1997, did Roth et al. present a semi-empirical approach to the problem where a numerical solution of the flow equation was used to derive the generalized covariance and cross-covariance models under prescribed head conditions [59]. But, contrary to Dong's, their work was done in the broader context of the inverse problem.

8 Inverse problems, kriging, and cokriging

The inverse problem, i.e. the identification of the transmissivity field from the head field by "inverting" the flow equation, has been the object of numerous publications in the groundwater literature and the reader is referred to a recent review by [37]. In a nutshell, Matheron's contribution was to provide a method to constrain the resulting transmissivity field to be a "regionalized variable" characterized by a mean, a variance and a covariance. Before Matheron's input, the inverse problem was known to be ill-posed, and to require additional constraints to obtain stable and robust transmissivity field estimates (insensitive to small variations in the input head data, and "plausible", as defined by Neuman, [53]. These constraints were seen as "regularization constraints", see [19], such as smoothness, zoning (zones of constant values), upper and lower bounds, etc. By showing that transmissivity fields could be seen as regionalized variables, Matheron opened the way for better stability, robustness and plausibility constraints. One approach [53] was to first krige the transmissivity field based on the local measurements, and then to minimize, in the objective function, a weighted sum of the quadratic errors in head and of the "distance" between the initial transmissivity field given by kriging and the final transmissivity field calculated by the inverse problem. Another approach [35, 20, 5], was to define the unknown transmissivity field by kriging, based on a known variogram, the local known tranmissivity values given by well tests, and on additional unknown values assigned to fictitious wells, which had to be identified by solving the inverse problem. This method, known as the "pilot point method", automatically generated conditioned transmissivity fields, with a prescribed variogram; the number and location of the pilot points (i.e. of the unknowns) was the tool to constrain the regularity of the transmissivity field. This method was later extended to Monte-Carlo conditionally simulated fields (i.e. multiple solutions of the inverse problem) by [57] and [33].

Intrinsically, however, the inverse problem can be seen as a cokriging problem, if the flow equation can be linearized, i.e. if the variance of $\ln T$ is small enough, lower than e.g. 1, and in the Gaussian case. The problem is to identify the tranmissivity based on the local measurements of transmissivity, and on an additional variable, the hydraulic head. These two variables are linked by the linearized flow equation, and, based on the analytical solution of the stochastic flow equation, the knowledge of the covariance of the transmissivity field is enough to calculate the cross-covariance of transmissivity and head, and the covariance of the head (see [17], as mentioned above). The estimation of the transmissivity field by cokriging is then straightforward. Matheron did not suggest cokriging as a means to solve the inverse problem but, in 1983, Kitanidis proposed to use an approach based on cokriging, first in 1D with Vomvoris [30] and, one year later, in 2D with Hoeksema. The first step was to evaluate the cross-covariance, C_{YH}, and the covariance, C_H, from C_Y, the unconditional covariance of Y, taking advantage of the linearization; in fact, at the same time, these authors also identified C_Y from the available data on both $Y = \ln T$ and H; the second step was cokriging itself. Rubin and Dagan (in [60, 61] further developed the approach. But, contrary to the Pilot Point method, this approach requires the variance of $\ln T$ to be small, which is a strong limitation in practice.

9 Conclusion

To summarize, Matheron's contributions to flow and transport in natural media are much greater than the number of papers related to fluid flow that he wrote or was a co-author of. In order of importance, we consider that the introduction of geostatistics into the definition, estimation and simulation of the properties and variables used in hydrogeology and hydrocarbon reservoir engineering was a major breakthrough. Without Matheron's work, the concept of regionalized variables in these disciplines may have been delayed by a number of years, maybe never applied. It made it possible to estimate with greater accuracy aquifer or reservoir properties, and to develop the concept of Monte-Carlo simulations, conditioned on the available information. It also impacted tremendously on the methods used to address the inverse problem.

Although not discussed in this chapter (see Delfiner and Haas, this volume, page 89), an additional major contribution to fluid flow in natural media is the definition of reservoir architecture by the facies simulation method (HERESIM®[3]). Proceeding from the estimation of properties (e.g. permeability) to that of the geologic formation geometry in space made it possible to include much more geological knowledge in aquifer or reservoir studies, and this is a line of research still very active today; it is being combined with the genetic approach, where the aquifers and reservoirs are represented by methods simulating the physical processes of the rock genesis [38]. Boolean methods, as applied e.g. to fractured media, and genetic methods are also an outcome of Matheron's work (see Delfiner and Haas, this volume).

Finally, Matheron's work at the pore scale and on upscaling, although it brought rigor to the definition of the fundamental laws of flow and transport in porous media, was less of a breakthrough in its applications as it remained rather theoretical and still slightly apart from practical applications. But it had a major influence on other fields of physics, e.g. for the basis of the study of abnormal diffusion and more generally for the homogenization of physical properties of random media.

References

1. Bear, J. (1972) Dynamics of fluids in porous media, American Elsevier, New-York.

---

[3] Reservoir modeling tool, registered trademark of Institut Français du Pétrole

2. Besbes, M., Delhomme, J.P., Hubert, P., Ledoux, E., Marsily, G. de (1978) Etude sur modèle mathématique des débits de fuite du réservoir d'Origny Sainte-Benoite., Internal Report, Paris School of Mines, CIG, 78/52, 50 p. plus figures.

3. Bouchaud, J.P. (1990a) Anomalous diffusion in random media: statistical mechanisms, models
and physical applications. Physical Report, 195, 127

4. Bouchaud, J.P., J Koplik, A Provata, S Redner (1990b) Superdiffusive Transport due to Random Velocities, Physical Review Letters, 64, 2503

5. Certes, C., Marsily, G. de (1991) Application of the pilot point method to the identification of aquifer transmissivity, Advances in Water Resources, vol.14, n° 5, p.284-300.

6. Chiles J.P. and Delfiner P. (1999) Geostatistics: Modeling spatial uncertainty" (Chapter 8: Scale effects and Inverse problems p.593-635), John Wiley and Sons, New York .

7. Dagan G. (1989) Flow and Transport in Porous Formations, Springer-Verlag, Berlin.

8. Dagan, G. (1993) High-order correction of effective permeability of heterogeneous isotropic formations of lognormal conductivity distribution. Transp. Porous Media, 12, 279-290.

9. Dagan, G. (2001) Effective, equivalent and apparent properties of heterogeneous media, Proceed. 20th Intern.Congress of Theoretical and Applied Mechanics (H. Aref and J.W. Phillips, eds). Kluwer Academic Press, Dordrecht, The Netherlands.

10. Delhomme, J.P. (1976) Applications de la théorie des variables régionalisées dans les sciences de l'eau, Doctoral thesis, Université de Paris VI.

11. Delhomme, J.P. (1978) Kriging in the hydrosciences, Advances in Water Resources, vol.1, n° 5, p.251-266.

12. Delhomme, J.P. (1979a) Spatial variability and uncertainty in groundwater flow parameters: a geostatistical approach, Water Resources Research, vol.15, n° 2, p.269-280.

13. Delhomme, J.P. (1979b) Kriging under boundary conditions, Oral communication, American Geophysical Union Fall Meeting, San Francisco (figures later reprinted in: Chiles J.P. and Delfiner P. (1999) p.322-323).

14. Deutsch, C.V., Journel, A.G. (1992) GSLIB, Geostatistical Software Library and User's Guide. Oxford University Press, New York.

15. de Wit, A. (1995) Correlation structure dependence of the effective permeability of heterogeneous porous media. Phys. Fluids, 7 (11), 2553-2562.

16. Dieulin, A., Marsily, G. de, Matheron, G. (1981) Growth of the dispersion coefficient with the mean traveled distance in porous media. Int. Symp. on the Quality of Groundwater, Amsterdam, March 1981. In: Quality of Groundwater, Elsevier, 1981.

17. Dong A. (1990) Estimation géostatistique des phénomènes régis par des équations aux dérivées partielles, Doctoral thesis, Ecole des Mines, Paris.

18. Eisntein, A. (1905) Über die von den Molekularkinetischen Theorie der Wärme gefordete Bewegung von in ruhenenden Flüssigkeiten suspendierten Teinchen. Ann. Phys. 17, 539-560.

19. Emsellem, Y., Marsily, G. de (1971) An automatic solution for the inverse problem. Water Resources Research vol.7, n° 5, p.1264-1283.

20. Fasanino, G., J.E. Molinard, G. de Marsily, V. Pelce (1986) Inverse modelling in gas reservoirs. Soc. of Petr.Eng. J, SPE 15592, 15 p.

21. Freeze, R.A. (1975) A stochastic-conceptual analysis of one-dimensional groundwater flow in non-uniform homogeneous media, Water Resources Research, vol.11, n° 5, p.725-741.

22. Fried, J.J., Combarnous, M.A. (1971) Dispersion in porous media. Adv Hydrosc. 7, 169-282.

23. Gelhar L.W. (1993) Stochastic Subsurface Hydrology, Prentice-Hall, Englewood Cliffs NJ.

24. Gelhar, L.W., Gutjahr A.L., Naff, R.L. (1979) Stochastic analysis of macrodispersion in a stratified aquifer. Water Resources Research Vol.15, n° 6, p.1387-1397.

25. Gutjahr A.L. and Gelhar L.W. (1981) Stochastic models for subsurface flow: Infinite versus finite domains and stationarity, Water Resources Research, vol.17, n° 2, p.337-350.

26. Hoeksema R.J. and Kitanidis P.K. (1984) An application of the geostatistical approach to the inverse problem in two-dimensional groundwater modeling, Water Resources Research, vol.20, n° **7**, p.1003-1020.

27. Indelman, P., Abramovich, B. (1994) A higher order approximation to effective conductivity in media of anisotropic random structure. Water Resources Research, vol.30, n° 6, p.1857-1864.

28. Journel, A.G.,(1989). Fundamentals of Geostatistics in Five Lessons, Short Courses of Geology, **8**, AGU, Washington D.C., 1989.

29. King, P. (1987) The use of field theoretic methods for the study of flow in heterogeneous porous media. J. Phys. A. Math. Gen., 20, 3935-3947.

30. Kitanidis P.K. and Vomvoris E.G. (1983) A geostatistical approach to the inverse problem in groundwater modeling, steady state, and one-dimensional simulations, Water Resources Research, vol.19, n° 3, p.677-690.

31. Kolmogorov, A.N. (1931) Über die analytischen Methoden in der Wahrscheinlichkeitsrechnung. Math. Ann. 104, 415-458.

32. Landau, L.D., Lifschitz, E.M. (1960) "Electrodynamics of continuous media". Pergamon Press, Oxford, United Kingdom.

33. Lavenue, A.M., Ramarao, B.S., Marsily, G. de, Marietta, M.G. (1995) Pilot point methodology for automated calibration of an ensemble of conditionally simulated transmissivity fields : Part 2 - Application. Water Resources Research, vol.31, n° 3, p.495-516.

34. Mantoglou, A., Wilson, J.L. (1982) The turning bands method for the simulation of random fields using line generation by a spectral method. Water Resources Research, vol.18, n° 5, p.645-658.

35. Marsily, G. de (1978) De l'identification des systèmes hydrologiques. Doctoral Thesis, Université Paris VI.

36. Marsily, G. de (1986) Quantitative Hydrogeology. Groundwater Hydrology for Engineers. Academic Press, New-York, 440 p.

37. Marsily G. de, Delhomme J.P., Coudrain-Ribstein A. and Lavenue A.M. (2000) Four decades of inverse problems in hydrogeology, in Zhang D. and Winter C.L. eds, Theory, Modeling, and Field Investigation in Hydrogeology, Geological Society of America Special Paper 348, p.1-17.

38. Marsily, G. de, Delay, F., Goncalvez, J., Renard, Ph., Teles, V., Violette, S. (2004) Dealing with spatial heterogeneity. Hydrogeology J., to appear.

39. Matheron, G. (1965) Les variables régionalisées et leur estimation. Masson, Paris.

40. Matheron, G. (1966, 1967, 1968) Structure et composition des perméabilités. April 1966. Genèse et signification énergétique de la loi de Darcy. November 1966. Méthode de Schwydler et règles de pondération. January 196. Composition des perméabilités en milieu poreux hétérogène : critique de la règle de pondération géométrique. March 1968. Revue de l'Institut Français du Pétrole, Rueil Malmaison.

41. Matheron G. (1967) Elements pour une théorie des milieux poreux, Masson, Paris.

42. Matheron G. (1969) Le krigeage universel, Cahiers du Centre de Morphologie Mathématique de Fontainebleau, Fasc.1, Ecole des Mines de Paris.

43. Matheron, G. (1971) La théorie des fonctions aléatoires intrinsèques généralisées. Note Géostatistique N° 117, Technical Report N-252, Centre de Géostatistique, Fontainebleau, France.

44. Matheron G. (1973) The intrinsic random functions and their applications, Advances in Applied Probability, vol. 5, p.439-468.

45. Matheron G. (1974) Effet proportionnel et lognormalité ou: Le retour du serpent de mer, Report N-374, Centre de Géostatistique, Ecole des Mines de Paris.

46. Matheron G. (1979) L'émergence de la loi de Darcy, Report N-592, Centre de Géostatistique, Ecole des Mines de Paris.

47. Matheron, G., Marsily, G. de (1980) Is transport in porous media always diffusive? A counter-example. Water Resources Research, vol.16, n° 5, p.901-917.

48. Matheron, G. (1984) L'émergence de la loi de Darcy. In : Marsily, G. de (Ed.) "Ecoulement dans les milieux poreux ou fissurés", Annales des Mines, 5-6, 1984, p. 11-16.

49. Matheron, G. (1985) Calcul de la dispersion macroscopique dans le cas à une seule dimension. Intern. Symposium on The Stochastic Approach to Subsurface Flow, Montvillargenne, June 3-6, 1985. Intern. Assoc. for Hydraulic Research, G. de Marsily, Ed., Paris School of Mines, Fontainebleau, France.

50. Matheron, G. (1993) Quelques inégalités pour la perméabilité effective d'un milieu hétérogène. Cahiers de Géostatistique, Fascicule 3, Compte Rendu des Journées de Géostatistiques, 25-26 mai 1993, p.1-20. Paris School of Mines publication.

51. Meier, P.M., Carrera, J., Sanchez-Vila, X. (1998) An evaluation of Jacob's method for the interpretation of pumping tests in heterogeneous formations. Water Resources Research, vol.34, n° 5, p.1011-1025.

52. Mizell S.A., Gutjahr A.L. and Gelhar W. (1982) Stochastic analysis of spatial variability in two-dimensional steady groundwater flow assuming stationary and nonstationary heads, Water Resources Research, vol.18, n° 6, p.1053-1067.

53. Neuman, S.P. (1973) Calibration of distributed parameter groundwater flow models viewed as a multiple objective decision process under uncertainty. Water Resources Research, vol.9, n° 4, p. 1006-1021.

54. Neuman S.P. (1980) A statistical approach to the inverse problem of aquifer hydrology, Part 3: Improved solution method and added perspective, Water Resources Research, vol.16, n° 2, p.331-346.

55. Neuman, S.P. (1984) Role of geostatistics in subsurface hydrology. In Geostatistics for Nature Resources Characterization, NATO –ASI, Verly et al, Eds, Part 1, 787-816, Reidel, Dordrecht, The Netherlands.

56. Noetinger, B. (2000) Computing the effective permeability of log-normal permeability fields using renormalization methods. C.R. Acad. Sci. Paris / Earth and Planetary Sciences, 331, 353-357.

57. Ramarao, B.S., Lavenue, A.M., Marsily, G. de, Marietta, M.G. (1995) Pilot point methodology for automated calibration of an ensemble of conditionally simulated transmissivity fields : Part 1 - Theory and computational experiments. Water Resources Research, Vol.31, n° 3, p.475-493.

58. Renard P. and Marsily G. de (1997) Calculating equivalent permeability: A review, Advances in Water Resources, vol.20, n° 5-6, p.253-278.

59. Roth C., Fouquet C. de, Chiles J.P. and G. Matheron (1997) Geostatistics applied to hydrology's inverse problem: Taking boundary conditions into account, in: Geostatistics Wollongong '96, Baafi E.Y. and Shofield N.A. eds, Kluwer, Dordrecht, vol.2, p.1085-1097.

60. Rubin, Y., Dagan, G. (1987a) Stochastic identification of transmissivity and effective recharge in steady groundwater flow. 1. Theory. Water Resources Research, vol. 23, n° 7, p. 1185-1192.

61. Rubin, Y., Dagan, G. (1987b) Stochastic identification of transmissivity and effective recharge in steady groundwater flow. 2. Case study. Water Resources Research, vol. 23, n° 7, p. 1193-1200.

62. Rubin, Y., Gomez-Hernandez, J. (1990) A stochastic approach to the problem of upscaling the conductivity in disordered media : Theory and unconditional numerical simulations. Water Resources Research, vol.26, n° 3, p.691-701.

63. Schwydler, M.I. (1962a) Les courants d'écoulement dans les milieux hétérogènes. *Izv. Akad. Nauk. SSSR, mekh. I mas,* n° 3, 185-190.

64. Schwydler, M.I. (1962b) Courants d'écoulement plans dans les milieux à hétérogénéité aléatoire. *Izv. Akad. Nauk. SSSR, mekh. I mas,* n° 6, -5-71.

65. Schwydler, M.I. (1963a) Sur les caractéristiques moyennes des courants d'écoulement dans les milieux à hétérogénéité aléatoire. *Izv. Akad. Nauk. SSSR, mekh. I mas,* n° 4, 127-129.

66. Schwydler, M.I. (1963b) Sur la précision de la prévision du débit des puits. *Izv. Akad. Nauk. SSSR, mekh. I mas,* n° 5, 148-150.

67. Schwydler, M.I. (1963c) Sur les calculs hydrodynamiques des écoulements de filtration dans les milieux poreux hétérogènes. *Dobyca nefti, teorija i praktica, Annuaire 1963,* 107-108, VNII.

Over Thirty Years of Petroleum Geostatistics

Pierre Delfiner and André Haas

Total

1 Introduction

If someone made a survey at the three major annual petroleum conferences, the AAPG for geologists, the SEG for geophysicists, and the SPE for reservoir engineers, and asked the participants if they knew who Georges Matheron was, the most likely answer would be "no". Yet the words variograms, kriging, multi-realizations, object models, etc., are ubiquitous in the technical sessions of these meetings. Geostatistical concepts and tools have become so widespread in the industry that people tend to treat them as common knowledge, as if they had always been around. But at the origin of these ideas there is Georges Matheron, an exceptional man who had a profound impact on the diffusion of probabilistic thinking in the Geosciences.

We will give a short overview of the history of Petroleum Geostatistics, highlighting its most significant milestones, the breakthroughs that made this approach essential for geology, geophysics, and reservoir. We will illustrate the account with a few representative examples, emphasizing why they were innovative at the time and how we see their limitations, and also their extensions, today in light of our current knowledge. Since this volume is a tribute to Georges Matheron we will deliberately focus on his work or work that he initiated, thus leaving out important topics such as Bayesian methods for example. To conclude we will outline the current trends in Petroleum Geostatistics and say a word about the future. The presentation will remain non-mathematical.

2 Kriging Techniques

Statistical techniques have long been used to estimate mining deposits. Geostatistics as such has been set on firm probabilistic ground during the sixties by Georges Matheron [40, 39] first at the French Geological Survey (BRGM) and then at the Center for Geostatistics of the Ecole des Mines de Paris, in Fontainebleau. One key feature was the introduction of a covariance or a variogram function to quantify spatial variability.

The first applications to petroleum took place in the early seventies and made use of kriging to interpolate structural or petrophysical properties in 2D [18]. Traditional methods such as inverse distance weighting only consider the distances between the data point and the estimated point and take no account of lateral continuity. The analysis of seismic data shows a great difference in variability between travel times to a given seismic marker, which vary continuously, and stacking velocities, which tend to be very noisy. Kriging uses this information and 'adapts' to the interpolated variable, performing as an exact interpolator or as a smoothing operator, whichever is more appropriate. As a bonus it also delivers a variance quantifying the uncertainty attached to the estimate.

Kriging provided a consistent framework for integrating data of different kinds –exact or uncertain measurements, slope information, trends– and also for combining different variables in order to compute hydrocarbon accumulations from grids of thickness, net-to-gross, porosity, saturation, and so on [19].

Commercial computer packages such as Krigepack (Elf, Total, Center for Geostatistics) and Bluepack (Center for Geostatistics) were developed and made geostatistical techniques readily accessible to the petroleum industry. They enabled the construction of reservoir models regarded as a stack of 2D maps of continuous variables. This elementary representation is perhaps sufficient to compute in-place volumes but nowadays appears too simplistic to account for heterogeneities governing underground fluid movement, and therefore the production of hydrocarbons. This led to the development of stochastic methods discussed below.

It may be interesting to step back for a moment and recall that the application of geostatistics to petroleum problems faced two apparently insurmountable difficulties. First, there was the number of wells. At the appraisal or early development stage of a field very few wells are available, typically less than ten, not enough for a meaningful statistical inference of the variogram. Second, there was the stationarity issue. Simple and ordinary kriging and the variogram itself assume the stationarity of the variable or at least of its increments, while most parameters of a petroleum reservoir show spatial trends. The geometry of an oil or gas reservoir must have a shape that ensures trapping of hydrocarbons, for example an anticline. Petrophysical parameters such as porosity or net-to-gross thickness often also show patterns that reflect the depositional mechanism. And of course all geological variables are in some relationship with depth. The keys that unlocked these two doors were the development of a new form of kriging capable of handling the presence of a trend, and the external drift method.

Kriging with a trend

This form of kriging takes into account the presence of a trend function described by a linear combination of space-dependent functions, such as a polynomial in the X- and Y-coordinates. Matheron (1971) called it "Universal Kriging", not because it is universally applicable but because it ensures that the error has zero mean whatever the true unknown polynomial coefficients. This technique is now often called by the more suggestive name of "Kriging with a trend", but Matheron took great care initially to distinguish the notion of 'trend', which is generic and vague, from the notion of 'drift' defined mathematically as the (space-dependent) expected value of the parent random function. In the Universal Kriging model the parent random function $Z(\mathrm{x})$ is modeled as the sum of two components

$$Z(x) = m(x) + Y(x) \tag{1}$$

where $m(x)$ is a polynomial (deterministic) function representing the drift and $Y(x)$ is the 'residual', a zero-mean random function with appropriate stationarity properties. This model is very appealing for geological applications because it accounts both for an interpretable large scale trend and for correlated small scale fluctuations. With Universal Kriging an estimate of $Z(x)$ is obtained without an explicit determination of the two components, and the uncertainty on the drift is reflected in a large kriging variance in extrapolated areas. The only difficulty, a serious one from a theoretical point of view, is the statistical inference of the variogram of $Y(x)$ because the experimental variogram of residuals tends to be biased. Much work has been devoted in the early seventies to the solution of this problem and Matheron [44] even developed a new theory that weakened stationary assumptions to higher-order increments. From a practical point of view, however, this inference problem is not really blocking.

Kriging with an external drift

A modification to Universal Kriging made it possible to address mapping problems with a small number of wells [9]. The idea is to use indirect but densely sampled data to make up for the scarcity of well information. Seismic is the most common and most useful source of additional information, thanks to spectacular advances in acquisition and processing technologies, going from 2D sections to $3D$ cubes, and now even 4D (time-lapse seismic) showing the modifications of reservoir fluid saturations through time. Seismic data allow the geophysicist to define the geometry of reservoirs, but also and increasingly to interpret seismic images in terms of lithology, petrophysics and fluid content by analysis of seismic attributes [17].

Seismic data are introduced in the Universal Kriging model through the drift. To take a specific example, suppose that $Z(x)$ is the depth to a given geological horizon measured at a few well locations and $T(x)$ the seismic travel time to that horizon sampled virtually everywhere. In the simplest case we can write the model as

$$Z(x) = a_0 + a_1 T(x) + Y(x) \tag{2}$$

where a_1 is the average velocity and a_0 a datum plane correction. This looks like a standard regression equation with the important difference that $Y(\mathrm{x})$ is a spatially correlated residual. As a consequence the interpolated map honors the well data, unlike the traditional time-depth functions used by geophysicists. By and large it may be considered that seismic measurements describe the *shape* of the reservoir, whereas borehole data provide *depth control*. This approach is also used extensively to map petrophysical variables using seismic attributes as an external drift.

Collocated cokriging

Another popular technique to estimate the value of a target variable, such as the depth of a geological horizon, using seismic data is the so-called 'collocated cokriging' [55]. This is a particular implementation of co-kriging (multivariate kriging) in which seismic only contributes one value co-located with the estimated point. The underlying idea is that once the seismic value is known at the target point, seismic data at other locations bring little additional information. The cross-covariance between the two variables is assumed proportional to the covariance of the target variable. With this model the co-kriging system is simplified considerably and only requires the specification of a correlation coefficient between seismic and target data at the same point. The approach has been used for mapping porosity from well porosity data and seismic attribute values [12].

An alternative, and more effective, implementation of collocated cokriging is one that also uses the seismic values co-located with the well data [22, 6]. Rivoirard [54] showed that this implementation does not result in any loss of information, i.e. coincide with the full co-kriging solution, provided that the cross-covariance is proportional to the covariance of seismic (which is generally smoother than the target variable). Whatever its implementation, the collocated version of co-kriging is by far the most used in the petroleum industry.

Factorial kriging

Geostatistical techniques are routinely used for Quality Control of seismic data, and in particular velocity analyses [52], [37]. The principle is to distinguish several components in the $3D$ variogram and interpret them either as processing artifacts or as genuine geological features. For example a variogram of stacking velocity may display three components: a nugget effect representing white noise, an anisotropic short-range component attributed to an in-line acquisition artifact, and a long-range isotropic component carrying true geological information. Once these ranges are identified the technique of 'Factorial Kriging' [46] allows the geophysicist to extract from the data the component representing the geological signal. When compared with filtering in the spectral domain, this spatial filtering technique has the advantage of not requiring data in a regular grid, not requiring tapering or padding to make the input grid periodic, and also of remaining local (whereas the Fourier transform smears a local anomaly over all frequencies).

3 Stochastic Methods

Despite its probabilistic setting kriging is a deterministic method in the sense that it provides a single answer, for example the most accurate, or the most likely, map given the data. Such map, however, is a far too smooth representation of reality. Most geological variables display erratic spatial variations at a small scale, which is the motivation for using a probabilistic model in the first place. The kriged map is not a possible map but rather the average of all possible maps. Stochastic methods provide these possible maps.

The need for realistic maps appeared initially in mining with the search for optimum selective mining methods, which required simulations of average grades over blocks the size of the selection unit. In petroleum, multiple realizations of top reservoir were first used to determine the distribution of oil-in-place despite the nonlinearity introduced by the truncation of volumes below an oil-water contact [19]; [8]. The need to reproduce reservoir heterogeneity, which is known to largely control the flow of fluids, also led practitioners to walk away from kriging and turn to stochastic simulation.

The Fourier transform or moving averages are classical methods for simulating a stationary Gaussian random function with a given covariance. In the early seventies Matheron developed another general algorithm, 'Turning Bands', simplifying the simulation in n-dimensional space to a simulation in $1D$. However there remained the problem of constraining the realizations to honor the data. Matheron's brilliant idea, mentioned only incidentally in his 1973 paper[1] [44], was to take advantage of the orthogonality of the kriging error and the kriging estimator, equivalent to independence in the Gaussian case, to pick the kriging error from a non-conditional simulation of the field and add it to the kriging estimate. This conditioning kriging was the breakthrough that launched stochastic simulation as a credible modeling tool.

The Sequential Gaussian Simulation (SGS) algorithm developed at Stanford University [31] made direct conditioning possible. Because of its simplicity it has become so popular as to make other methods look obsolete. However,

---

[1] Originally Matheron had submitted two papers but unfortunately was forced to condense them into one, which became difficult to read.

this algorithm has its own problems. As simulated points become data points, the size of the kriging system keeps growing until it becomes necessary to use moving neighborhoods; this can generate numerical artifacts or require large computation times. Non-conditional simulations techniques have gained renewed interest especially due to the extremely efficient Fast Fourier Transform (FFT) now also available in $3D$.

In any case an essential requirement is the possibility of transforming physical quantities into Gaussian variables and the other way around. Even so, the above methods are suited to the simulation of continuous variables and are inadequate to represent major heterogeneities. This led to the development of specific methods to simulate categorical variables such as lithofacies or rock types. We will review only three. In the early nineties these approaches, advocated by different geostatistical schools, seemed to be in competition. A few years later it has become obvious that they are in fact complementary. The choice of the most appropriate method depends on the type and scale of the deposit. The combination of different methods can even prove useful to represent different levels of heterogeneity (Begg et al., 1996; Dubrule, 1998).

Indicator simulation

The simplest method, developed primarily at Stanford University, is to code the presence/absence of the facies of interest with a binary indicator function. The variogram of this indicator is supposed to capture the spatial distribution of the facies. It has a geometric interpretation as the probability that two points h apart belong to different facies. The algorithm used most is the Sequential Indicator Simulation algorithm (SIS) which is a variant of the sequential simulation method [1]; [30, 29]. This algorithm provides great flexibility since the facies can be characterized by variograms with different ranges and specific anisotropies. When there are several facies its main limitation is to ignore the relations between the facies. On the other hand, taking these into account would involve full indicator co-kriging, a formidable task requiring the determination of all indicator cross-covariances. Furthermore, there are no theoretical models available for representing a set of indicators in a consistent manner (this led Matheron, [45], to develop 'Disjunctive Kriging', but this approach is not used in the petroleum industry and will not be discussed here).

Truncated Gaussian

Matheron objected to the use of indicator covariances to describe the geometry of geological facies. He pointed out that a covariance is not a powerful tool for describing the geometry of a random set model. For example, in a two-phase medium made of grains and pores, the connectivity of the pores is generally very different from that of the grains, but their indicator covariances are identical. So, indicator covariances cannot give information on connectivity. In addition there are complex mathematical conditions that a covariance function must satisfy to serve as a valid model for an indicator covariance [47]. In view of these difficulties and in order to ensure mathematical consistency, Matheron preferred to start from a $3D$ model and derive indicators. He proposed the truncated Gaussian simulation method.

This method defines the different facies by applying multiple thresholds to a continuous Gaussian random function. To give a simple example in $1D$, consider a gamma-ray log in a vertical well in a sandstone reservoir. A high value of gamma-ray indicates shale, a low value indicates sand, and intermediate values indicate shaly sand. If suffices to simulate a continuous curve representing the gamma-ray log and apply the thresholds to obtain at once the vertical arrangement of all lithofacies.

This approach was implemented in the HERESIM ® package (IFP, Center for Geostatistics) and enabled the construction of facies models in very realistic fluvio-deltaic environments, taking into account variations in vertical and horizontal facies proportions [48]; [53]. In a $3D$ model the thresholds are adapted so as to respect vertical facies proportion curves. Likewise the parameters of the continuous Gaussian curve are selected to match the facies indicator variograms. In this model facies transitions necessarily take place between facies i to $i + 1$ or to $i - 1$, which requires the facies to be rank-ordered. This limitation is meaningful only in the case of simple sequences such as shoreface environments –but is more questionable in the case of complex environments. A generalization of the model is to use two or more Gaussians, for example the first Gaussian may represent the geometry of fluviatile channels and a second one that of eolian systems [34].

Boolean Objects

Indicator simulation and truncated Gaussian construct simulations at every point and are often called 'pixel-based' methods, by opposition with 'object-based' methods which are more global. Object-based simulation populates the space with geological objects such as channels or sand bars, levees, etc. themselves sampled from relevant statistical distributions. In the Boolean model, objects are placed at random independently and are amalgamated if they overlap (Boolean union). This model has two main parameters: the mean number of objects per unit of volume and the shape, size and orientation of the objects, which are to match observed facies proportions and shapes of sedimentary bodies.

Matheron developed the general theory of Boolean models in his 1967 landmark book on porous media. He derived the relations between the parameters of the model and the statistical properties of the resulting process, such as the distribution of the number of primary objects intersecting an arbitrary finite set B (this was an early use of multi-point statistics, even with an infinite number of points!). At the time, however, Matheron's aim was not simulation but description (mathematical morphology) and he did not consider the critical problem of conditioning.

[23] used a Boolean model to simulate rectangular stochastic shales in sandstone reservoirs. These discontinuous shales set a tortuous environment for fluid flow and the authors computed the permeability of large grid blocks from the disposition of the shales within the block. It was a breakthrough in the oil industry because it introduced heterogeneity as a major factor controlling permeability. Subsequent developments considered more complex shapes such as channels [51]. The main difficulty in this approach is the evaluation of the numerous parameters defining the shapes of the channels (a sinusoidal strip for example has six parameters: amplitude, wavelength, width, thickness, orientation, and length). Once these parameters are known the simulation of a Boolean model is straightforward. However, conditioning a simulation by observations is not so easy and remained a hurdle for quite some time. Some commercial packages used the brute force method of trial and error: generate simulations until finding one that satisfies the constraints. Another approach relied on simulated annealing. Following a suggestion by Matheron in 1990, Lantuéjoul [32], [33] proposed a rigorous conditioning algorithm based on birth-and-death processes (the trick is the possibility to *delete* previous objects).

The geological realism of the Boolean model opens the door to a variety of petroleum applications. A good model of meandering channels, for example, can be used to determine the percentage of channels connected to a given well, and therefore the remaining potential for in-fill drilling. On this subject, it is interesting to note the existence of a *percolation threshold*. There is a critical value of the density of objects above which all objects are connected. A similar effect also occurs with the truncated Gaussian model [2].

Uncertainty studies

The development of free geostatistical software such as GSLIB [11] popularized geostatistics among oil companies and familiarized production geologists with the notion of conditional simulation. However, the ability to produce a large number of realizations remained little used for quite some time, in part due to computing limitations and more importantly to the difficulty of exploiting the results. With the advent of fast computers and the development of $3D$ geomodelers, multiple realizations of reservoir characteristics can be generated easily and combined into realistic $3D$ stratigraphic models (e.g. [13]. Hundreds or thousands of realizations of reservoir geometry, geological facies distribution at different scales, and petrophysical variables, are routinely produced to determine the probability distribution of hydrocarbon in place volumes, as well as the sensitivity of the results to model parameters. Caveat: experience with such studies indicates that uncertainty tends to be underestimated, in particular because realizations are generated under the same geological model [38].

Multiple realizations studies are now mainstream in the petroleum industry. Methodologies have been defined to integrate static and dynamic uncertainties in a comprehensive uncertainty assessment workflow [7]. The difficulty is to pass from static uncertainty, typically a distribution of oil originally in place (OOIP), to recoverable reserves and production profiles, which are obtained by running a flow simulator. In practice the number of runs has to be limited (a typical run may take from 2 to 4 days on a single CPU). This is achieved by careful selections of realizations –which are *not* equiprobable– and by use of Experimental Design techniques.

4 Geostatistics as an Integral Part of Geosciences

Looking back at the evolution of Petroleum Geostatistics we see a trend toward specialization of the methods and integration of the geostatistical approach and algorithms into domain-specific applications. Let us illustrate this trend in each of the three traditional métiers of petroleum geosciences: geology, geophysics, and reservoir.

4.1 Geology

Indicator simulation, truncated Gaussian, or object models methods attempt to capture the complex shapes and arrangements of geological bodies by way of a purely descriptive approach. In order to improve geological realism the next step is to incorporate the natural processes governing deposition and erosion into the simulation model itself.

Early process-based stochastic models

Early work used Markov chain analysis to model facies sequences along a vertical. Matheron [42] and Jacod and Joathon [27] developed process-based random models where the sedimentation process is governed by a differential equation relating sediment thickness to influx of sedimentary material and sea depth. This model had the advantage to be in $3D$ and have geologically meaningful control parameters: rate of influx of sedimentary material, compaction, subsidence, meandering. In some cases these models could also be conditioned on well data [28]. For some reason it seems that Matheron and his group lost interest in the subject for about thirty years until it was revived by the simulation of channels.

Current research

In meandering channelized reservoirs the natural meandering of the channel controls the deposition of sandy point bars and other sedimentary bodies. Meandering rivers have long been of interest to scientists, and hydraulic studies dating back to the eighties [25] have enabled the development of 2D equations that reproduce in a realistic manner the evolution of the channel migrating on its floodplain. These equations capture physical laws such as mass and momentum conservation and are deterministic, but still rapidly capable of reproducing varied shapes, even starting from an initial channel represented by a plain broken line. A recent thesis [35, 36] combines this genetic approach with a stochastic model introducing randomness through the variations of erodibility within the floodplain, and the occurrence, intensity or size of the different elements.

Lopez extended the process to $3D$ and added associated sedimentary bodies, which in turn can be used to model meandering channelized reservoirs at the scale of the reservoir. Migration of the channel leads to the deposition of point bars, occasional cut-offs create oxbow lakes filled in with mudplugs, while levee breaching leads to crevasse splays and possibly avulsions. In addition, occasional overbank floods lead to the deposition of silt and shales over the aggradating floodplain, while wetland facies may be deposited in lowlands.

The direct construction of a realization is only one aspect of a process-based stochastic approach. For the model to be operational the construction must be controlled by a limited number of key parameters such as channel section or overbank flood frequency. Various architectures can be reproduced by varying the parameters. In practice these parameters can be selected to honor data statistics (e.g. vertical proportion curves of facies from wells). Given a set of parameters, different realizations can be generated. The final challenge, still at the research stage, is to condition these realizations, either on soft regional data such as seismic time slices, or on well data.

4.2 Geophysics

Because of its large volume of data seismic has long been a domain of choice for the application of geostatistics. New challenges are posed by the fast pace of innovation in acquisition and processing technology. High resolution surveys can provide images of stunning quality, allowing the identification of 'architectural elements' in turbiditic systems, such as channels, levees, lobes, bars, etc. Dubrule [15] gives a comprehensive account of the contribution of geostatistics to the integration of seismic data in Earth Models. We will highlight two important developments.

Geostatistical inversion

Inversion of seismic data to acoustic impedance (the product ρv of density by velocity) is an efficient tool for the integration of geological and geophysical data. Well logs provide geological data with a vertical resolution of 0.5 to 1 m whereas seismic data have a vertical resolution of the order of 30 m (sometimes better under exceptional conditions). Standard optimization-based deterministic inversion produces acoustic impedance values at the scale of seismic and calibrated to the measured well log impedances. Geostatistical inversion, also named 'stochastic inversion', aims at doing the opposite, namely providing acoustic impedance values at the scale of logs and constrained by seismic.

The algorithm [20] is an extension of SGS. At a given grid node a number of vertical acoustic impedance traces are generated by geostatistical simulation conditioned on well log data. Each of these traces is convolved with a seismic wavelet and compared to the actual seismic trace observed at that location. The acoustic impedance trace providing the best match is selected and used as conditioning data for the next simulated point. In the end the whole space is filled with traces, thus providing a realization of acoustic impedance in $3D$ with the high frequencies present in the log data. This technique which was developed ten years ago is still not routine in the industry, perhaps because it is one step short of the final goal of reservoir characterization, which is the translation of seismic attributes into geological and petrophysical properties.

Velocity fields

An essential step in the processing of seismic reflection data is repositioning reflectors observed in time sections at their correct horizontal and vertical position. This process, called 'migration', requires a model of the velocity field. In the simplest case the velocity field is assumed constant, so that the seismic rays are straight, and velocity is estimated by a deterministic method. The problem with this approach is that it only captures the large scale component of the velocity field. Even weak perturbations of this field can result in important migration errors. Matheron [49] developed a probabilistic model in which he could relate the migration errors to the covariance of the velocity perturbations. This research note initiated and inspired three theses at the Ecole des Mines de Paris (Touati 1996, Iooss 1998, Geraets 2002) and a number of publications listed in the bibliography of Iooss et al. [26]. Their focus shifted from the study of migration errors to the characterization of heterogeneities in the velocity field. In particular, an inversion method has been developed allowing estimation of the velocity covariance from the covariance of the observed travel times, thus giving information on the size and orientation of velocity heterogeneities. This literature is very technical and geared toward an audience of geophysicists. Here the power of geostatistics lies in its blending with physics.

4.3 Reservoir

Upscaling

In 1967 Matheron published a highly original book entitled 'Elements for a theory of porous media'. It was only 166 pages long but it laid both the foundations of Mathematical Morphology and the foundations of the relationship between geostatistics and fluid mechanics. After all these years this book is still a bible for researchers in the field of porous media.

On the dynamic part the book discussed the problem of upscaling, that is, the derivation of physical properties of heterogeneous media as the scale increases. At the microscopic scale of the pores the flow is governed by the Navier Stokes equation; at the macroscopic scale of a core Darcy's law appears, defining a key parameter, the permeability k; at the megascopic scale of a large grid block in a dynamic model a global permeability K appears. In reservoir engineering applications, upscaling concerns the change from macro to megascopic scale, computing K from the spatial distribution of k. Matheron established that geometric averaging is correct for lognormally distributed (scalar) permeabilities but only in 2D flow. In $3D$, and contrary to common practice, the upscaled permeability cannot be the geometric mean; it is closer to the arithmetic mean, approximately at two-thirds of the distance separating the arithmetic and harmonic means. Matheron conjectured an exact upscaling formula that is still being discussed today (see paper by J.P. Delhomme and G. de Marsily in this same volume, page 69). He always kept an interest in the subject and published a last note on it in 1993, two years before retirement.

Integration of production data

Integration of production data in petroleum reservoir models is a major objective for petroleum geostatistics. A dynamic model of the reservoir matching the data and the available production history is expected to provide reliable forecasts of the quantity of hydrocarbon produced at the various stages of field development. Such integration, however, is an extremely difficult challenge because the relations between production data, either scalar or function of time, and petrophysical fields are very complex and highly nonlinear.

Well tests are the easiest data to integrate and much work has been devoted to this subject. Using simple methods one can take into account apparent permeabilities deduced from well tests and regarded as spatial averages. Simulated annealing appears as an interesting solution, the method being very versatile and capable of handling diverse constraints [10]; however it should be used with care because imposing inconsistent constraints may lead to artifacts. Simpler methods based on kriging constrained by spatial averages are also available [21]. Finally, optimization methods using dynamic simulations with gradients make it possible to carry out an inversion of the complete time-pressure curve [5]. Estimated parameters may be local (limits, shapes) or global (facies proportions, averages, etc.)

These inversion methods can be generalized for production history matching. Stochastic models are updated using algorithms such as pilot points or Markov Chain Monte Carlo (MCMC). The results obtained in the scope of the EC-sponsored PUNQ project seem extremely interesting but are difficult to compare and to generalize to real life cases involving a large number of parameters to be fitted [16]. It is difficult at this time to tell which method will emerge as the best among all those tested.

Today the limiting factor for the integration of production data is the speed of dynamic reservoir simulators. Fast simulators, such as streamline simulators, provide a partial solution [3]. Their combination with an iterative calibration of stochastic models using a gradual deformation approach looks promising [24].

5 The Future of Petroleum Geostatistics

We have insisted on the need for geostatistics to be integrated with the disciplines where it is applied. To make meaningful contributions geostatisticians must understand the problems and the context of the problems, and therefore become geoscientists themselves. They are encouraged to do so by another factor, the lack of attractive career paths lined up for them. In oil companies geostatistics is usually not recognized as part of "core business" but more as a support activity. As a result many (if not most) very competent geostatisticians recycle themselves as geophysicists, reservoir engineers, sometimes geologists, and actually do well in their new métier.

Professional mobility is a good thing, provided that the outflow of talents is compensated by an inflow of recruits. Unfortunately that is not necessarily the case. The perception that geostatistics is a competency, as opposed to a discipline such as geology or geophysics, and therefore can be acquired if needed by proper on-the-job training, makes it difficult to justify hiring academically trained geostatisticians. They come in competition with much sought for reservoir engineers for example. In conclusion, the future of petroleum geostatistics lies in the willingness of oil companies to open the door to young talented geostatisticians, and to fund with research contracts the academic departments that train them.

Acknowledgement. This paper borrows from a presentation to the IAMG in Trondheim in 1999 entitled: "Petroleum Geostatistics, from stone age to industrial times", by André Haas and Olivier Dubrule. The authors are indebted to Jacques Rivoirard for his help in writing the section on process-based models.

References

1. Alabert, F. (1987). Stochastic Imaging Of Spatial Distributions Using Hard and Soft Information. M. Sc. thesis, Stanford University, California.
2. Allard, D., and HERESIM Group (1993). On the connectivity of two random set models: the truncated Gaussian and the Boolean. In Geostatistics Tróia '92, A. Soares, ed. Kluwer, Dordrecht, Netherlands, Vol. 1, 467–478.
3. Batycky, R.P., Blunt, M.J. and Thiele, M.R.(1997). A 3D Field-scale Streamline-based Reservoir Simulator, SPE Reservoir Engineering, **12**, 246-254.

4. Begg, S.H., Kay, A., Gustason, E.R. and Angert, P.F. (1996). Characterisation of a Complex Fluvio-Deltaic Reservoir for Simulation, SPE Formation Evaluation, 147-153.
5. Blanc, G., Guérillot, D., Rahon, D. and Roggero, F. (1996). Building Geostatistical Models Constrained by Dynamic Data - A Posteriori Constraints, SPE Paper # 35478.
6. Chilès, J.P. and Delfiner, P. (1999). Geostatistics- Modeling Spatial Uncertainty. Wiley & Sons, New York, 695 p.
7. Corre, B., Thore, P., de Feraudy V. and Vincent, G. (2000). Integrated Uncertainty Assessment for Project Evaluation and Risk Analysis. SPE paper # 65205.
8. Delfiner, P., and J.P. Chilès (1977). Conditional simulation: A new Monte-Carlo approach to probabilistic evaluation of hydrocarbon in place. SPE Paper # 6985.
9. Delfiner, P., J.P. Delhomme, and J. Pélissier-Combescure (1983). Application of geostatistical analysis to the evaluation of petroleum reservoirs with well logs. In Proceedings of the SPWLA 24th Annual Logging Symposium, Calgary, June 1983.
10. Deutsch, C. (1992). Annealing techniques Applied to Reservoir Modeling and the Integration of Geological and Engineering (Well Test) Data. PhD Thesis, Stanford University, California .
11. Deutsch, C.V. and Journel, A.G. (1992). "GSLIB: Geostatistical Software Library and User's Guide", Oxford University Press , New York.
12. Doyen, P.M., Den Boer, L.D. and Pillet, W.R. (1996). Seismic Porosity Mapping In The Ekofisk Field using a New Form of Collocated Cokriging, SPE Paper # 3649..
13. Dubrule, O., Basire, C., Bombarde, S., Samson, Ph., Segonds, D. and Wonham, J. (1997). Reservoir Geology Using 3D Modelling Tools, SPE Paper # 38659.
14. Dubrule, O. (1998). Geostatistics in Petroleum Geology, A.A.P.G. Continuing Education Course Notes Series #38.
15. Dubrule, O. (2003). Geostatistics for Seismic Data Integration in Earth Models. EAGE 2003 Distinguished Instructor Short Course, No. 6.
16. Floris, F.J.T., Bush, M.D., Cuypers, M., Roggero, F. and Syversveen, A.R.(2001). Methods for quantifying the uncertainty of production forecasts: a comparative study. In Petroleum Geoscience, Vol. 7, 2001, S87-S96.
17. Fournier, F. (2004). Interprétation orientée réservoir des attributs sismiques: une contribution à la lithosismique. Rapport 58 145. Institut Français du Pétrole.
18. Haas, A. and Viallix, J.R (1976). Krigeage Applied to Geophysics: the Answer to the Problem of Estimates and Contouring, Geophysical Prospecting, 24, 49-69.
19. Haas, A., and C. Jousselin (1976). Geostatistics in the petroleum industry. In Advanced Geostatistics in the Mining Industry, M. Guarascio, M. David, and C. Huijbregts, eds. Reidel, Dordrecht, Holland, 333–347.
20. Haas, A. and Dubrule, O. (1994). Geostatistical Inversion of Seismic Data, First Break, 12, 561569 (1994).
21. Haas, A. and Noetinger, B. (1996). Stochastic Reservoir Modeling Constrained by Well Test Permeabilities. In Geostatistics Wollongong '96, E.Y. Baafi and N.A. Schofield, eds. Kluwer, Dordrecht, Netherlands, Vol. Publ., Dordrecht, Vol 1, 501-511.
22. Haas, A. and Biver, P. (1998). Simulations stochastiques en cascade, In Cahiers de Géostatistique: Ecole des Mines de Paris, 6, p.31.

23. Haldorsen, H.H. and Lake, L. (1984). A New Approach to Shale Management in Field Scale Simulation, SPE Paper # 10976.

24. Hu, L.Y., Le Ravalec M and Blanc G. (2001). Gradual deformation and iterative calibration of truncated Gaussian simulations. In Petroleum Geoscience, Vol. 7, 2001, S25-S30.

25. Ikeda S., Parker G., and Sawai K. (1981). Bend theory of river meanders. Part 1. Linear development, *Journal of Fluid Mechanics*, vol. 112, pp. 363-377.

26. Iooss B., Geraets D., Mukerji T., Samuelides Y., Touati M., and Galli A. (2003). Inferring the statistical distribution of velocity heterogeneities by statistical traveltime tomography. Geophysics, 68, 1714-1730.

27. Jacod J. and Joathon P. (1971). Use of random-genetic models in the study of sedimentary processes. Journal of the International Association for Mathematical Geology, 3(3), 219-233.

28. Jacod J. and Joathon P. (1972). Conditional simulation of sedimentary cycles in three dimensions. In Proceedings of the International Sedimentary Congress, Heidelberg, August 1971, D.F. Merriam, ed. Plenum Press, 139-165.

29. Journel, A.G. and Gomez-Hernandez, J.J. (1989), Stochastic Imaging of the Wilmington Clastic Sequence, SPE Paper # 19857.

30. Journel, A.G., and F. Alabert (1989). Non-Gaussian data expansion in the Earth Sciences. Terra Nova, 1(2), 123–134.

31. Journel, A.G.,(1989). Fundamentals of Geostatistics in Five Lessons, Short Courses of Geology, 8, AGU, Washington D.C., 1989.

32. Lantuéjoul, C. (1997). Iterative algorithms for conditional simulations. In Geostatistics Wollongong '96, E.Y. Baafi and N.A. Schofield, eds. Kluwer, Dordrecht, Netherlands, Vol. 1, 27–40.

33. Lantuéjoul, C. (2001). Geostatistical Simulation. Models and Algorithms. Springer Verlag.

34. Le Loc'h, G., and A. Galli (1997). Truncated plurigaussian method: theoretical and practical points of view. Proceedings of the 5th International Geostatistics Congress, E. Baafi (Ed.), Wollongong, Australia, 22-27 September 1996.

35. Lopez S. (2003). Modélisation de reservoirs chenalisés méandriformes, approche génétique et stochastique. PhD Thesis, Ecole des Mines de Paris.

36. Lopez S., I. Cojean I., Rivoirard J., Fouche O., Galli. A.(2004). Process-based stochastic modelling in the example of meandering channelized reservoirs, EAGE, Paris.

37. Magneron, C., Leron, A. and Sandjivy, L. (2003). Spatial quality control of seismic stacking velocities using geostatistics. EAGE 65th Conference & Exhibition, Stavanger, Norway.

38. Massonnat, G.J. (2000). Can We Sample the Complete Geological Uncertainty Space in Reservoir-Modeling Uncertainty Estimates? SPE Journal, Vol. 5, No. 1.

39. Matheron, G. (1962-1963). Traité de géostatistique appliquée, Tome I; Tome II: Le krigeage. I: Mémoires du Bureau de Recherches Géologiques et Minières, No. 14 (1962), Editions Technip, Paris; II: Mémoires du Bureau de Recherches Géologiques et Minières, No. 24 (1963), Editions B.R.G.M., Paris.

40. Matheron, G. (1965). Les variables régionalisées et leur estimation. Une application de la théorie des fonctions aléatoires aux Sciences de la Nature. Masson, Paris, 306p.

41. Matheron, G. (1967). Eléments pour une théorie des milieux poreux.. Masson, Paris.

42. Matheron, G. (1969). Les processus d'Ambartzumian et leur application en géologie. Technical Report N-131, Centre de Géostatistique, Fontainebleau, France.

43. Matheron, G. (1970). La théorie des variables régionalisées et ses applications. Cahiers du Centre de Morphologie Mathématique de Fontainebleau, Fasc. 5, Ecole des Mines de Paris. Translation (1971): The Theory of Regionalized Variables and Its Applications.

44. Matheron, G. (1973). The intrinsic random functions and their applications. Advances in Applied Probability, 5, 439–468.

45. Matheron, G. (1976). A simple substitute for conditional expectation: The disjunctive kriging. In Advanced Geostatistics in the Mining Industry, M. Guarascio, M. David, and C. Huijbregts, eds. Reidel, Dordrecht, Holland, 221–236.

46. Matheron, G. (1982). Pour une analyse krigeante des données régionalisées. Technical Report N-732, Centre de Géostatistique, Fontainebleau, France.

47. Matheron, G. (1987). Suffit-il, pour une covariance, d'être de type positif ? Sciences de la Terre, Série Informatique Géologique, 26, 51–66.

48. Matheron, G., Beucher, H., De Fouquet, C. and Galli, A. (1987) Conditional Simulation of the Geometry of Fluvio-Deltaic Reservoirs. SPE Paper # 16753.

49. Matheron G. (1991). Géodésiques aléatoires. In Cahiers de Géostatistique: Ecole des Mines de Paris, 1, 1-18.

50. Matheron G. (1993). Quelques inégalités pour la perméabilité effective d'un milieu poreux hétérogène. In Cahiers de Géostatistique : Ecole des Mines de Paris, 3, 1-20.

51. Petit, F.M., Biver, P., Calatayud, P.M., Lesueur, J.L. and Alabert, F. (1994). Early Quantification of Hydrocarbon in Place through Geostatistical Object-Modelling and Connectivity Computations, SPE Paper # 28416.

52. Piazza J.L., Sandjivy, and L. Legeron S. (1997). Use of geostatistics to improve seismic velocities: case studies : 67th Annual Mtg. SEG expanded abstracts , 1293-1296.

53. Ravenne, C. and Beucher, H. (1988). Recent Development in Description of Sedimentary Bodies in a Fluvio Deltaic Reservoir and Their 3D Conditional Simulations, SPE Paper # 18310

54. Rivoirard, J. (2001). Which Models for Collocated Cokriging? Mathematical Geology, Vol 33, No 2.

55. Xu, W., Tran, T.T., Srivastava, R.M. and Journel, A.G. (1992). Integrating Seismic Data in Reservoir Modelling: The Collocated Cokriging Alternative, SPE Paper # 24742.

The expansion of environmental geostatistics

Roberto Bruno[1] and Chantal de Fouquet[2]

[1] Dipartimento di Ingegneria Chimica, Mineraria e delle Tecnologie Ambientali,
Università di Bologna
[2] Centre de Géostatistique, Ecole des Mines de Paris

1 Introduction

Environment, and more precisely pollution, appeared very early among the
initial applications of geostatistics [15, 19, 33], but is seldom mentioned in
Matheron's writing. Nevertheless, it is indeed in reference to the pollution
context that he examines, at the end of his essay on the practice of probabilities, "Estimating and choosing", the operational character of non-linear
estimators [25, 31].

For a mathematician, the conclusion may seem disappointing. Indeed,
Matheron suggests regarding the different expressions of the conditional expectation given by various probabilistic models as a mere set of algorithms,
each one depending on a limited number of parameters. The practician has
then only to choose the algorithm best suited to his problem [31], p. 137.

Drawing a guideline between the pragmatism imposed by practice, and
mathematical rigor, Matheron stipulates for never granting a blind confidence to the model. Indeed geostatistical modeling consists in constructing
a simplified but operational representation of the studied phenomenon, which
nevertheless imposes to take into account all the available information, qualitative as well as quantitative. The model relevance is its aptitude to describe
the phenomenon correctly.

Most geostatistical concepts and models were first created to solve estimation and selection problems for mining exploitation. But the operational
concepts forged by Matheron were soon extended to a wide range of domains.
Indeed the priority given to the physical phenomenon, the modeling adapted
to each specific case, i.e. the "monoscopic" modeling according to Matheron's
terminology, explains the expansion of geostatistics in disciplines as varied as
pollution studies, fisheries, agronomy, biodiversity. . .

With just a slight adaptation of the terminology, pollution problems in
different environments (soils, water or air) can thus be formulated using mining concepts such as estimation and especially selection, developed more than
thirty years ago [20, 21, 22]. Typically, the studied problems deal with:

- characterization and quantification of spatial or space-time variability, generally in a multivariate context;
- estimation, to map a quantity or to assess the accuracy of the obtained maps;
- optimization of a sampling pattern, to reach a given accuracy, or to find an acceptable compromise between accuracy and sampling cost;
- comparison of the concentrations with quality threshold, taking into account the different spatial or time "supports" of measurements and "selection" units;
- identification of volume / tonnage of soil involved in pollution removal, as a function of the threshold and of the reference support. More generally, "selectivity curves" should be determined;
- evaluation of the information effect, that is the consequences of estimation errors when delimitating a polluted area;
- "risk assessment", such as the evaluation of the population exposed to concentrations exceeding a threshold in air pollution, or the first arrival time of a pollutant to drinking water pumping.

At the beginning of environmental geostatistics, the theoretical solution of a wide range of problems was then already available and, from a shallow point of view, "environment" could have appeared as a new field of application or adaptation for well-known methods. For example, multivariate problems gave an opportunity to reconsider the cokriging. The very convenient external drift method has been widely used to introduce some *a priori* knowledge in the estimation process. The time component of phenomena has been taken into account by widening a little the class of usual covariance models. Nevertheless, estimating pollutant concentrations in rivers, as well as the need for specific space-time models are making new developments necessary.

Before presenting some geostatistical approaches to environmental problems, let us clarify what this term means here.

In the broad sense, "environment" appoints a natural or more often an anthropised media, and environmental studies aim at quantifying its state or assessing its evolution. The studied questions come from soil sciences (cartography of depth of different horizons, quantification of physical-chemical parameters for "precision" agriculture); from agronomy (forest inventory, counting of vegetal or animal species); from pluviometry or climatology (namely characterizing temperature fields), from ecology... Applications of geostatistics to epidemiologic studies aim at detecting or quantifying the regionalization of disease risk, trying to split the influence between the surrounding medium and individual parameters.

The time evolution of relief and seismic risk are also studied. The very interesting field of stochastic hydrology, in which Matheron did pioneer work, is dealt with in another chapter of the present book.

In a quite restricted sense, environmental geostatistics deals with relationships between variables depicting the various media (population density, actual or previous land use, altitude) and concentrations of diverse substances in the air, in superficial waters like oceans and rivers or in aquifers, in soils and sediments. The questions are treated at various scales: atmospheric pollution can be characterized on a continental or a national scale, or in the immediate surrounding of a factory chimney. Pollutant concentrations in aquifers are studied on the regional scale, or locally near a pollution "source".

As for all ranking, the "environmental" qualifier remains somewhat arbitrary. Thus for agriculture, nitrates are first nutriments; when streaming toward rivers they become pollutants. Apparently analogous problems, such as exposure quantification to dust or noise, fall within industrial hygiene for professional nuisances or "environment" in urban context.

Therefore, for more than a decade the development of environmental geostatistics has been reflecting the evolution of our society concerns. In the absence of specialized journals, the quadrennial international geostatistical conferences make it possible to summarize this evolution: the "environment" theme, missing in the Avignon conference in 1988, is present among "other applications" in Troia (1992); autonomous in Wollongong (1996), it represents more than 25% of the presentations in Cape Town (2000). Meanwhile, since 1996, numerous case studies were presented at the biennial "Geoenv" conferences. Most of the previously mentioned subjects are treated in the acts of these congresses.

We do not intend to build here a state of the art of environmental geostatistics. In the first part, we investigate some characteristics common to many environmental problems, and examine their consequences on the models choice. Conversely, the setting of some technical aspects, namely normative ones, would gain in clarity by using geostatistical concepts. Looking at the major classes of problems, we show the adaptation of geostatistical models to the environmental context. In a second section, original examples, relative to problems met in various media, illustrate the power of geostatistical concepts. Thus, "simple" methodological transpositions (change of workspace, variations on kriging choosing additional constraints or using *ad hoc* covariances) bring conceptual solutions to a wide variety of problems. In conclusion, methodological development prospects are presented.

2 Characteristics and methods of environmental studies

Environmental studies presents some common characteristics, such as the variables properties, the multivariate or space-time context, and the important question of exceeding a threshold. Before presenting some useful or widely used models, let us first return to "the mean", an apparently simple concept.

2.1 The mean: operational mode or physical quantity?

The distinction between physical quantity and statistical parameter shows its relevance in the regulation context of water or air quality, as introduced in the following example.

In order to assess river water quality, nitrate (NO_3) and other fertilizer concentrations are regularly measured at some monitoring stations. To compare concentrations in different areas or examine their evolution, a few synthetic quantitative indicators are calculated, based on the water framework European directive. These calculations use the classical statistical inference of the distribution parameters for independent and identically distributed variables: the expectation of the distribution is obtained as the arithmetical average of experimental data measured during one year, and the variance of the associated estimation error is taken equal to $\frac{\sigma^2}{n-1}$, with n the number of data and σ^2 their experimental variance.

In temperate climates, the concentrations of most substances in rivers show an annual periodicity due to seasonal flow and tributaries variations. For example, in France, surface streaming carrying the fertilizers toward rivers is more important in winter. Therefore, nitrates concentrations are higher in winter and lower in summer, a ratio between these values being commonly larger than 10. In many stations, for a better survey, sampling frequency is doubled during the periods of high nitrates concentrations, from November to April. The "statistical" annual mean and the computed quantiles of the distribution necessarily increase when the 12 regular monthly measurements are completed by 6 additional winter measurements. Indeed preferential sampling obviously induces a bias on the estimation of these "synthetic indicators".

If the underlying model of successive and independent drawings from the same random variable is acceptable for the successive values obtained using a die, it is no more valid for the successive concentration measurements at a stream station. Why?

First, assuming an identical distribution of concentrations all year long is physically unrealistic in agricultural areas, because of the seasonal variations of nitrate concentrations. In addition, due to the time correlation of concentrations that has been effectively demonstrated, the hypothesis of independence is all the less relevant as the measurements are close in time.

Thus the underlying probabilistic model, usual in classical statistics, is here not suitable. Moreover what quantity do we really want to estimate?

Let z(t) denotes the concentration during time at one given station. The yearly mean z_T of concentration z during year T, is the time integral $z_T = \frac{1}{T} \int_T^z (t) dt$. The physical quantity to be estimated is thus defined regardless any model, whether stochastic or deterministic, chosen to depict the concentrations, and does not require any hypothesis of time stationarity. The geostatistical estimation of z_T consists in computing an approximate value of this integral by an "optimal" weighted average of the experimental data.

For this, the concentration z(t) at this station is considered as a realization of a one dimensional Random Function (a stochastic process) Z(t). The unknown yearly mean concentration is then estimated, for example using a linear combination of the data. "Kriging" corresponds to the optimal weighting, ensuring the minimum of the estimation variance.

Let us now consider a large time interval, for example thirty years. In the absence of major changes during this period (which is not the case for nitrates), the "expectation" parameter $E[Z(t)]=m(t)$ is supposed to be a constant m, whereas the thirty annual means of the concentrations will all be different. For an ergodic model of Stationary Random Function, the expectation parameter m is the limit of the time average of concentrations over larger and larger time intervals, or equivalently, the limit of the average of all "annual means of the concentrations" on these intervals.

What we want to calculate in practice is indeed the annual average concentration for each year, and not the expectation of these values.

To summarize, the yearly mean $z_T = \frac{1}{T} \int_T^z (t)dt$ represents a physical quantity, whereas the arithmetical mean of the measurements during one year and the "statistical" calculation of the associated variance rather represent an operational algorithm. Matheron pointed out other confusions between concept and operational mode, for example concerning the "drift" [21, 7].

2.2 "Environmental" context

Environmental problematic present specific characteristics mostly linked to some classes of variables and properties. Their joint recurrence defines what we can call the "environmental context". In the following, three general characteristics are examined: additivity, multivariability and temporality.

Additive variables or not?

Many "environmental" variables do not verify the important property of additivity.

Most of the parameters which characterize soils are intrinsically non-additive. For example, pH is equal to the logarithm of the H^+ ions concentration, up to the sign. The "cationic exchange capacity", which gives the total quantity of exchangeable cations that the soil can adsorb, is expressed in milliequivalents per 100g of soil or clay, and is a function of pH.

For non additive variables kriging remains pertinent as long as only one support is considered: the quantity to be estimated must then be defined on the same support as the data, supposed to be identical for all samples. To estimate non additive quantities on supports different from that of the sampling, a numerical approach, possibly based on simulations, can sometimes be used. Most often additional information about the phenomenon remains indispensable.

An example: on a field, the biodiversity can be quantified by the number of different species on it this number is not an additive variable and cannot be reached from the number of species counted on sample plots (or "placettes") of areas **s** of few square meters. Indeed, without any other information, we only know that the number of species existing in the union of two sample plots stands between the maximum and the sum of the number of species counted for each plot. This depends on whether the species found in the less diversified plot are present in the other, or no species is common to both sample plots. "Block" kriging of the number of species present in the whole field would radically underestimate the wanted quantity. Moreover, in an area **S**, the number of species estimated in that way would be quite often lower than the number of species counted in some of the sample plots located within **S**. In reality, "block" kriging gives an estimator of the average number of species existing in the sample plots **s** whose union gives the area **S**.

Without further information, the number of species can be estimated only for areas having the same geometry as the sample plots. In the absence of "phenomenological" information, simulations cannot help solve the problem.

A blind application of kriging, which does not take into account the properties of the studied variables, can obviously lead to absurd results. A detailed investigation of the variable properties is hence necessary, before applying kriging to a new domain.

Variables such as concentrations are not always additive, and sometimes have to be weighted or complemented with other variables to provide additive quantities suitable for a linear estimation. Let us for example consider a thin horizontal layer of variable thickness, sampled by vertical boreholes with a unique sample measured over the thickness of the layer. To estimate the concentration, one uses the thickness and the "accumulation", defined as the product of the thickness by the associated mean concentration. In soil sciences, grades (of organic material, of clay...) are usually given with reference to the mass of fine materials in the samples. The estimation is made coming back to additive quantities such as the mass of clay or organic material, and the mass of fine materials within the sample.

Multivariate analysis

A second characteristic is the great number of factors occurring in "environmental" problems, and then the multivariate modeling. Let us leave aside the classification problems [31]; it is however possible to distinguish, though not exhaustively, various kinds of relationships.

Relationships between homologous variables

As an example, let us consider the concentrations of various pollutants in the soils of former industrial sites. A given activity generates a "pollutant plume", but diverse activities may be simultaneously or successively present on a site. For instance, a petrochemical plant can simultaneously generate a metallic pollution (from lead or other added products) and an organic one (PAH or polycyclic aromatic hydrocarbons, among others). The evolution of each product depends on its physic-chemical properties, mainly solubility, but also on the soils type, the history of the site... Spatial correlations between the concentrations reflect the relationships between the various substances.

As regards air pollution, complex photochemical reactions can be induced between ozone and nitrogen oxides, reactions that need to be taken into account for short term previsions.

Relationships between "explicative variables" and "concentrations"

Auxiliary variables such as type and density of ground occupancy partially depict the urban medium or indicate some economic activity (density of the buildings, characteristics of transport or industry infrastructures). One aims at assessing the "predictive" degree of these "explicative" variables for some pollutant concentrations. Putting in evidence and modeling these relationships help improve the estimation of the concentrations when the medium is known with enough details at the studied scale. It is the case for air pollution at the scale of an urban area, or for agricultural pollution of rivers or aquifers at a regional scale.

Relationships between "markers" and "concentrations"

"Organoleptic" observations (presence of filling materials, visible tar traces, smell...) do not explain the pollution, but, as its consequence, can help detecting it. If their correlation with concentrations is high enough, these qualitative or semi-quantitative measurements, imprecise but easy to acquire, can be used to improve the precision of the estimations, at low additional sampling cost. An example in organic soil pollution is presented below.

Variables linked by partial derivative equations

In hydrology, for example, the diffusivity equation links transmissivity and piezometric head. Joint modeling of head and transmissivity must consider these phenomenological relationships. First order linearization of this equation leads, for a macroscopic flow with a constant head gradient, to a linear equation between the transmissivity perturbation and the Laplacian of the head perturbation. From initial works by Matheron [23], Dong [9] deduced relationships between generalized covariances of perturbations terms, as well as the corresponding degree of stationarity.

Numerical approaches based on simulations allow breaking some restrictive hypotheses on one hand but necessitate an accurate specification of the whole limit conditions on the other hand (see the paper on hydrogeology in this book, page 69).

Variables describing a complex dynamic system

When calculating possible climatic evolutions or making short-term predictions for air quality on urban areas, it is necessary to take into account the dynamics of the system. In these models, the number of parameters returns a very complex system, and it seems illusory to directly deduce a consistent set of simple and cross space-time covariances between the variables. Quite often, only some of these variables are "interesting" for the studied problem: in ecology for example, we are interested in pluviometry, and not in pressure field, which is nevertheless necessary to predict the rain height. The geostatistical modeling of the interest variables concentrates then on the differences between the predictions from a model (usually deterministic) and the data. These "data assimilation" methods are presented in the following.

The classical set of multivariate models is shown to be very rich. Its flexibility allows adapting it in order to model most of the previous relationships. In particular, the modeling of a non-linear regression of a concentration Z on some "explicative variables" $y = \{y_1, ..., y_n\}$, offers several variants from a deterministic to a stochastic depiction of the links between these variables. We detail some typical examples:

- Deterministic relationships using models "with residual" of the type $Z(x) = f(y(x)) + R(x)$, where R is a stationary Random Function. When the variability of the residuals depends on the local characteristics of the environment, a model of the type $Z(x) = f(y(x)) + g(y(x))R(x)$ is preferred.
  Some "dexterities" allow simplifying the previous models. For example, the linearization of the relationships between concentration and auxiliary variables is often acceptable with a good approximation using the translated logarithm $\log\left(1 + \frac{y}{m}\right)$ of these variables.
- Deterministic relationships that are not completely specified, as the external drift model. The relationship $Z(x) = a_0 + \sum_i a_i f_i(y_i(x)) + R(x)$ is then valid locally, that is in a kriging neighborhood; the coefficients a_i, not available in practice, are filtered by the non-bias conditions of the kriging.
- Stochastic relationships of a linear type, as the linear coregionalisation model. Within this model, simple and cross-covariances are written as linear combinations of elementary structures: $\gamma_{ij}(h) = \sum_\ell c_{ij}^\ell \gamma_\ell(h)$. The variables are linear combinations of space (or time) components corresponding to different variability scales.
- Stochastic shifted relationships. These models deal with relations of type: $Z(x) = Y(x - x_0) + R(x)$. This class of models can be widened to the correlation between a random function Y and its derivatives $\frac{\partial Y}{\partial x_i}$, null when the variables Y and $\frac{\partial Y}{\partial x_i}$ are taken at the same point. Other coregionalisation models are built by regularization, derivation, substitution etc.

- Non-linear stochastic relationships. This includes relationships between transformed variables such as indicators, translated logarithms, anamorphosis, or truncated variables such as concentrations in the presence of a "detection limit".

This class of multivariate models requires that more than the first two moments of the random functions are specified.

Time component

The time component constitutes a third important characteristic of environmental context. Schematically, several classes of modeling can be distinguished, depending on the way time is treated.

Phenomena where time can be taken out of the stochastic modeling

Time discretization may allows returning a space-time problem to a multivariate modeling. In air pollution, seasonal measurements can be modeled as "winter" and "summer" concentrations. The multivariate approach offers great flexibility, by allowing a fine modeling of time non-stationarity. This can be useful when the variability irregularly evolves with time, or in the presence of seasonal effects. However the time component of the estimation variance should not be neglected, when only a part of the considered period is monitored.

In some decompositions of the type "deterministic prediction plus innovation term", the system dynamics is taken into account by the deterministic mode and the stochastic model works only on "residuals". Concentration Z at date t_i is split in two terms, a prediction, calculated by a deterministic dynamic model, and a random residual, spatially correlated: $Z(x, t_i) = f(x, t_i) + R_i(x)$. The main spatial non-stationarities are taken into account by the deterministic model, whilst stationarity hypotheses can be stricter on the residuals than on the concentration Z.

In the advanced version of "data assimilation" methods, the covariance matrix of the errors between the predictions of the phenomenological model and the "observations" at some measurement stations evolves with time. Some simplified prototypes (ecological modeling of a population evolution) or operational ones (prediction of the pollution level at the scale of urban areas) are used to constrain the predictions by the last available measurements [1;37;4] (see further on).

Phenomena without specific modeling of the dynamics

The space-time models first presented in the literature were often a simple extension of usual spatial models, time being treated as an additional coordinate. The covariances $C(h, \tau)$, h and τ being respectively the space and the time increment, were using a geometric or a zonal anisotropy, respectively of the form $C\left(\sqrt{\frac{h^2}{a^2}+\frac{\tau^2}{b^2}}\right)$ or $C_s(h)+C_t(\tau)$. The often-used factorized covariances, product of a spatial and a time component, correspond to the following markovian property (or "screen effect" in the geostatistical terminology): conditionally to Z(x,t), the variables Z(x',t) and Z(x,t') are independent. This model, previously proposed by Matheron for sedimentary facies simulations [10], has very efficient Markov properties: when measurements are synchronous and always located at the same stations, the estimation of the concentration Z at any measurement date depends for any point only on the measurements at this same date. Hence, this very specific model is not adapted for depicting a propagation phenomenon, whose covariance in the propagation direction (at constant velocity v) would admit a component of type $C(h-v\tau)$. Several authors [for example 12] have examined the separability of space and time in geostatistical modeling.

Working on the cartography of the bathymetry near a coast from measurements taken during almost three decades, [6] gave a nice example of covariance factorization. The covariance is decomposed in three terms: two exponential terms, respectively for the time and for the spatial component parallel to the coast (notation ‖), and a cardinal sine covariance for the spatial component perpendicular to the coast (notation ⊥). This attenuated periodic component represents the location of the various sand bars met when moving away from the coast (in fact, it is the covariance of the residuals normalized by the 28 years average depth). This covariance is written:

$$C\left(h_\parallel, h_\perp, \tau\right) = \sigma^2 . \exp\left(-\frac{|h_\parallel|}{a_\parallel}\right) . \frac{a_\perp}{h_\perp} \sin\left(\frac{h_\perp}{a_\perp}\right) . \exp\left(-\frac{\tau}{a_\tau}\right)$$

This model of time component is convenient only when discretised with a mesh of one year. In fact, it does not take into account the sand movement, and specifically the bars migration at short time intervals; their modeling would necessitate the fitting of a propagation component at small space and time distances.

Generalizing the anamorphosed Gaussian model, Matheron [24] proposed a space-time modeling of air pollutant concentrations, which allows taking into account the spatial non-stationarity. In this model, detailed in the following, time intervenes just as a coordinate.

Dynamic phenomena

A full space-time modeling is necessary, for example, to make short-term predictions of air pollution level, or to study a pollutant transfer from the soil to an aquifer and evaluate the first arrival time at drinkable water pumping. Specific probabilistic modeling is only beginning.

The theoretical solutions of some very general equations can be used to enlarge the class of available models. In 1992, Matheron presented generic models of space-time intrinsic random functions, built as solutions of some physical equations; unfortunately, these models have never been published.

Another approach, purely numerical, consists in associating geostatistical simulations of one or several "parameters", such as permeability, concentrations, etc. and "phenomenological" softwares to deduce the associated flows, ecological state or even meteorological fields.

So a great number of geostatistical simulations of transmissivity fields have been used as entry for flow simulators that compute the associated head in transient flow. Simple and cross-covariances of flow and transmissivity, numerically computed on the whole set of joint realizations of flow and transmissivity, are automatically consistent with the border constraints and the flow equations. These bivariate covariances, space-time concerning the head and space only concerning the transmissivity, are used for conditioning the transmissivity from the head values.

In oil reservoir modeling, the calculated permeability values are constrained by the results of well tests. Flow simulators are used to establish the strongly non-linear relationships between the variables. This procedure associates geostatistics and phenomenological modeling, thanks to the present calculation capacities. If this purely numerical approach appears to be very useful, it still holds some limitations, namely the definition of "border constraints", sometimes poorly known. Sensitivity studies can help identify the influence of this non controllable part of the modeling.

However this approach should be extended to other disciplines, such as hydrogeology or geotechnics.

2.3 A typical non-linear problem: comparison to a threshold

Non linear problems are not specific to environment, selectivity constituting a major part of mining geostatistics. Comparing concentrations to a quality threshold, a recurrent question for pollution studies, is in fact essentially a selectivity problem. This question includes different aspects.

- What does a threshold represent, when the support of the concentrations is not specified?
- How to foresee the global (i.e. on the entire site) or local (i.e. on a given dense sampled area) proportion of values exceeding a threshold? Usually the selection support differs from the support of the experimental data.
- What are the consequences of the unavoidable selection errors when delimitating polluted areas on estimated concentrations instead of real ones, which in practice always remain unknown?

Support effect

The notion of support seems more intuitive for time than for space phenomena. Then, let us go back to time means in the case of air pollution.

A daily concentration is the mean of twenty-four hourly concentrations. So it is necessarily included between their minimum and their maximum values.

For a given observation period, a month for example, the mean of all daily concentrations is equal to the mean of the hourly concentrations, but the daily minimum is necessarily superior to the hourly minimum, and conversely, the daily maximum is necessarily inferior to the hourly maximum. During the month, the experimental variance (the "dispersion" variance) of the hourly concentrations is then greater than the variance of the daily concentrations. Hence, the proportion of concentrations exceeding a given threshold differs between hourly and daily values. Hourly and daily concentrations have not the same histogram.

The European regulation for air quality systematically specifies the time support, namely distinguishing between yearly and hourly limit values. For limit values concerning the protection of human health, defined by quantiles on hourly values, the field is specified: it is the civil year. This regulation, very precise for time values, is muted concerning the spatial support. Now, spatially the support effect has the same consequences: in a given area, when the volume or the surface of the considered "units" increases, the proportion of their high or low values decreases.

For soil pollution, the volume of the samples, generally about a few liters, is much smaller than the volume of the selection blocks during remediation, usually units ranging from a few to dozens of cubic-meters, or more. Deducing the tonnage of selection blocks exceeding a threshold concentration from the histogram of the samples concentrations gives a boorishly erroneous evaluation of the volumes to remediate (Fig. 1(a)).

The choice of the pertinent support should depend on the future usage of the site. In case of strong quality constraints, as for setting up schools or lots, retaining the support of the experimental data consists in imposing that after remediation none of the samples taken on the site will exceed the threshold. For the same threshold, a larger support can be chosen when industrial implants are foreseen on the site.

In soil pollution regulation is still deficient concerning the spatial support related to threshold values.

Matheron has proposed several " change of support models", to forecast the histogram of block values knowing the histogram of samples values (see Rivoirard in this book, page 17).

2.4 Information effect

The "information effect" refers to the consequences of selection errors. The delineation of polluted zones, comparing estimated values to quality threshold, leads to two types of error, corresponding in statistics to the so-called risks of first and second type (Fig. 1(b) cases C and D):

- When the estimated value is higher than the threshold, whereas the actual concentration is lower (case D), the selection of such zones generates unnecessary additional costs. In case of repeated too pessimist predictions, the credibility of these alerts can eventually be questioned.

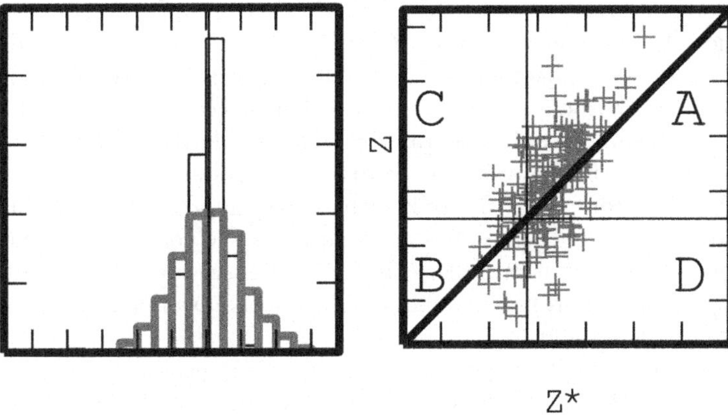

(a) Support effect: histogram of samples (thick grey line) and histogram of block values (fine line) superimposed. The vertical line indicates the mean, indentical for both support

(b) Information effect: scatter diagram of estimated concentrations (abscissa) and actual ones (ordinate)

Fig. 1. Support and information effect

- The case when the estimation is lower than the threshold, whereas the real concentration is higher (case C), is more problematic, because it can produce a sanitary risk. After a site remediation, controls will show the necessity of additional works, making in balance the economy of the project. Furthermore a too important discrepancy between prediction and reality can lead, for soil pollution, to an erroneous choice of a remediation process.

Unavoidable, selection errors are reduced when the scatter diagram between the real value $Z(x)$ in ordinate and its estimation $Z^*(x)$ in abscissa is close to the first bisector line. In particular, it is the case when the accuracy of the estimation is improved: the variance of the estimation error indicates the distance between $Z(x)$ and the first bisector line at point with abscissa $Z^*(x)$. Estimators such as kriging or disjunctive kriging, which ensure "best accuracy" among a class of estimators, usually have a conditional bias: regression of Z on Z^* goes away from the first bisector line. It is often possible to reduce this conditional bias by choosing an appropriate kriging neighborhood.

2.5 Probability of exceeding a threshold

Once the selection support chosen ("points" or "blocks" of a few to several quadrate-meters or cubic-meters) the polluted zone is delineated by "marking" the blocks (or the points) with concentrations exceeding a given threshold. This is done using an indicator variable. The regularization of the indicator is the local proportion of values exceeding the threshold.

As the indicator variable is not known except on the few experimental data, it has to be estimated. The "best estimator" is the conditional expectation which, for an indicator variable, coincides with the conditional probability. Thus, mapping this probability should indicate the local proportion of values exceeding the threshold.

In practice, from the experimental data, it is impossible to verify the validity of the whole probabilistic model used for this estimation. For example the bivariate spatial distribution (the distribution in two points simultaneously) is only inferred for a few distances, according to the sampling schema. The conditional expectation can then be replaced by a less requiring estimator, namely disjunctive kriging , but without guaranteeing that the estimated probability remains within the $[0, 1]$ interval.

The algorithm for calculating the conditional probability within a "reasonable" probabilistic model can also be used. This probability should rather be interpreted as a conventional measure of the uncertainty when classifying the block (or the point) as polluted or not polluted.

In practice, it is wise to examine this probability map when comparing an estimated map to a threshold value, in addition to the map of estimation variance: the "polluted zone" will sometimes be difficult to precisely delineate.

1. *Calculation of the "recoverable" pollution*

In mining geostatistics, the "selectivity" of a distribution represents the capacity to recover as much "metal" as possible by extracting as little ore as possible. Namely, the selectivity can be characterized by the following recovering functions [29]:

- The $Q(T)$ curve, giving the mass of "metal" or pollutant recovered when selecting the T volume of "ore" or soil containing the highest concentrations;
- The curve $B(s)$, expressing the conventional benefit, defined as the mass of recovered pollutant beyond the threshold s.

These curves supply a synthetic information on the lost of selectivity due to support or information effect. Still little used in environment, they are useful to choose the threshold or the selection support.

The anamorphosed multigaussian model

Frequently calculations are made using the anamorphosed multigaussian model, which often proves to be well suited to pollution phenomena. In this model the concentration Z(x) is written as the transformation by an anamorphosis function φ, of a stationary random function Y(x) with standard multinormal spatial distribution: $Z(x) = \varphi(Y(x))$. We suppose here for simplicity that φ is strictly increasing on its definition domain. Within this classic model the conditional distribution at a given location can be directly calculated, without any simulation process.

Conditional distribution

For a random function with standardized multigaussian spatial distribution, the distribution of $Y(x)$ conditionally to the data $Y_\alpha = Y(x_\alpha)$ is simply a normal variable, with expectation equal to kriging (with known mean) $Y^*(x)$ from the Y_α, and with residual variance equal to kriging variance.

Knowing the experimental data Z_α being equivalent knowing the Gaussian transform $Y_\alpha = \varphi^{-1}(Z_\alpha)$, the conditional random function Z(x) can be written:

$$Z(x) = \varphi(Y^*(x) + \sigma_K(x)W(x))$$

where $Y^*(x)$ denotes the kriging of $Y(x)$ from the Y_α, and $\sigma_K^2(x)$ the kriging variance; W is a random function with normal spatial distribution, with zero mean and a unit variance, spatially independent from Y^*. The covariance of W is not stationary, but for many calculations, this does not intervene.

With this conditional distribution, the calculation of different quantities is immediate.

Probability of exceeding a threshold

Thanks to the bijectivity of the anamorphosis, the threshold value on Gaussian transforms becomes simply $\varphi^{-1}(s)$. The conditional probability is easily derived at any point:

$$P(Z(x) > s) = P\left(Y^*(x) + \sigma_k(x)W(x) > \varphi^{-1}(s)\right)$$
$$= 1 - G\left(\frac{\varphi^{-1}(s) - Y^*(x)}{\sigma_K(x)}\right)$$

G denoting the distribution function of the reduced normal distribution.

Kriging being an exact estimator, estimated values are equal to the data at any experimental point $Y^*(x_\alpha) = Y_\alpha$ and the kriging variance is null. The probability value is then 0 or 1 whether Z_α is inferior or superior to the threshold. Far from the experimental points, kriging converges toward the expectation of Y, which is null in the model, as well as the kriging variance converges toward the a priori variance, which equals one in the model. Then, we retrieve the a priori probability $1 - G(\varphi^{-1}(s))$.

This probability indicates then the reliability of the selection for a given threshold. With well-differentiated results near to 0 or 1 an imperfect but realistic selection can be envisaged.

Quantiles

Let q^ϖ be the quantile of order ϖ of a Gaussian variable, i.e. the value exceeded with probability 1-ϖ. The anamorphosis being increasing, it is easy to deduce from the normal conditional distribution that the quantile of order ϖ of Z(x) is $\varphi\left(Y^*(x) + \sigma_K(x)q^\varpi\right)$: mapping the quantiles of the local distribution of the concentrations is very easy.

Probability interval is deduced in the same way. The anamorphosis being non-linear, this interval is generally no longer symmetrical around the estimation of Z(x).

This model is easy to generalize to take into account some auxiliary information.

A space-time model

To represent some pollution phenomena, supposed to be stationary in time but not in space, Matheron proposed a very flexible model [Matheron, 1974], which is a generalization of the previous one. The concentrations of various pollutants, possibly completed by some other information, are supposed to be measured at some experimental points x_α, supplying for each substance i, a long enough time serie $Z\left(i, x_\alpha, t_\beta\right)$. The phenomenon being supposed to be stationary in time, the $Z\left(i, x, t\right)$ are considered point by point as the transformed of a standard Gaussian by an anamorphosis function spatially non-stationary: $Z\left(i, x, t\right) = \varphi_{ix}\left(Y_{ixt}\right)$. In addition, let us suppose that the Y_{ixt} have a bigaussian distribution. The Hermitian development of the anamorphosis $\varphi_{ix}(y) = \sum_k \psi_k(i, x)\eta_k(y)$ defines non-stationary fields of coefficients $\psi_k(i, x)$ which have to be estimated. For example, the non-stationary expectation of the variable i is given by the $\psi_0(i, x)$, and its non-stationary variance by $\sum_{k=1}^{\infty} \left(\psi_k(i, x)\right)^2$.

The estimation is then split in two phases:

- The time series at the experimental point x_α, allow fitting the anamorphosis that is the estimation of the $\psi_k(i, x_\alpha)$ coefficients, by assuming time stationarity for each variable at each experimental point. Out of the experimental points, the $\psi_k(i, x)$ coefficients are then estimated by kriging or cokriging.
- The calculation and the fitting of the space-time variograms of the Gaussian variables $Y_{ix_\alpha t}$ built at each experimental point x_α give a multivariate model of stationary covariance $\rho_{ij}\left(h, \tau\right)$. These covariances make it possible to estimate $Z\left(i, x, t\right)$, or to calculate the probability that the concentration exceeds a given limit value, using non linear estimators such as disjunctive kriging or conditional expectation.

How to check the validity of this model?

The time variograms at the experimental points x_α will or will not invalidate the hypothesis of time stationarity. The spatial variograms of the $\psi_0(i,.)$ coefficients allow controlling their regionalization. We can also verify the positivity of the calculated anamorphosis of the concentration apart from the experimental points, or compute a low order quantile of this concentration. Finally, several criteria allow checking the bigaussian character of the Y_{ixt}.

This model type can be extended to a spatially stationary phenomenon with time non-stationarity, due for example to seasonality. An anamorphosis varying with time will be inferred and then interpolated between the various measurement dates.

As for most approaches, these space-time models are compatible with the classical covariance models.

3 The potentialities of the geostatistical approach

In the second part, we examine –not exhaustively of course– the diversity of problems that geostatistics can help solve. The chosen examples, either applicative or methodological, are taken in various contexts.

3.1 Mapping using auxiliary variables

Bobbia et al [5] were among the first to map the concentrations of urban air pollutants with the help of auxiliary variables as emissions inventories or population density, in order to improve the precision of the estimation and obtain more realistic maps.

For mapping of annual median of the NO_2 daily concentration over the Paris urban area within the period 1997-1999, measurements are available at only 20 sites located mainly in Paris and the near suburbs. Kriging gives then a poorly contrasted map due to the small number of monitored sites. The NO_2 concentration is linked with the nitrogen oxides emission, locally known from inventories (fig. 2(a)) and notated NO_x in the following. Due to chemical reactions in the air, and to the diffusion and transport process of the nitrogen oxides from the sources, the relation between the NO_2 concentration median and the NO_x inventory is not linear. But the linearity with the logarithm of the emissions is quite acceptable, with a high correlation coefficient equal to 0.93 (fig 2(b)). As emission inventories are not available on the whole of the studied area, a cokriging is then performed. In comparison to kriging, cokriging emphasizes the increase in concentrations close to the main road axes (fig. 2(c).).

3.2 Changing the workspace for pluviometry

Quite often environmental characteristics rather than geographic proximity account for the similarities between the observed values of a variable. This

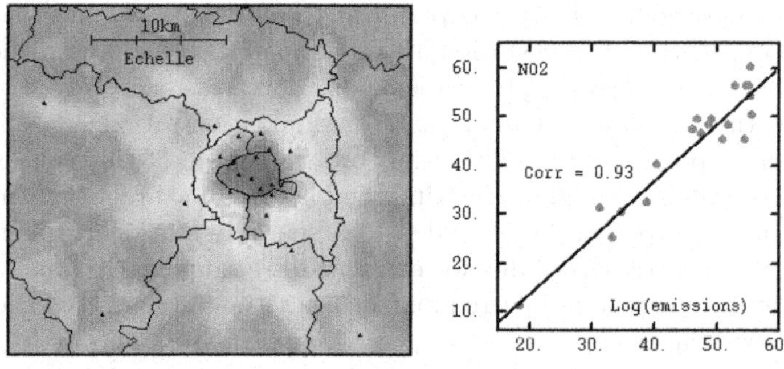

(a) Logarithm of annual NO_x emissions (kriging map).

(b) Scatter diagram between NO_2 median measurements (ordinate) and NO_x emisions logarithm (abscissa).

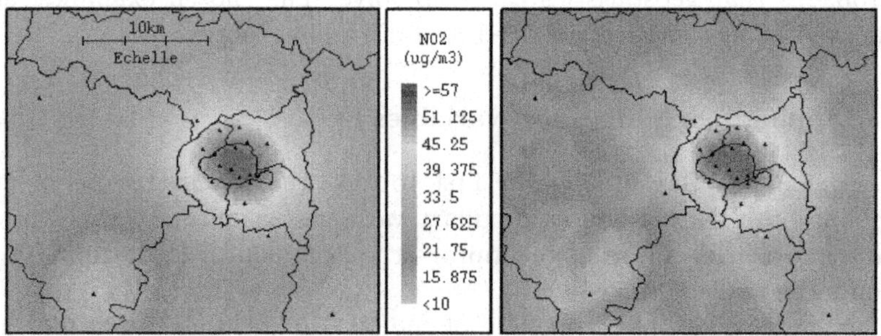

(c) Kriging (left) and Cokriging (right) of NO_2 using emissions logarithm

Fig. 2. Estimation of NO_2 concentrations on Paris urban area.

is why several authors have tried to estimate some variables of interest (rain height, air pollution...) using a deformation of the geographic space. But the criteria defining this deformation are partly arbitrary, and consistency conditions are tricky to ensure, such as the bijectivity of the transformation. Dealing with a bivariate problem, the relationship between rain height and atmospheric pressure in south-west Europe, Biau [2] proposes an elegant solution: the change of working space by means of explicative variables.

A Principal Component Analysis made on pressure measurements during several decades shows that the two first factors explain an important part of the pressure variability. The mapping of these factors indicates that the first one corresponds to the gradient from low pressure above Iceland to high pressure above Azores, while the second one shows the West-Atlantic "abnormality" [3]. At any point, these first two factors are known from the climatologic models.

Let us now consider a fixed location. Similar pressure situations then correspond to close points in the first two factors space. Since rainfall height is largely determined by the pressure, the differences between rainfalls heights at two dates are better described by the "distance" of the corresponding pressure fields, measured in the plane of the first two factors, than it is by the time interval between the two events.

At each location, the 1D modelling of rainfall in time is therefore replaced by a 2D modelling in the factors space. The rainfall variogram is then computed and kriged in this 2D space.

The authors show that most of the variograms thus obtained at different measurement stations present a structure and are consistent with the local characteristics of the rainfall. Besides, this description takes into account the strongly marked contrasts observed between local rainfalls due to the contrasted geography of southern Europe.

In this case, authors are interested only by the temporal and not by the spatial interpolation of rainfalls heights.

3.3 Using qualitative observations in soil pollution

The soils on former industrial sites present sometimes high organic pollution by "tar" or more precisely polycyclic aromatic hydrocarbons (PAH). Too often the areas to be processed are delimited from estimations based on a poor number of chemical samples. As a high pollution by PAH is partly detected on the field by the smell, tar traces, etc., can these qualitative observations (or organoleptic drawings) be exploited to improve the estimates precision at a low cost?

On the site of an old coking plant, Jeannée et al. [16] have qualitative observations and a factorial analysis shows their redundancy. Given the target of analyzing the benzo(a)pyrene concentrations, which is a PAH often sampled because of its pathogen properties, the authors try to improve the estimation by using the correlations between concentration and organoleptic observations.

None of these observations allows detecting exhaustively the whole set of the sampled high concentration values. The organoleptic observation best correlated with concentrations is the presence of tar traces. All the traces correspond to high concentrations even though some high concentrations samples do not present tar traces (fig. 3).

The modality coding of a qualitative variable is arbitrary. But it is known that the qualitative function best statistically correlated with a quantitative variable is the mean per class, i.e. the regression of concentration on the qualitative observation. With only two modalities, the correlation is obviously

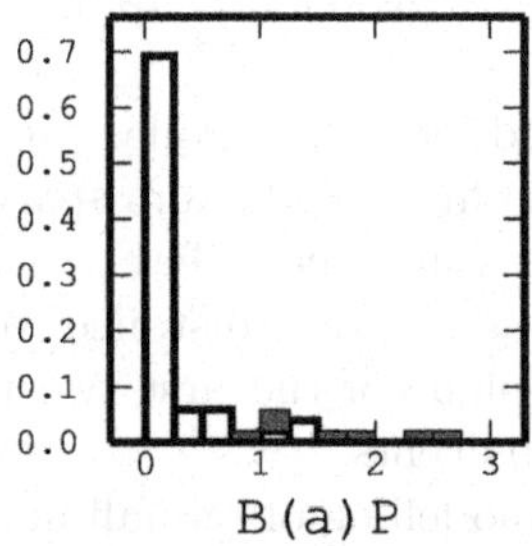

Fig. 3. Histogram of B(a)P concentration, with indication of the presence of tar traces (in dark grey).

the same with the concentration, using the indicator variable or the mean by class.

The experimental variograms and their modeling show the good spatial correlation of these variables (fig. 4).

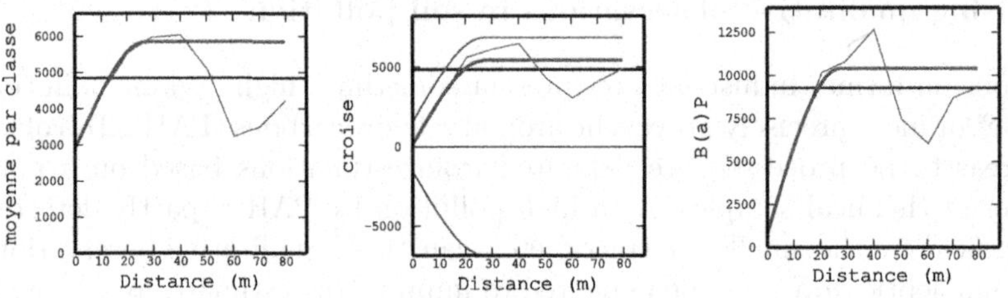

Fig. 4. Fitting simple variograms of B(a)P concentration (right), of the mean per class (left), and of the cross-variogram (middle).

Cross-validation is used to quantify the utility of these observations when estimating the concentrations. For this, 50% of measurements are eliminated and estimated again by cokriging, the observations being considered known in every point. Beside the scatter plot analysis (not shown), the synthetic criterion of the mean squared estimation error shows that, compared to the kriging, cokriging using the auxiliary data lowers the estimation errors. The results obtained by exploiting the tar presence prove to be better than those obtained using the other qualitative observations.

An economic sampling can thus improve the precision. But large enough sampling for chemical analyses remains essential.

3.4 Modeling directional data by complex random Functions

Wind speed, flow in an aquifer, dip or direction, gradient of a scalar quantity... all are vectorial variables which can be described by complex variables. Lajaunie et al. [17], citing results obtained by Matheron examined the interest of a specific treatment of these variables: is the kriging of a complex variable better than the cokriging of its components?

The covariance of a stationary complex variable Z, supposed to have a zero mean, is a Hermitian form $C(h) = E\left[Z(x + h)\bar{Z}(x)\right]$ such as:

$$ Var\left(\sum_i \lambda_i Z(x_i)\right) = \sum_{i,j} \lambda_i \bar{\lambda}_j C(x_i - x_j) \geq 0 \,. $$

It is easy to see that the real part C^R of the complex covariance is a real covariance.

The stationarity of the variable $Z = X + iY$ is not equivalent to the stationarity of the real part X and the imaginary part Y respectively. When both these variables are mutually stationary, the real part of the complex covariance is the sum of both simple covariances: $C^R(h) = C_X(h) + C_Y(h)$, and the imaginary part $C^I(h) = C_{XY}(-h) - C_{XY}(h)$ corresponds to the odd part of the cross-covariance of the X and Y components. The bivariate spatial structure of the pair (X,Y) then holds more precise information than the covariance of the complex variable.

When the cross-covariance C_{XY} of the components is symmetrical, the imaginary part of the complex covariance is zero and C(h) is then identical to its real part. The kriging weights are then real numbers. Therefore the complex kriging can bring some improvement compared to monovariate kriging of real and imaginary part only when their cross-covariance C_{XY} is not symmetrical.

Cokriging of the components usually gives a more accurate estimation than the complex kriging of Z. Indeed the latter minimizes the sum of the estimation variances for real and imaginary parts, while cokriging minimizes each term separately, for a wider class of covariances.

Is the "performing" modeling by complex Random Functions then to be rejected in favour of a more "classical" bivariate modeling? The authors show that the gain brought by complex kriging is precisely due to the economy of this model, in particular when a consistent fitting of the simple and cross-covariances is problematic. It is namely the case for a non symmetrical cross-covariance C_{XY}.

3.5 Estimating or simulating variables linked by linear partial differential equations

Derivative or gradient data can be used in a simpler way considering the covariance models deduced from the differential relationships between variables. If the derivability properties of Random Functions of order two were known well before the beginning of geostatistics (see for example [18]), Matheron has generalized the results, using the theory of the Intrinsic Random Function of order k (IRF-k) [23].

He demonstrated namely the following result: if Z is IRF-k p times derivable, and D an order p derivative operator, then DZ is an IRF-$(k-p)$ if $k \leq p$ and it is a stationary Random Function if $k > p$. In particular, the solution Z of the Poisson equation $\Delta Z = Y$ is an IRF-1 if Y is a Stationary RF, and an IRF-$(k+2)$ if Y is an IRF-k.

Theory shows that the relationships between covariances are simply obtained by exchanging expectation and derivation relatively to each term: if Z is a stationary RF with a covariance $K(h)$, its derivative are stationary and with zero mean; the simple and cross-covariances between Z and its gradient components can be written

$$E\left[\frac{\partial Z}{\partial x_i}(x).Z(x')\right] = \frac{\partial}{\partial x_i} E\left[Z(x).Z(x')\right] \text{ and}$$

$$E\left[\frac{\partial Z}{\partial x_i}(x).\frac{\partial Z}{\partial x_{j'}}(x\prime)\right] = \frac{\partial^2}{\partial x_i \partial x_j'} E\left[Z(x).Z(x\prime)\right] \text{ which gives}$$

$$E\left[\frac{\partial Z}{\partial x_i}(x).Z(x+h)\right] = -\frac{\partial}{\partial h_i} K(h) \text{ and}$$

$$E\left[\frac{\partial Z}{\partial x_i}(x).\frac{\partial Z}{\partial x_j'}(x+h)\right] = -\frac{\partial^2}{\partial h_i \partial h_j} K(h) .$$

When covariance K is decreasing with h, $Z(x+h)$ is positively correlated to the derivative $\frac{\partial Z}{\partial x_i}$ at point x. These relationships remain valid for IRF-k. In particular, as each derivation reduces the IRF-k degree by one unit, the degree of the derivative covariance is reduced by 2 units, in consistency with the polynomial indetermination of the generalized covariances.

The integration of the stationary RF offers in fact a quite intuitive approach to the IRF-k theory. In this model, only some "allowable linear combinations" also called "generalized increments" are supposed to be stationary with finite variance. Indeed these linear combinations mean filtering some polynomial integration constants.

These relationships allow estimating [9] or simulating [11] some variables linked by linear partial differential equations. For example, estimating or simulating a gradient can be done independently of any discretisation, thus avoiding the smoothing effect of methods such as finite differences.

For example, let us examine the gradient estimation from the data $Z(x_\alpha)$ within a stationary Random Function model. λ_i^α denotes the weight attributed to $Z(x_\alpha)$ when estimating the derivative $\frac{\partial Z}{\partial x_i}$ at point x, and λ_α the weight when estimating the variable Z itself. When the mean of Z is unknown, the non-bias condition $E\left[\frac{\partial Z^*}{\partial x_i}(x) - \sum_\alpha R\lambda_i^\alpha Z(x_\alpha)\right] = 0$ imposes that the sum of the weights attributed to all experimental values of Z is zero when estimating one derivative : $\sum_\alpha \lambda_i^\alpha = 0$, instead of one when estimating Z itself. As usual, the first term of the kriging system depends on the covariance between the data, and the second term on the covariances between the data and the variable to be estimated, the system being completed according to the non bias conditions. From the linearity of the covariance with regard to each variable, it is easy to show that the kriging weights for the estimation of the derivative are equal to the derivative of the kriging weights for the estimation of the variable Z: $\lambda_i^\alpha = \frac{\partial}{\partial x_i}\lambda_\alpha$. Hence, any realization verifies: $\frac{\partial(z^*)}{\partial x_i} = \left(\frac{\partial z}{\partial x_i}\right)^*$, as long as the kriging of Z and its derivatives are performed using a covariance model consistent with the derivation and using the same set of experimental points.

These relationships allow estimating the (perturbation of) transmissivity in a way consistent with the (perturbation of) head. In practice, for complex flows, purely numerical solutions based on simulation techniques are sometimes unavoidable.

3.6 Data assimilation

Data assimilation allows incorporating ("assimilating") some measurements in a space-time numerical model which depicts some phenomena as complicated as the evolution of meteorological or oceanographic conditions, in order to improve the predictions. Schematically, the system state Z_n at discretized time t_n is supposed to be entirely given by the previous step n-1 via a non linear function f depicting the system evolution, associated to a correction of modeling errors. This can be written: $Z_n = f_n(Z_{n-1}) + \varepsilon_n^m$ or more generally $Z_n = f_n(Z_{n-1} + \varepsilon_n^m)$. The measurements, supposed to be known at some of these instants n, can be written as a function of the system, associated with some measurement errors: $Y_n = h(Z_n) + \varepsilon_n^o$. The function h, non necessarily linear, takes into account among others the often huge support difference existing between the mesh of the phenomenological numeric model on which Z_n is regularized (for example one or more km) and the quasi-punctual volume of the measurements.

There exists a great number of variations on the data assimilation techniques. A review can be found in [1]. The sequential methods make the spatial correlation matrixes of the "prediction" errors evolve, by propagating the uncertainties on the successive steps. The advantage is that the error matrixes obtained that way are consistent with the modeling of the phenomenon. For example a non linear prediction method with a linear correction can be split in two stages:

- "forecasting" which consists in letting the system evolves numerically following its own dynamics, from the estimation Z_{n-1}^* obtained at the previous step: $Z_n^f = f_n(Z_{n-1}^*)$;
- correcting, by cokriging of $Z_n - Z_n^f$ from the difference between this "sketch" and the observations: $Z_n^* = Z_n^f + K_n\left(Y_n - h\left(Z_n^f\right)\right)$; K_n then refers to a linear operator.

The difficulty consists in computing the covariance used for this cokriging. For example the evolution equation are linearized, and the dimension of the covariance matrix is reduced by keeping only the higher eigenvalues (filter called "RRSQRT" short for "reduced rank square root"); or a purely numerical approach is used, based on simulations ("En KF" short for "Ensemble Kalman filter"). The operational versions of these methods are in fact based on simplifications.

These methods finely associate geostatistical techniques and phenomenological equations.

4 Conclusion

The vast majority of papers concerning environmental geostatistics are thus based on concepts and methods created prior to the 1980's. Matheron's target in *Estimating and choosing* got a final answer: the validation is the "sanction by the practice".

Which direction will the new methodological developments take? Using auxiliary variables as coordinates, as well as changing the work space (for modeling concentration along the river flows or on large domains that require taking into account the earth roundness) should significantly widen the validity field of existing models.

Coupling geostatistics to phenomenological models (short term forecast of air quality, modeling of basin shedding pollutants, modeling of reactive transport of pollutants ...) should improve the relevance of the estimations. Taking into account the physical context and the current capacities of modeling is essential for integrating geostatistics in many "predicting" processes, avoiding thus the risk of misevaluating some first order effects.

Let us finish with some words by Matheron [31, p 7]: [in this book]

> *"there will not be found any "world view", whether explicit or implicit, but only methodological advice for practitioners of probability. One can, if necessary, distinguish between the prime mover and the code. The prime mover, i.e. the "dialectic", is always implicitly present. It incites us to progress, to start new endeavors, without ever remaining at rest. It enables us to "understand" all of the previous developments in a vast retrospective synthesis, but only with hindsight, like the Minerva bird which rises after sunset. It may be that this is just an anthropomorphic illusion. At any rate, this point of view is only suitable for an historian or a philosopher (...). Not so to the practitioner: in his work he rather needs a code, a sort of plumb line to help him build straight walls. This does not mean that he subscribes to any particular philosophy of the plumb line, but only that he likes his job".*

Acknowledgement. The authors thank J. Rivoirard and P. Chauvet for carefully reviewing the paper and F. Poirier for her help with the translation.

References

1. Bertino L. 2001. Assimilation de données pour la prédiction de paramètres hydrodynamiques et écologiques : cas de la lagune de l'Oder. Thèse de Doctorat de l'Ecole des Mines de Paris.
2. Biau G. 1999. Downscaling of precipitation combining kriging and empirical orthogonal function analysis. Proceedings of the Second European Conference on Geostatistics for Environmental Applications held in Valencia, Spain, November 18-20,1998. Kluwer Academic Publishers.
3. Biau G., Zorita E., von Storch H., Wackernagel H. 1999. Estimation of precipitation by kriging in the EOF Space of the Sea level pressure Field. Journal of Climate, Volume 12. pp 1070-1085.
4. Blond N. 2002. Assimilation de données photochimiques et prévision de la pollution atmosphérique. Thèse de doctorat de l'Ecole Polytechnique, Palaiseau, France.
5. Bobbia M. Mietlicki F., Roth C. 2000. Surveillance de la qualité de l'air par cartographie : l'approche géostatistique. Prix du meilleur poster INRETS, 5-8 juin Avignon, France.
6. Bourgine B.,Chilès J.-P.,Watremez P. 2001. Space-time modelling of sand beach data. Proceedings of the Third European Conference on Geostatistics for Environmental Applications held in Avignon, France, November 22-24, 2000. Kluwer Academic Publishers.
7. Chilès, J.-P. 1979. La dérive à la dérive. Note du Centre de Géostatistique et de Morphologie Mathématique, Ecole des Mines de Paris, Fontainebleau.
8. Chilès J.-P., Delfiner P. Geostatistics: modelling spatial uncertainty. 1999. N.Y. Wiley (Wiley series in probability and statistics: applies probability and statistics section).
9. Dong A. 1990. Estimation géostatistique des phénomènes régis par des équations aux dérivées partielles. Thèse de docteur en géostatistique. Ecole des Mines de Paris.
10. Fouquet, C. de, Beucher, H., Galli, A. et Ravenne C. 1989. Conditional simulation of random sets: Application to an argillaceous sandstone reservoir. In: M. Armstrong (ed.), *Geostatistics*, Dordrecht : Kluwer, vol.2.
11. Fouquet, C. de. 2001. Joint simulation of a random function and its derivatives. In: W. Kleingeld and D.G. Krige, Geostats 2000 Cape Town: proceedings of the sixth international geostatistics congress held in Cape Town, South Africa, in April 2000, Geostatistical Association of Southern Africa, vol. 1, pp. 84-93.
12. Gneiting T. 2002. Nonseparable stationary covariance functions for space-time data. J. American Statistical Association, 97: 560-600.
13. Gómez-Hernandez J., Soares A., Froidevaux R. (Eds), 1999. GeoENVII- Geostatistics for Environmental Applications. Proceedings of the Second European Conference on Geostatistics for Environmental Applications held in Valencia, Spain, November 18-20, 1998. Kluwer Academic Publishers.
14. Grzebyk M., Dubroeucq D. 1994. Quantitative analysis of distribution of soil types: existence of an evolutionary sequence in Amazonia. Geoderma, 62, pp285-298.

15. Guibal D., 1973. L'estimation des okoumés du Gabon- problèmes méthodologiques. Note du Centre de Géostatistique, Ecole des Mines de Paris, Fontainebleau.

16. Jeannée N., Fouquet C. de. 2003. Apport d'informations qualitatives pour l'estimation des teneurs en milieux hétérogènes : cas d'une pollution de sols par des hydrocarbures aromatiques polycycliques (HAP). *Comptes rendus Geoscience*, Vol. 335, no. 5, p. 441-449.

17. Lajaunie C., Bejaoui R. 1991. Sur le krigeage de fonctions complexes. Note du Centre de Géostatistique, Ecole des Mines de Paris, Fontainebleau.

18. Lévy P. 1965. Processus stochastiques et mouvement brownien. Gauthiers-Villars. Paris.

19. Marbeau J.-P., 1974. Pollution des huîtres du bassin du Morbihan. Note du Centre de Géostatistique, Ecole des Mines de Paris, Fontainebleau.

20. Matheron G. 1965. Les variables régionalisées et leur estimation. Editions Technip, Paris.

21. Matheron G., 1970. La théorie des variables régionalisées, et ses applications. Les Cahiers du Centre de Morphologie mathématique, fascicule 5. Ecole des Mines de Paris, Fontainebleau.

22. Matheron G. 1973. Le krigeage disjonctif. Note du Centre de Géostatistique, Ecole des Mines de Paris, Fontainebleau.

23. Matheron G. 1973. The intrinsic random functions and their applications. Adv. Appl. Prob., N.5, pp439-468.

24. Matheron G. 1974 Les fonction de transfert des petits panneaux. Note du Centre de Géostatistique, Ecole des Mines de Paris, Fontainebleau.

25. Matheron G. 1978. Estimer et choisir. Les cahiers du centre de morphologie mathématique de Fontainebleau. Ecole des Mines de Paris.

26. Matheron G. 1978. L'estimation globale des réserves récupérables. Note du Centre de Géostatistique, Ecole des Mines de Paris, Fontainebleau.

27. Matheron G. 1978. Le krigeage disjonctif et le paramétrage local des réserves. Note du Centre de Géostatistique, Ecole des Mines de Paris, Fontainebleau.

28. Matheron G. 1984. The selectivity of the distributions and the "second principle of geostatistics". Proc. 2nd NATO ASI, "geostatistics for natural resources characterization". G. Verly et al. eds. Dordrecht, Holland. Reidel Publ.

29. Matheron G. 1985. Comparaison de quelques distributions du point de vue de la sélectivité. Sciences de la terre n.24, Série Informatique géologique.

30. Matheron G. 1989. The internal consistency of models in geostatistics. In: M. Armstrong (ed.), *Geostatistics*, Dordrecht: Kluwer, vol. 1.

31. Matheron G. 1989. Estimating and choosing: an essay on probability in practice, Springer. Traduction A.M. Hasofer.

32. Monestiez P., Allard D., Froidevaux R. (Eds), 2001. GeoENVIII- Geostatistics for Environmental Applications. Proceedings of the Third European Conference on Geostatistics for Environmental Applications held in Avignon, France, November 22-24, 2000. Kluwer Academic Publishers.

33. Orfeuil J.-P. 1977. Etude, mise en œuvre et test d'un modèle de prédiction à court terme de pollution atmosphérique. Note du Centre de Géostatistique, Ecole des Mines de Paris, Fontainebleau.

34. Rivoirard J. 1991. Introduction au krigeage disjonctif et à la géostatistique non linéaire. Cours C-139. Centre de géostatistique, ENSMP, Fontainebleau.

35. Saporta G. 1990. Probabilités, analyse des données et statistique. Editions Technip, Paris.

36. Rivoirard J. 1994. Introduction to Disjunctive Kriging and Non-linear Geostatistics. Clarendon press, Oxford.

37. Sénégas J., Wackernagel H., Rosenthal W., Wolf T. 2001. Error covariance modelling in sequential data assimilation. – *Stochastic environmental research and risk assessment 1, pp 65-86.*

38. Soares A., Gómez-Hernandez J., Froidevaux R. (Eds), 1997. GéoENVI- Geostatistics for Environmental Applications. Proceedings of the Geostatistics for Environmental Applications Workshop, Lisbon, Portugal, November 18-19, 1996. Kluwer Academic Publishers.

39. Wackernagel H., Webster R., Oliver M. 1988. A geostatistical method for segmenting multivariate sequences of soil data. – In H.H. Bock, ed. Elsevier.

40. Webster R., Oliver M.A. 1989. Disjunctive kriging in agriculture. In: M. Armstrong (ed.), *Geostatistics*, Dordrecht: Kluwer, vol. 1.

Part II

Random Sets

Random Closed Sets

I. Molchanov

Department of Mathematical Statistics and Actuarial Science, University of Bern

1 Early years of random sets

Concepts and results involving random sets appeared in probabilistic and statistical literature long time ago. The origin of the modern concept of a random set goes as far back as the seminal book by A.N. Kolmogorov [22] (first published in 1933) where he laid out the foundations of probability theory. He wrote [22, p. 46]

> Let G be a measurable region of the plane whose shape depends on chance; in other words let us assign to every elementary event ξ of a field of probability a definite measurable plane region G.

In modern terminology, G is said to be a random set, which is not necessarily closed, see [37, Sec. 2.5]. It should be noted also that even before 1933 statisticians worked with confidence regions that can be naturally described as random sets.

The next major contribution was due to H.E. Robbins [43, 44] who discovered the formula

$$\mathbf{E}\mu(X) = \int \mathbf{P}\{x \in X\}\mu(\mathrm{d}x)\,, \tag{1}$$

which relates expectation of a σ-finite measure of a random set with the integral of its coverage function. Despite being a simple application of Fubini's theorem, Robbins' formula marked the first rigorous result concerning random sets. Actually, the special case of this formula appears already in [23] (published in 1933 in a physical journal).

For a while results concerning random sets remained scattered in the literature. A rapid development and growing interest in geometric probabilities in the late sixties called for formalisation of the concept of a random set. Developments in mathematical theories of cones and capacities by G. Choquet and the growing literature on set-valued functions greatly facilitated this task. On the other hand, advances in microscopy and image analysis stimulated appearance of new models for random sets and subsequent developments of statistical tools suitable for their formal analysis.

2 The definition of a random set

The crucial breakthrough done by G. Matheron was to concentrate on random sets with closed values and formally define them as random elements whose values belong to the family $\mathcal{F}$ of closed subsets of a given space E. The formal definition required endowing $\mathcal{F}$ with a σ-algebra generated by the topology on $\mathcal{F}$ that is now commonly known under the name of the Fell topology, see [4] and [9] for discussion of this and many other topologies on the space of closed sets. The idea behind the definition of a random closed set is that a random closed set is accessible through the knowledge of the fact whether or not it hits any given compact set. Matheron's definition of a random closed set in [30] is formulated as follows.

Definition 1. *A map* $X : \Omega \to \mathcal{F}$ *from a probability space* $(\Omega, \mathfrak{F}, \mathbf{P})$ *to the family* $\mathcal{F}$ *of closed subsets of a locally compact separable Hausdorff space* E *is called a* random closed set *if* $\{X \cap K \neq \emptyset\} \in \mathfrak{F}$ *for every* K *from the family* $\mathcal{K}$ *of compact subsets of* E.

The main idea behind Definition 1 is that it is observable if X hits or misses given deterministic sets. Because of this reason the underlying topology on $\mathcal{F}$ is sometimes called the hit-or-miss topology. Note that the book [30] rests on several previous publications of G. Matheron. In particular, in [26] the measurability issues for random closed sets are considered together with the first characterisation theorem for their distributions using the avoidance functional. The report [27] contains the concept of the hit-or-miss topology and the characterisation results for distributions of random closed sets using the capacity functionals.

From the contents of the book [30] it appears that G. Matheron was unaware of the parallel work in set-valued analysis, where E.G. Effros [7] studied the Borel σ-algebra on $\mathcal{F}$ for the case when the carrier space E is a complete separable metric space that is not necessarily locally compact. Since then it is typical to speak about Effros-measurable set-valued functions (also known as multivalued functions or correspondences). Another serious breakthrough was due to C. Himmelberg [19] who proved the following Fundamental Measurability theorem that established equivalence of various definitions of measurability for set-valued functions in Polish spaces.

Theorem 1. *Let* E *be a separable metric space and let* X *be a function on* $(\Omega, \mathfrak{F}, \mathbf{P})$ *with values in the family of closed subsets of* E. *Consider the following statements.*
(1) $\{\omega : X \cap B \neq \emptyset\} \in \mathfrak{F}$ *for every Borel set* $B \subset E$.
(2) $\{\omega : X \cap F \neq \emptyset\} \in \mathfrak{F}$ *for every* $F \in \mathcal{F}$.

(3) $\{\omega :\ X \cap G \neq \emptyset\} \in \mathfrak{F}$ for every open $G \subset E$ (in this case X is said to be Effros-measurable).
(4) $\varrho(y, X) = \inf\{\varrho(y, x) :\ x \in X\}$ is a random variable for each $y \in E$.
(5) There exists a sequence $\{\xi_n\}$ of E-valued random elements (measurable selections of X) such that X almost surely coincides with the closure of $\{\xi_n,\ n \geq 1\}$.
(6) The graph of X, i.e. the set $\{(\omega, x) \in \Omega \times E :\ x \in X(\omega)\}$, belongs to the product σ-algebra of $\mathfrak{F}$ and the Borel σ-algebra on E.
Then the following results hold.
(i) $(1) \Rightarrow (2) \Rightarrow (3) \Leftrightarrow (4) \Rightarrow (6)$
(ii) If E is a Polish space (i.e. E is also complete) then $(3) \Leftrightarrow (5)$.
(iii) If E is a Polish space and the probability space $(\Omega, \mathfrak{F}, \mathbf{P})$ is complete, then (1)–(6) are equivalent.

Although it is possible to deduce numerous measurability results concerning operations with random sets from Theorem 1, it was G. Matheron who first realised the importance of semicontinuity concept for random closed sets. Many operations with sets are not continuous but only semicontinuous, so that measurability can be deduced by establishing semicontinuity of the corresponding maps. The semicontinuity concept also relates random closed sets to problems in stochastic optimisation [45, 46], where random closed sets naturally appear as epigraphs of lower semicontinuous functions.

3 Distributions of random sets

The next issue dealt by G. Matheron after formally defining a random closed set was to describe its distribution in an "economical" way. This is a highly important question since the Borel σ-algebra on $\mathcal{F}$ is so rich that it is infeasible to explicitly allocate probabilities to every event that belongs to it. This was a typically probabilistic question which is not usually dealt with in the set-valued analysis literature.

G. Matheron followed the traditional approach of constructing a probability measure by extending its values from a semi-algebra of sets to the corresponding σ-algebra. This semi-algebra consists of finite unions of the events $\{X \cap K \neq \emptyset\}$ for all $K \in \mathcal{K}$. Note that the families $\{F \in \mathcal{F} :\ F \cap K \neq \emptyset\}$ also generate the Borel σ-algebra on $\mathcal{F}$. In other words, G. Matheron found necessary and sufficient condition for a functional

$$T_X(K) = \mathbf{P}\{X \cap K \neq \emptyset\}, \quad K \in \mathcal{K},$$

to be extendible to a probability measure on $\mathcal{F}$. The functional T_X is called a *capacity functional* of X. It is sometimes called the hitting (or trapping) functional or plausibility functional. An immediate observation is that T_X is sub-additive, but typically non-additive (unless X is a random singleton), i.e.

$T_X(K_1 \cup K_2)$ is less but not necessarily equal to $T_X(K_1) + T_X(K_2)$ for disjoint K_1 and K_2.

The capacity functional has several basic properties

(i) $T_X(\emptyset) = 0$ and $0 \leq T_X(K) \leq 1$ for every $K \in \mathcal{K}$;

(ii) T_X is upper semicontinuous on $\mathcal{K}$, i.e. $T_X(K_n) \downarrow T_X(K)$ if $K_n \downarrow K$;

(iii) T_X is completely alternating (also called alternating of infinite order), i.e. the following recurrently defined functionals

$$\Delta_{K_1} T_X(K) = T_X(K) - T_X(K \cup K_1)$$

$$\cdots$$

$$\Delta_{K_n} \cdots \Delta_{K_1} T_X(K) = \Delta_{K_{n-1}} \cdots \Delta_{K_1} T_X(K) - \Delta_{K_{n-1}} \cdots \Delta_{K_1} T_X(K \cup K_n)$$

are non-positive for every $n \geq 1$ and $K, K_1, \ldots, K_n \in \mathcal{K}$.

It is easy to see that

$$\Delta_{K_n} \cdots \Delta_{K_1} T_X(K) = -\mathbf{P}\{X \cap K = \emptyset, \ X \cap K_1 \neq \emptyset, \ldots, X \cap K_n \neq \emptyset\},$$

so that condition (iii) generalises the monotonicity concept for multivariate cumulative distribution functions. Note that the above notation for the successive differences is taken from the harmonic analysis literature [5] and so differs from the notation used in [30].

A function ϕ on the family of all subsets of E with values in the extended real line is called a capacity if it is monotone, $M_n \uparrow M$ implies $\phi(M_n) \uparrow \phi(M)$ for arbitrary sets $M, M_n \subset E$, and $\phi(K_n) \downarrow \phi(K)$ if $K_n \downarrow K$ are compact sets. The above properties single out those capacities (obtained by extending T onto the family of all subsets of E) that correspond to distributions of random closed sets. The key result in random sets theory says that every functional T satisfying (i)-(iii) above corresponds to the distribution of a unique random closed set.

Theorem 2 (Choquet-Kendall-Matheron theorem). *Let* $T : \mathcal{K} \mapsto [0,1]$. *There exists a unique random closed set X with capacity functional T such that $\mathbf{P}\{X \cap K \neq \emptyset\} = T(K)$ for every $K \in \mathcal{K}$ if and only if T satisfies conditions (i)-(iii).*

G. Matheron [30] attributed this theorem to G. Choquet [6], where it indeed appears but not in such an explicit form. It was observed by D.G. Kendall [21] that the trapping probabilities define the distribution of a random set with not necessarily closed values provided the family $\mathcal{K}$ is replaced by another appropriately chosen family of trapping sets. G. Matheron gave a clear formulation of this theorem given above and provided an independent proof based exclusively upon the first principles of extending a probability measure from a semi-algebra to a σ-algebra. Later on with advances of harmonic analysis on semigroups [5] and the theory of lattices [11, 40] new proofs of this result appeared. Indeed, the family of closed sets is a semigroup and a lattice with the main operation being union. Within both of these frameworks, defining a measure on a semigroup or lattice is one of the key issues. The original

Choquet's proof of Theorem 2 is based on a representation of positive definite functions on cones and is similar to the harmonic analysis proof outlined below.

Proof. The family $\mathcal{K}$ of compact sets is an Abelian semigroup with respect to the union operation. Let $\mathcal{I}$ be the set of all sub-semigroups I of $(\mathcal{K}, \cup)$, which satisfy

$$K, L \in I \;\Rightarrow\; K \cup L \in I \quad \text{and} \quad K \subseteq L, \; L \in I \;\Rightarrow\; K \in I.$$

Define $\tilde{K} = \{I \in \mathcal{I} : K \in I\}$ and equip $\mathcal{I}$ with the coarsest topology in which the sets $\tilde{K}$ and $\mathcal{I} \setminus \tilde{K}$ are open for all $K \in \mathcal{K}$. It is possible to prove that the map

$$c(I) = E \setminus \bigcup_{K \in I} \mathrm{Int}K$$

is continuous on $\mathcal{I}$ ($\mathrm{Int}K$ is the interior of K), and

$$c^{-1}(\mathcal{F}^K) = \bigcup_{L \in \mathcal{K}, \; K \subset \mathrm{Int}L} \tilde{L},$$

where $\mathcal{F}^K = \{F \in \mathcal{F} : F \cap K = \emptyset\}$. Indeed, $c(I) \cap K = \emptyset$ if and only if there exists $L \in I$ such that $K \subset \mathrm{Int}L$. It follows from (ii) that

$$Q(K) = \sup\{Q(L) : \; L \in \mathcal{K}, \; K \subset \mathrm{Int}L\},$$

where $Q(K) = 1 - T(K)$.

Note that $\mathcal{I}$ (with identical involution) is isomorphic to the set of semi-characters on $(\mathcal{K}, \cup)$, i.e. complex-valued maps on $\mathcal{K}$ satisfying $\chi(\emptyset) = 1$ and $\chi(K \cup L) = \chi(K)\chi(L)$. Property (iii) implies that T is a completely alternating function on $(\mathcal{K}, \cup)$. It is possible to prove that the corresponding function Q is negative definite on $\mathcal{K}$, i.e.

$$\sum_{i,j=1}^{n} a_j \bar{a}_k Q(K_j \cup K_k) \leq 0$$

for any complex numbers $a_1, \ldots, a_n$, $n \geq 1$. By [5, Prop. 4.17], there exists a measure ν on $\mathcal{I}$ such that $Q(K) = \nu(\tilde{K})$. Now the continuity property of (Radon) measures ($\sup_\alpha \mu(G_\alpha) = \mu(\cup_\alpha G_\alpha)$ for upward filtering family of open sets G_α) yields

$$\nu(\cup_{L \in \mathcal{K}, \; K \subset \mathrm{Int}L} \tilde{L}) = \sup\{\nu(\tilde{L}) : \; L \in \mathcal{K}, \; K \subset \mathrm{Int}L\} = \nu(c^{-1}(\mathcal{F}^K)),$$

so that $Q(K) = \mu(\mathcal{F}^K)$, where μ is the image measure of ν under the continuous mapping $c : \mathcal{I} \mapsto \mathcal{K}$.

The uniqueness part follows from the fact that the families $\{F \in \mathcal{F} : F \cap K = \emptyset, F \cap K_1 \neq \emptyset, \ldots, F \cap K_n \neq \emptyset\}$ generate the Borel σ-algebra on $\mathcal{F}$.

The lattice-theoretic proof [40] is even more powerful, since it is applicable for a non-Hausdorff space E. However, it is generally unknown how to characterise distributions of random closed sets in a space E that is either not locally compact or not separable, see [38].

4 Further developments

When random closed sets have been properly defined, their distributions characterised and measurability properties of some operations established, the theory of random sets was brought to a stage when it was desirable to obtain results parallel to those well-known in probability theory for random variables and stochastic processes. This was not easy however as the space of all closed (or compact) sets is not a linear space, while most of conventional techniques in probability theory are adapted to studies of random elements in linear spaces.

4.1 Special random sets and models

The capacity functional $T_X(K)$ defined for all compact sets K is a complicated object. In simple cases it is possible to define it using direct probabilistic arguments. For example, if $X = \{\xi\}$ is a random singleton, then $T_X(K) = \mathbf{P}\{\xi \in K\}$ is a probability measure that can be efficiently defined. More complicated examples of random sets appear from stochastic processes as excursion sets, e.g. $X = \{x \in E : \ \xi(x) \geq t\}$, where t is a real number and ξ is a real-valued random process indexed by E with upper semicontinuous paths (in order to ensure that X is closed). However, it remains an important task to develop new models for random sets, provide manageable expressions for their capacity functionals and relate properties of capacity functionals to those of the corresponding random closed sets.

For instance, it is possible to characterise random closed sets with almost surely convex realisations in terms of their capacity functionals. A random closed set X is almost surely convex if and only if

$$T_X(K) + T_X(K \cup K_1 \cup K_2) = T_X(K \cup K_1) + T_X(K \cup K_2) \qquad (2)$$

for every convex compact sets K, K_1 and K_2 such that K_1 and K_2 are separated by K in a sense that the segment joining any two points of K_1 and K_2 hits K, see [30].

G. Matheron introduced one extremely important model for random sets called the Boolean model. The basic idea follows the concept of a point process, which is a collection of points in a carrier space. The principal new feature is that the carrier space becomes the family of compact sets, so that one works with a collection of sets instead of collection of points, see Chap. 1.2 of this volume for an in-depth survey of this important model.

A Poisson point process $\{F_1, F_2, \ldots\}$ on $\mathcal{F}$ is determined by a measure on $\mathcal{F}$, that is not necessarily finite. Similarly to the Choquet theorem, it is possible to show that every such measure ν uniquely corresponds to a function $\Psi(K) = \nu(\{F \in \mathcal{F} : \ F \cap K \neq \emptyset\})$, which is upper semicontinuous, completely alternating, satisfies $\Psi(\emptyset) = 0$, but not necessarily is bounded by 1 from above and may even be infinite. Then

$$T_X(K) = 1 - \exp\{-\Psi(K)\} \tag{3}$$

is the capacity functional of a random closed set X that is the union of the sets $F-1, F-2, \ldots$ that form the underlying Poisson process on $\mathcal{F}$. If the functional Ψ satisfies (2), then X is called the semi-Markov random closed set. It was shown by G. Matheron [28] that these sets include Boolean models with convex grains and also unions of Poisson flats.

Many other important random closed sets are related to paths of a Brownian motion W_t, $t \geq 0$, or other stochastic processes with values in $\mathbb{R}^d$. Assume that W_t starts at x from a compact set D and is killed whenever it leaves D. Denote by X the path of W_t (the set of points visited at least once). This is an example of a random *fractal* set [8, Chap. 15,16]. The corresponding capacity functional is related to hitting probabilities of the Wiener process. For instance, if the initial position x is distributed according to the equilibrium probability measure on D, then $T_X(K)$ is the ratio $C(K)/C(D)$, where $C(\cdot)$ stands for the Newton capacity of the corresponding set. If $X_t = \{W_s : \ s \leq t\}$ is the part of the path up to time t, then its r-neighbourhood $X_t^r = \{x : \ \varrho(x, X_t) \leq r\}$ is called the Wiener sausage. Results on volumes of the Wiener sausage are closely related to the probability that the Wiener process hits obstacles that form a Boolean model, see [50].

Further recent results concern such concepts like capacity equivalence for random sets. Two random closed sets X and Y are called capacity equivalent if there are positive constants c_1 and c_2 such that

$$c_1 T_X(K) \leq T_Y(K) \leq c_2 T_X(K)$$

for every $K \in \mathcal{K}$. It is shown in [41] that the path of the Wiener process is capacity equivalent to a sequence of sets related to some branching processes on a tree generated by successive partitions of the unit square. This concept if closely related to Radon-Nikodym derivatives of capacities considered in [15].

4.2 Expectation

The concept of averaging for random sets was not mentioned at all in [30], although the relevant ideas in set-valued analysis (concerning integration of set-valued functions) appeared well before 1975 in R.J. Aumann's pioneering work [2]. Aumann defined the integral of a set-valued function F as the set of integrals of all measurable functions f such that $f(t) \in F(t)$ for all parameter values t. In application to random sets the idea of the corresponding expectation was first explicitly spelt out in [1]. The crucial step was to consider all random singletons ξ that almost surely belong to a random closed set X. Such ξ is called a selection of X. It is well-known that an almost surely non-empty random closed set possesses a selection under rather mild conditions on the carrier space E.

A random closed set X in a Banach space E is called integrable if it has at least one integrable selection. The selection expectation (also called Aumann or set-valued expectation) of an integrable random closed set X is defined as the closure of the set of expectations of all integrable selections of X

$$\mathbf{E}X = \mathrm{cl}\{\mathbf{E}\xi : \ \xi \in X \text{ a.s., } \xi \text{ integrable}\} .$$

Taking closure in the right-hand side is essential as the family of all $\mathbf{E}\xi$ is not necessarily closed if the carrier space E is infinite dimensional.

Numerous results concerning the selection expectation include dominated convergence theorem and the Fatou lemma. It is possible to define the conditional expectation that leads to the concept of set-valued martingales [17]. However, the selection expectation has a serious drawback that reduces the range of its practical applications for averaging of sets. On a non-atomic probability space it always returns convex results, i.e. $\mathbf{E}X$ coincides with the expectation of the convex hull of X. Furthermore, if X is bounded, then $\mathbf{E}h(X, u) = h(\mathbf{E}X, u)$, where

$$h(K, u) = \sup\{u(x) : \ x \in K\}$$

is the support function of K, and u is a linear continuous functional on E.

Alternative definitions of expectation [34, 36] make it possible to work with non-convex set, although these expectations do not have so nice mathematical properties as the selection expectation.

4.3 Minkowski sums

Despite Minkowski addition and related morphological operations with sets were described and the corresponding measurability results established in [30], the corresponding limit theorems remained beyond the scope of G. Matheron's attention. These limit theorem were derived first for random compact sets in Euclidean spaces and then generalised for random closed sets in Banach spaces without the compactness and boundedness assumptions, see e.g. [14, 18]. For simplicity we consider here only the case of random compact sets in the Euclidean space $E = \mathbb{R}^d$. Recall that $K \oplus L = \{x + y : \ x \in K, \ y \in L\}$ denotes the Minkowski sum of K and L.

Let $X, X_1, X_2, \ldots$ be a sequence of independent identically distributed random compact sets in $\mathbb{R}^d$. Assume that X is integrably bounded, i.e. $\|X\| = \sup\{\|x\| : \ x \in X\}$ is an integrable random variable. In this case all selections of X are integrable. The strong law of large numbers for random compact sets [1] establishes

$$n^{-1}(X_1 \oplus \cdots \oplus X_n) \to \mathbf{E}X , \tag{4}$$

where the convergence is understood with respect to the Hausdorff metric ϱ_H. Recall that the Hausdorff distance between two sets K and L is the smallest positive r such that K is contained in the r-neighbourhood of L and L is contained in the r-neighbourhood of K.

The central limit theorem concerns the speed of convergence in (4). Since it is not possible to subtract $\mathbf{E}X$ from the Minkowski average in the left-hand side of (4), one usually formulates the central limit theorem for square integrable (i.e. $\mathbf{E}\|X\|^2 < \infty$) random sets as the following convergence in distribution property

$$\sqrt{n}\varrho_H(n^{-1}(X_1 \oplus \cdots \oplus X_n), \mathbf{E}X) \xrightarrow{\mathrm{d}} \sup_{\|u\|=1} |\zeta(u)|, \tag{5}$$

where ζ is a Gaussian random field on the unit sphere with the covariance explicitly determined by the distribution of the support function $h(X, \cdot)$ of X, see [51]. Related works include characterisation of stable and infinite divisible for Minkowski addition random sets [12, 13]. It should be noted that the limiting random field (5) does not have an explicit geometrical meaning, and it is an open problem to provide a sensible geometric interpretation of ζ.

4.4 Weak convergence

Weak convergence of random closed sets is a special case of weak convergence of probability measures. Along the same line with the Choquet-Kendall-Matheron theorem, it is possible to show that a sequence of random closed set $\{X_n, \ n \geq 1\}$ converges weakly to a random closed set X if and only if $T_{X_n}(K) = \mathbf{P}\{X_n \cap K \neq \emptyset\}$ converges to $T_X(K) = \mathbf{P}\{X \cap K \neq \emptyset\}$ as $n \to \infty$ for each $K \in \mathcal{K}$ satisfying $T_X(K) = T_X(\mathrm{Int}\,K)$. These sets K correspond to the families of closed sets $\{F \in \mathcal{F} : F \cap K \neq \emptyset\}$ that are continuous with respect to the probability measure on $\mathcal{F}$ that describes the distribution of X, see [39]. In an internal report [29] G. Matheron suggested an equivalent definition that relies on the convergence of the capacity functionals $\limsup T_{X_n}(K) \leq T_X(K)$ for all compact sets K and $\liminf T_{X_n}(G) \geq T_X(G)$ for all open sets G.

It is well-known that the weak convergence of random variables is metrised by the Lévy metric. The weak convergence of random compact sets can be also described using the Lévy distance between their distributions. For random closed sets X and Y define

$$\mathfrak{L}(X, Y) = \inf\{r > 0 : T_X(K) \leq T_Y(K^r) + r, \ T_Y(K) \leq T_X(K^r) + r, \ K \in \mathcal{K}\}.$$

It is shown in [33] that X_n weakly converges to a random compact set X if and only if $\mathfrak{L}(X_n, X) \to 0$ as $n \to \infty$.

4.5 Unions

While the Minkowski addition of random sets generalises conventional sums of random vectors, unions of random sets have their parallel interpretation in the studies of extremes of random variables. Consider a random half-line $X = (-\infty, \xi]$ and independent realisations $X_n = (-\infty, \xi_n]$, $n \geq 1$, of the random closed set X. Then $X_1 \cup \cdots \cup X_n$ is the half-line bounded by $\max(\xi_1, \ldots, \xi_n)$, while $X_1 \oplus \cdots \oplus X_n$ is the half-line bounded by $\xi_1 + \cdots + \xi_n$.

A random closed set X is infinite divisible for unions if, for every n, X can be represented as a union of n independent identically distributed random closed sets. Infinite divisible random closed sets were characterised by G. Matheron in [30] as having the capacity functional of the form (3), where Ψ is a capacity that satisfies the same conditions (i)-(iii) as T with the only exception of the range of values that can now be $[0, \infty]$. The infinite values of Ψ appear due to the fixed points x that belong to X almost surely. The modern proof of (3) for infinitely divisible sets rests on the theory of lattices, see [40].

If, for every $n \geq 1$, there is a number $a_n > 0$ such that $a_n X$ has the same distribution as $X_1 \cup \cdots \cup X_n$ for $X_1, \ldots, X_n$ being independent identically distributed realisations of X, the random set X is called union-stable. Union-stable sets are infinite divisible, and it is possible to show that (3) holds with an additional requirement that $\Psi(sK) = s^\alpha \Psi(K)$ for some $\alpha \neq 0$ and every $s > 0$, $K \in \mathcal{K}$. The case of X without fixed points [30] is much easier to handle than the general case. This is due to the fact that non-trivial random sets with fixed points (e.g. the set of zeroes for the Wiener process or a randomly rotated cone) may satisfy $T_X(sK) = T_X(K)$ for every $s > 0$ and so it is difficult to turn the union-stability condition into a functional equation for T_X, see [33].

The characterisation of union-stable sets is naturally accompanied by a spectrum of limit theorems where union-stable random closed sets appear as weak limits, see [33]. These limit theorems are typically formulated in terms of capacity functionals of random sets, e.g. using the function $f(x) = T_X(xK)$ that should be regularly varying at infinity (or zero) for a sufficiently large family of compact sets K.

4.6 Functionals of random sets

A measurable functional of a random closed set automatically becomes a random variable. The earliest result concerning functionals of random sets is Robbins' formula (1) that is applicable to relate the expectation of $\mu(X)$ for a general σ-finite measure μ to the covering probabilities $\mathbf{P}\{x \in X\}$ of X. While the assumption of σ-finiteness of μ is absolutely essential, it was apparently overlooked in [30]. However, many interesting functionals of X can be represented as values $\mu(X)$ for a not necessarily σ-finite μ. The most well-known examples of such functionals are the surface area and the cardinality of X. As the capacity functional is the ultimate characteristics of a random closed set, it is quite natural to conjecture that the expected value of $\mu(X)$ for a general measure μ can be expressed using the capacity functional of X. Although this problem remains open, some preliminary results can be found in [3].

Another family of functionals of random sets is closely related to problems that appear in stochastic optimisation. For a real-valued random function

$\xi(x)$, $x \in E$, with almost surely lower semicontinuous realisations, its epigraph

$$\operatorname{epi} \xi = \{(x,t) : \ \xi(x) \le t\}$$

becomes a random closed set in $E \times \mathbb{R}$. The crucial point is to notice that the infimum of $\xi(x)$ for x from a compact set K is a random variable and the points, where this infimum is achieved, form another random closed set. This observation sparked a considerable activity in stochastic optimisation literature, see [45, 46]. For instance, to ensure weak convergence of infima, it suffices to prove that the sequence of the corresponding epigraphs converges weakly as random closed sets.

Further results on functionals of random sets rely on assuming particular models for random sets. Examples of these results are integral geometrical formulae for Boolean models of random sets [49], Boolean random functions [20, 48] and results for convex hulls of random points [47].

4.7 Statistics

Despite the fact that G. Matheron's book [30] does not deal explicitly with any statistical issue concerning random sets, the developed probabilistic tools formed a platform for further developments of statistical techniques. While statistical issues for point processes had been in the focus of attention of statisticians for quite a while before 1975, the first statistical paper on random sets [42] appeared later in 1977. It concerned estimation of a domain accessed though Poisson points inside it. The natural estimator is the convex hull of these Poisson points. This estimator is however biased and has to be rescaled to eliminate the bias. Modern developments in this problem are surveyed in [24].

Realisations of random sets are available through values of some functionals or numerical measurements. Determining of sets using values of functionals or measurements was initiated in [25] and further put into the framework of mathematical morphology in [16].

For general random sets, it is possible to build the empirical capacity functional in the same manner as empirical measures are defined. Let X_n, $n \ge 1$, be i.i.d. realisations of a random closed set X. Define the empirical capacity functional as

$$T_n^*(K) = \frac{1}{n} \sum_{i=1}^{n} 1_{X_i \cap K \neq \emptyset}, \quad K \in \mathcal{K}.$$

The strong law of large numbers immediately implies that $T_n^*(K)$ converges almost surely to $T(K)$ for any given K. However, the uniform convergence may fail even over a simple family of sets K. For instance, let X be a random closed subset of $\mathbb{R}$ defined as $\xi + M$, where ξ is normally distributed (say) and M is a nowhere dense set of a positive Lebesgue measure. Then it is easy to see that

$|T_n^*(\{x\}) - T(\{x\})|$ does not converge uniformly to zero over $x \in [0,1]$. The uniform convergence properties have been explored in [31], where it was shown that the empirical capacity functional converges uniformly over the family of all compact sets if X coincides almost surely with the closure of its interior, $\mathrm{Int}X$, and $\mathbf{P}\{\mathrm{Int}X \cap K \neq \emptyset\} = T(K)$ for each $K \in \mathcal{K}$. The corresponding central limit theorem and applications to estimation of quantiles of random sets were discussed in [32]. Further results on weak convergence of families of probability measures dominated by empirical capacity functionals can be found in [10].

While statistical techniques for general random closed sets are still quite scarce, more is known for particular models of random sets. This concerns, in particular, the Boolean model, where a range of statistical tools exists [35, 49], and union-stable random sets [33]. Statistical techniques for random sets commonly rely on minimisation of minimum contrast or method of moments. This means that parameters are estimated by matching moments of some functionals of the sample with the moments calculated (theoretically, numerically or by simulations) for the underlying model. Approaches based on likelihood are understandably quite complicated to work out, since the complete likelihood function is very difficult to write even for models based on the Poisson assumption.

5 Final remarks

The range of citation of Matheron's random sets book [30] is extremely wide and stretches far beyond the literature specifically concerned random sets. Apart from a tremendous impact on mathematical morphology and image analysis, its random sets chapters have been cited by many authors who wrote on harmonic analysis on semigroups, sample paths properties of stochastic processes, set-indexed processes, set-valued analysis, stochastic optimisation and integral geometry. The up-to-date state of the random sets theory is presented in [37].

Matheron's book on random sets left enough open ends in random sets theory to ensure its fruitful development for nearly thirty years. I am pleased to note that this book was translated into Russian and published in 1978, very soon after its English edition appeared in 1975. The translator, V.P. Nosko, and the editor, V.M. Maksimov, did a great job going through the uneasy text and supplying their comments to occasional unclear or difficult places. When it came to writing my second year undergraduate project in 1980, I discovered the Russian translation of Matheron's book in a book shop and was impressed by the way topology, convex geometry and probability theory merge together. And I am still fascinated by it.

References

1. Artstein, Z. and Vitale, R. A. (1975). A strong law of large numbers for random compact sets. *Ann. Probab.* **3**, 879–882.
2. Aumann, R. J. (1965). Integrals of set-valued functions. *J. Math. Anal. Appl.* **12**, 1–12.
3. Baddeley, A. J. and Molchanov, I. S. (1997). On the expected measure of a random set. In D. Jeulin, editor, *Advances in Theory and Applications of Random Sets*, 3–20, Singapore. Proceedings of the International Symposium held in Fontainebleau, France (9-11 October 1996), World Scientific.
4. Beer, G. (1993). *Topologies on Closed and Closed Convex Sets*. Kluwer, Dordrecht.
5. Berg, C., Christensen, J. P. R. and Ressel, P. (1984). *Harmonic Analysis on Semigroups*. Springer, Berlin.
6. Choquet, G. (1953/54). Theory of capacities. *Ann. Inst. Fourier* **5**, 131–295.
7. Effros, E. G. (1965). Convergence of closed subsets in a topological space. *Proc. Amer. Math. Soc.* **16**, 929–931.
8. Falconer, K. J. (1990). *Fractal Geometry*. Wiley, Chichester.
9. Fell, J. M. G. (1962). A Hausdorff topology for the closed subsets of a locally compact non-Hausdorff space. *Proc. Amer. Math. Soc.* **13**, 472–476.
10. Feng, D.-J. and Feng, D. (2004). On a statistical framework for estimation from random set observations. *J. Theoret. Probab.* **17**, 85–110.
11. Gierz, G., Hofmann, K. H., Keimel, K., Lawson, J. D., Mislove, M. and Scott, D. S. (1980). *A Compendium of Continuous Lattices*. Springer, Berlin.
12. Giné, E. and Hahn, M. G. (1985). Characterization and domains of attraction of p-stable compact sets. *Ann. Probab.* **13**, 447–468.
13. Giné, E. and Hahn, M. G. (1985). M-infinitely divisible random sets. *Lect. Notes Math.* **1153**, 226–248.
14. Giné, E., Hahn, M. G. and Zinn, J. (1983). Limit theorems for random sets: application of probability in Banach space results. In A. Beck and K. Jacobs, editors, *Probability in Banach spaces, IV (Oberwolfach, 1982)*, volume 990 of *Lect. Notes Math.*, 112–135. Springer, Berlin.
15. Graf, S. (1980). A Radon-Nikodym theorem for capacities. *J. Reine Angew. Math.* **320**, 192–214.
16. Heijmans, H. J. A. M. and Molchanov, I. S. (1998). Morphology on convolution lattices with applications to the slope transform and random set theory. *J. Math. Imaging and Vision* **8**, 199–214.
17. Hess, C. (1999). Conditional expectation and martingales of random sets. *Pattern Recognition* **32**, 1543–1567.
18. Hess, C. (1999). The distribution of unbounded random sets and the multivalued strong law of large numbers in nonreflexive Banach spaces. *J. Convex Analysis* **6**, 163–182.
19. Himmelberg, C. (1974). Measurable relations. *Fund. Math.* **87**, 53–72.
20. Jeulin, D. (1997). *Morphological Models of Random Structures*. CRC Press, Boca Raton, Florida.
21. Kendall, D. G. (1974). Foundations of a theory of random sets. In E. F. Harding and D. G. Kendall, editors, *Stochastic Geometry*, 322–376. Wiley, New York.
22. Kolmogorov, A. N. (1950). *Foundations of the Theory of Probability*. Chelsea, New York.

23. Kolmogorov, A. N. and Leontovitch, M. A. (1992). On computing the mean Brownian area. In A. N. Shiryaev, editor, *Selected works of A. N. Kolmogorov, Volume II: Probability and mathematical statistics*, volume 26 of *Mathematics and its applications (Soviet series)*, 128–138. Kluwer, Dordrecht, Boston, London.

24. Korostelev, A. P. and Tsybakov, A. B. (1993). *Minimax Theory of Image Restoration*. Springer, New York.

25. Lyashenko, N. N. (1983). Statistics of random compacts in the Euclidean space. *J. Soviet Math.* **21**, 76–92.

26. Matheron, G. (1967). *Eléments pour une Théorie des Milieux Poreux*. Masson, Paris.

27. Matheron, G. (1969). Théorie des ensembles aléatoires. Technical report, Les Cahiers du Centre de Morphologie Mathematique, Fascicule 4, Paris School of Mines.

28. Matheron, G. (1972). Ensembles fermés aléatoires, ensembles semi-Markoviens et polyèdres poissoniens. *Adv. Appl. Probab.* **4**, 508–541.

29. Matheron, G. (1975). La convergence en loi des fermés aléatoires. Technical Report Internal Report N-409, Paris School of Mines, Fontenebleau.

30. Matheron, G. (1975). *Random Sets and Integral Geometry*. Wiley, New York.

31. Molchanov, I. S. (1987). Uniform laws of large numbers for empirical associated functionals of random closed sets. *Theory Probab. Appl.* **32**, 556–559.

32. Molchanov, I. S. (1990). Empirical estimation of distribution quantiles of random closed sets. *Theory Probab. Appl.* **35**, 594–600.

33. Molchanov, I. S. (1993). *Limit Theorems for Unions of Random Closed Sets*, volume 1561 of *Lect. Notes Math.*. Springer, Berlin.

34. Molchanov, I. S. (1997). Statistical problems for random sets. In J. Goutsias, R. Mahler and H. T. Nguyen, editors, *Applications and Theory of Random Sets*, volume 97 of *The IMA Volumes in Mathematics and its Applications*, 27–45, Berlin. Springer.

35. Molchanov, I. S. (1997). *Statistics of the Boolean Model for Practitioners and Mathematicians*. Wiley, Chichester.

36. Molchanov, I. S. (1998). Random sets in view of image filtering applications. In E. R. Dougherty and J. Astola, editors, *Nonlinear Filters for Image Processing*, chapter 10. SPIE.

37. Molchanov, I. (2005). *Theory of Random Sets*. Springer, New York.

38. Nguyen, H. T. and Nguyen, N. T. (1998). A negative version of Choquet theorem for Polish spaces. *East-West J. Math.* **1**, 61–71.

39. Norberg, T. (1984). Convergence and existence of random set distributions. *Ann. Probab.* **12**, 726–732.

40. Norberg, T. (1989). Existence theorems for measures on continuous posets, with applications to random set theory. *Math. Scand.* **64**, 15–51.

41. Peres, Y. (1996). Intersection-equivalence of Brownian paths and certain branching processes. *Comm. Math. Phys.* **177**, 417–434.

42. Ripley, B. D. and Rasson, J.-P. (1977). Finding the edge of a Poisson forest. *J. Appl. Probab.* **14**, 483–491.

43. Robbins, H. E. (1944). On the measure of a random set. I. *Ann. Math. Statist.* **15**, 70–74.

44. Robbins, H. E. (1945). On the measure of a random set. II. *Ann. Math. Statist.* **16**, 342–347.

45. Rockafellar, R. T. and Wets, R. J.-B. (1998). *Variational Analysis.* Springer, Berlin.
46. Salinetti, G. and Wets, R. J.-B. (1986). On the convergence in distribution of measurable multifunctions (random sets), normal integrands, stochastic processes and stochastic infima. *Math. Oper. Res.* **11**, 385–419.
47. Schneider, R. (1988). Random approximations of convex sets. *J. Microscopy* **151**, 211–227.
48. Serra, J. (1989). Boolean random functions. *J. Microscopy* **156**, 41–63.
49. Stoyan, D., Kendall, W. S. and Mecke, J. (1995). *Stochastic Geometry and its Applications.* Wiley, Chichester, second edition.
50. Sznitzman, A.-S. (1998). *Brownian Motion, Obstacles and Random Media.* Springer, Berlin.
51. Weil, W. (1982). An application of the central limit theorem for Banach-space-valued random variables to the theory of random sets. *Z. Wahrsch. verw. Gebiete* **60**, 203–208.

The Boolean Model:
from Matheron till Today

Dietrich Stoyan[1] and Klaus Mecke[2]

[1] Institut für Stochastik, TU Bergakademie Freiberg
[2] Institut für Theoretische Physik, Universität Erlangen-Nürnberg,

1 Introduction

Until the 1970s random sets were only a marginal or exotic part of probability theory. This situation has changed completely since the publication of the fundamental and seminal book by Matheron [43]. This book has laid the fundamentals of the theory of random closed sets, provided the suitable measure-theoretic machinery and offered the fundamental theorems. It also presented an excellent introduction to the theory of the Boolean model.

The Boolean model appeared early in applied probability, typically in the context of attempts to describe random geometrical structures of physics and materials science. In most of these papers, only the case of spherical grains was studied in *ad hoc* approaches. It was Matheron who created the general theory of the Boolean model. Already in [39] the stationary and non-stationary versions of the model were studied. Even the case of a Cox process for the germs was considered, and the grains were not necessarily convex. In [40] the Boolean model is introduced in its stationary form, for any kind of grains; the case of convex grains is studied in relation with the important semi Markov property. This property is developed further in [42], where the Boolean model for non convex and convex grains is studied, the intensity $\lambda(x)$ being a measure, so that the non stationary case is covered. Finally, in [43] the Boolean model appears in three different places: first, a Poisson point process on closed sets in considered (pp. 57–61), generating abstract Boolean models in general spaces (even non euclidean); then (pp. 61–62) the euclidean case in considered, and detailed in the stationary case; the convex case is specialized in connection with the semi Markov property (pp. 137–139).

The present paper describes in its first three sections briefly the classical results such as given in [43] and [61], beginning with some remarks on random closed sets, together with some modern concepts for the characterisation of their distribution. It then informs in section 4 about results for the Boolean model under reduced invariance conditions. While in the classical papers the model was stationary and isotropic, later also the non-isotropic and even non-stationary case has been investigated. In the physical literature the idea of a Boolean model on a lattice appeared, which is briefly sketched in section 5. Also physical problems have led to the study of percolation for the Boolean model. A very successful heuristic approach to this difficult problem is described in section 6. Finally, important statistical methods for the Boolean model are described in section 7. A lot of papers have been published on statistical problems for the Boolean model, but a good part of them presents practically unrealistic and instable methods. The authors go back to the sources, in particular to Serra's classical work, and describe briefly those methods they consider as stable and powerful.

2 Characterising the Distribution of a Random Closed Set

Matheron [41, 43] showed that the distribution of a random closed set Ξ is given by the probabilities

$$T(K) = P(\Xi \cap K \neq \emptyset) \qquad \text{for } K \in \mathbb{K},$$

where $\mathbb{K}$ denotes the system of compact subsets of the space in which Ξ exists; in the present paper this is the d-dimensional Euclidean space $\mathbb{R}^d$. Sometimes, T is called the *capacity functional*, because T is a so-called alternating capacity of infinite order in the sense of Choquet.

It is easy to show that invariance properties of random closed sets can be expressed in terms of the capacity functional. A random closed set Ξ is called *stationary* if its distribution is translation invariant, i.e. Ξ and $\Xi_h = \{y : y = x + h, x \in \Xi\}$ have the same distribution for all $h \in \mathbb{R}^d$. This is equivalent to

$$T(K) = T(K + h)$$

holding for all $K \in \mathbb{K}$ and all $h \in \mathbb{R}^d$. Isotropy is an analogous property connected with rotations around the origin. Also here, a characterisation by means of the capacity functional is possible.

The system $\mathbb{K}$ is so large that in non-trivial cases it is hopeless to specify $T(K)$ for all $K \in \mathbb{K}$. A natural idea is therefore to consider only K's belonging to subsets of $\mathbb{K}$. In particular, families of compact sets K are used which are parametrised (preferably) by real parameters. This leads to functions that can be presented graphically. Such families of compact subsets of $\mathbb{R}^d$ are e.g. the systems of all singletons $\{x\}$ for $x \in \mathbb{R}^d$, pairs $\{x, y\}$ for x and y in $\mathbb{R}^d$, triplets $\{x, y, z\}$ for x, y and z in $\mathbb{R}^d$, segments, and spheres.

The first case corresponds to the covering function defined by

$$p(x) = P(x \in \Xi).$$

If the random closed set is stationary, then $p(x)$ is a constant, which here is denoted by p. This value can be interpreted as the *volume fraction*, i.e.

$$p = E(\nu_d(\Xi \cap U))$$

where U is the unit cube, $U = [0,1]^d$. For a compact random set, the function $p(x)$ can be considered as a kind of mean or expectation of Ξ.

The function C given by

$$C(x,y) = P(x \in \Xi, y \in \Xi)$$

is called *covariance* or two-point probability function. In the stationary and isotropic case, the covariance only depends on the distance r between the two points x and y and is denoted as $C(r)$.

A generalisation to three or more points is natural — however, closed-form expressions for higher-order probability functions are known only for very special models. Third-order probability functions play an important role in the study of physical bulk properties of two-phase random materials, see Jeulin & Ostoja-Starzewski [29], Jeulin [28], and Torquato [66].

Also the case where K is the ball of radius r centred at the origin o, $K = b(o,r)$, or a segment of length r with one endpoint in the origin and some specified direction, $K = s(o,r)$, is practically important. However, for a stationary random closed set Ξ with positive volume fraction it is more natural to consider the intersections of the ball or segment with Ξ when the centre or endpoint does not belong to Ξ. The functions

$$H_s(r) = P(\Xi \cap b(o,r) \neq \emptyset \mid o \notin \Xi)$$
$$H_l(r) = P(\Xi \cap s(o,r) \neq \emptyset \mid o \notin \Xi) \qquad \text{for } r \geq 0$$

are called *spherical* and *linear contact distribution functions*. In the anisotropic case H_l depends on the direction of the segment $s(o,r)$. In most practically interesting cases, these functions are indeed distribution functions. They have a nice geometrical interpretation: $H_s(r)$ is the distribution of the random distance from o to the nearest point of the random set Ξ, under the condition that the origin o is not in Ξ. Analogously, $H_l(r)$ belongs to the directional distance from o to Ξ, where the direction is given by the segment direction.

Both functions H_s and H_l are closely related to volume fractions of dilated sets. For example, it is

$$p + H_s(r)(1 - p) = \text{ volume fraction of } \Xi \oplus b(o,r) \, .$$

Therefore, the following idea of generalisation introduced by Mecke [44, 46, 47, 49], Mecke et al. [51], and Jacobs et al. [26, 27] seems to be natural: Consider, for any stationary random closed set Ξ, the set $\Xi \oplus b(o,r)$ and use the corresponding intensities of the curvature or Minkowski measures w_k (see Stoyan, Kendall & Mecke, [63], p. 235; Mecke & Wagner[53]), treated as functions $w_k(r)$ of r, in order to describe the distribution of Ξ. Brodatzki & Mecke [10, 11] developed computer algorithms for the calculation of the $w_k(r)$. Additionally, it is useful to consider also $\Xi \ominus b(o, r)$ and so to use functions of a variable r taking both positive and negative values. This approach turned out to be very fruitful. While often $H_s(r)$ and equivalently $w_0(r)$ does not present sufficient information, the whole family of $w_0(r), \ldots, w_d(r)$ characterises the structure of Ξ very well, as demonstrated e.g. in Arns et al. [3, 4, 5] and Jacobs et al. [26].

Furthermore, the Minkowski measures w_k can directly be related to physical performance of heterogeneous materials where the spatial structure Ξ is essential for the material properties. In König et al. [35] it could be shown that for a fluid in an arbitrarily shaped container modelled, for instance, by a Boolean model the surface tension and other thermodynamic properties depend only linearly on the Minkowski measures $w_k(r)$ and not on other shape descriptors such as powers of Gaussian and mean curvatures (see also Mecke & Arns[50]).

3 Formulas for the Boolean Model

The *Boolean model* is the most famous and most frequently used random set model. It is a mathematically rigorous formulation of the idea of an infinite system of randomly scattered particles, see Hadwiger & Giger [21]. So it is a fundamental model for geometrical probability and stochastic geometry. The Boolean model has a long history; the first relevant papers appeared in the beginning of the 20th century in the physical literature, see the references in Stoyan, Kendall & Mecke [63]. The name 'Boolean model' appeared first in Matheron [40] to discriminate this set-theoretic model from (other) random field models, which appear in geostatistical applications.

The Boolean model is constructed by means of two components: a system of grains and a system of germs. The germs are the points x_1, x_2, $\ldots$ of a homogeneous Poisson process of intensity λ. The grains form a sequence $\{\Xi_n\}$ of i.i.d. random compact sets. Typical examples are spheres, discs, segments, and Poisson polyhedra. A further random compact set Ξ_0 having the same distribution as the Ξ_n is sometimes called the 'typical grain'. The Boolean model Ξ is the union of all grains shifted to the germs,

$$\Xi = \bigcup_{n=1}^{\infty} (\Xi_n + x_n) \, .$$

Its existence as a random closed set is given if

$$E(\nu_d(\Xi_0 \oplus K)) < \infty \qquad \text{for all } K \in \mathbb{K} \, ;$$

a sufficient condition is

$$ER^d < \infty$$

for the radius of the circumscribing sphere of Ξ_0, see Heinrich (1993). In the following it is always assumed that the typical grain Ξ_0 is convex. This does not mean that non-convex grains are unimportant. For example, the case where Ξ_0 is a finite point set corresponds to Poisson cluster point processes.

The basic parameters of a Boolean model are intensity λ and several parameters characterising the typical grain Ξ_0. While for simulations the complete distribution of Ξ_0 is necessary, for a statistical description it often suffices to know that the basic assumption of a Boolean model is acceptable and to have some mean values of geometrical characteristics of Ξ_0, e.g. its mean volume.

The capacity functional of the Boolean model Ξ is given by the simple formula

$$P(\Xi \cap K \neq \emptyset) = 1 - \exp(-\lambda E(\nu_d(\Xi_0 \oplus \check{K}))) \qquad \text{for } K \in \mathbb{K}, \qquad (1)$$

where $\check{K}$ is the set $\{-k \,:\, k \in K\}$. The derivation of this formula is given in Matheron (1975). Its structure is quite similar to the emptiness probability of the Poisson process or to the probability that a Poisson random variable does not vanish. It can perhaps be partially explained when applied to the particular case $\Xi_0 = \{o\}$. Then, the Boolean model is nothing else but the random set consisting of all points of the Poisson process of germs. Consequently,

$$P(\Xi \cap K \neq \emptyset) = 1 - \exp(-\lambda \nu_d(K)) \,.$$

The calculation of the capacity functional of a Boolean model poses a non-trivial geometrical problem, viz. the determination of the mean

$$E(\nu_d(\Xi_0 \oplus \check{K})) \,.$$

Here, integral geometry ([60], and [34]) helps if K is convex. If Ξ_0 is a ball, then the classical Steiner formula gives the result. If Ξ_0 is not spherical but isotropic (with distribution invariant with respect to rotations around the origin o), a generalisation of this formula found by Matheron leads to a formula in which the so-called Minkowski functionals W_k or intrinsic volumes appear. For example, in the three-dimensional case it is

$$E(\nu_3(\Xi_0 \oplus \check{K})) = \overline{V} + \frac{1}{\pi}M(K)\overline{S} + \frac{1}{\pi}\overline{M}S(K) + V(K)$$

where $M(K) = 3W_2(K)$, $S(K) = 3W_1(K)$ and $V(K) = W_0(K)$ are integral of mean curvature, surface area and volume of K, and $\overline{M}$, $\overline{S}$ and $\overline{V}$ are the corresponding means of Ξ_0. For the non-isotropic case see section 4 and for lattice configurations see section 5.

For non-convex K, the numerical determination of $T(K)$ is rather difficult. Already the numerical determination of the covariance $C(r)$, which belongs to the case where the set K is a pair of points of distance r, is a difficult problem unless Ξ_0 is a ball or a Poisson polyhedron, since it needs the set covariance of Ξ_0. For some particular cases, also formulas for three-point probabilities are given, i.e. probabilities of the form

$$P(x_1 \in \Xi,\ x_2 \in \Xi,\ x_3 \in \Xi),$$

namely in the planar case for circular, rectangular and Poisson polygonal grains and for the spatial case with spherical and Poisson polyhedral grains.

By means of the formula for the capacity functional and the generalised Steiner formula it is easy to give formulas for the spherical and linear contact distribution functions. The spherical contact distribution function satisfies

$$H_s(r) = 1 - \exp\left(-\lambda \sum_{k=1}^{d} \binom{d}{k} \overline{W}_k r^k\right) \qquad \text{for } r \geq 0$$

where the $\overline{W}_k$ are the expectations of the Minkowski functionals of Ξ_0. The linear contact distribution function $H_l(r)$ is an exponential distribution function with parameter $\lambda \frac{b_{d-1}}{b_d} \overline{W}_1$, where b_k is the volume of the unit sphere of $\mathbb{R}^k$.

Formula (1) also leads to the result that the intersection of a Boolean model with a linear subspace of $\mathbb{R}^d$ is again a Boolean model; formulas for the intensity and the mean Minkowski functionals of the grains of the induced lower-dimensional Boolean model are given in Matheron [43], p. 146.

Finally, for the densities of the Minkowski functionals or intensities w_k of the Minkowski or curvature measures of the Boolean model formulas are known (for variances see section 5, Mecke [48] for general grain shapes, and Kerscher et al. [33] for spheres in three-dimensional Euclidean space). The simplest characteristic of this type is volume fraction V_V, which is given by the formula

$$V_V = 1 - e^{-\lambda \overline{V}}, \tag{2}$$

where $\overline{V}$ is the mean d-dimensional volume of the typical grain. This is an easy consequence of formula (1) for the capacity functional.

Also for specific surface area S_V a nice general formula holds, namely

$$S_V = \lambda \overline{S} e^{-\lambda \overline{V}}, \tag{3}$$

where $\overline{S}$ is the mean $(d-1)$-dimensional surface area of the typical grain. In the other cases the Minkowski measures are signed measures, and thus the intensities can be negative. The following gives the intensities for $d = 2$ and 3 in a stereological notation, similarly as in Stoyan, Kendall & Mecke [63], pp. 76–77.

To $d = 3$ and $k = 1$ corresponds the specific mean curvature M_V, which satisfies

$$M_V = \lambda\left(\overline{M} - \frac{\pi^2}{32}\lambda \overline{S}^2\right) e^{-\lambda \overline{V}}, \tag{4}$$

where $\overline{M}$ is the mean integral of mean curvature of the typical grain.

The intensity related to the Euler-Poincaré characteristic is often called specific connectivity number and will be denoted here by $N_A(d = 2)$ and $N_V(d = 3)$. It holds

$$N_A = \lambda\left(1 - \frac{\lambda\overline{L}^2}{4\pi}\right)e^{-\lambda\overline{A}}$$
(5)

and

$$N_V = \lambda\left(1 - \frac{\lambda\overline{M}\,\overline{S}}{4\pi} + \frac{\pi}{384}\lambda^2\overline{S}^3\right)e^{-\lambda\overline{V}},$$
(6)

where $\overline{L}$ is the mean boundary length of the typical grain. These formulas were first given by Miles [55]. The general formulas for the d-dimensional case are best presented in Weil [72], his formulas (6) and (7). Brodatzki & Mecke [10, 11] developed computer algorithms for the calculation of Minkowski functionals $W_k(\Xi)$ for a configuration Ξ which allow in addition to the mean values w_k the numerical estimation of second order moments of $W_k(\Xi)$ for the Boolean model - analytically calculated by Mecke [48] (see Kerscher et al. [33] for variances in the Boolean model with spheres).

Since $\Xi_0 \oplus b(o, r)$ is convex and isotropic if Ξ_0 has these properties, the formulas above yield explicit expressions for the functions $w_k(r)$ for $r > 0$ introduced in section 2. Thus it is clear that for the Boolean model these functions for $r > 0$ depend only on the mean Minkowski functionals of Ξ_0 and λ.

There are also positive or absolute curvature measures, which were introduced by Matheron [43] and Schneider [59]. In the cases $d = 2$ and 3 those measures related to the Euler-Poincaré characteristic and to the integral mean curvature are of particular interest, see Stoyan, Kendall & Mecke [63], pp. 238–242. The corresponding intensity is called the specific convexity number N_V^+, which satisfies in all dimensions

$$N_V^+ = \lambda e^{-\lambda\overline{V}},$$
(7)

as shown by Matheron [43].

We mention here a famous application of the Boolean model which goes back to another hero of probability theory in the 20th century, Kolmogorov. He used it in the context of modelling of crystallisation processes, see Kolmogoroff [36] and Capasso et al. [12]. Consider a Poisson process system of germ points in which radial growth starts at time $t = 0$. The speed of growth α is the same for all germs. So at time t, each germ becomes a sphere of radius αt if enough space is available. The volume fraction of the corresponding Boolean model is

$$V_V(t) = 1 - \exp\left(-\lambda\frac{4}{3}\pi\alpha^3 t^3\right) \quad \text{for } t \geq 0.$$
(8)

One can assume that growth is stopped where two spheres come into contact, while otherwise growth into empty space still is possible. A nice visualisation

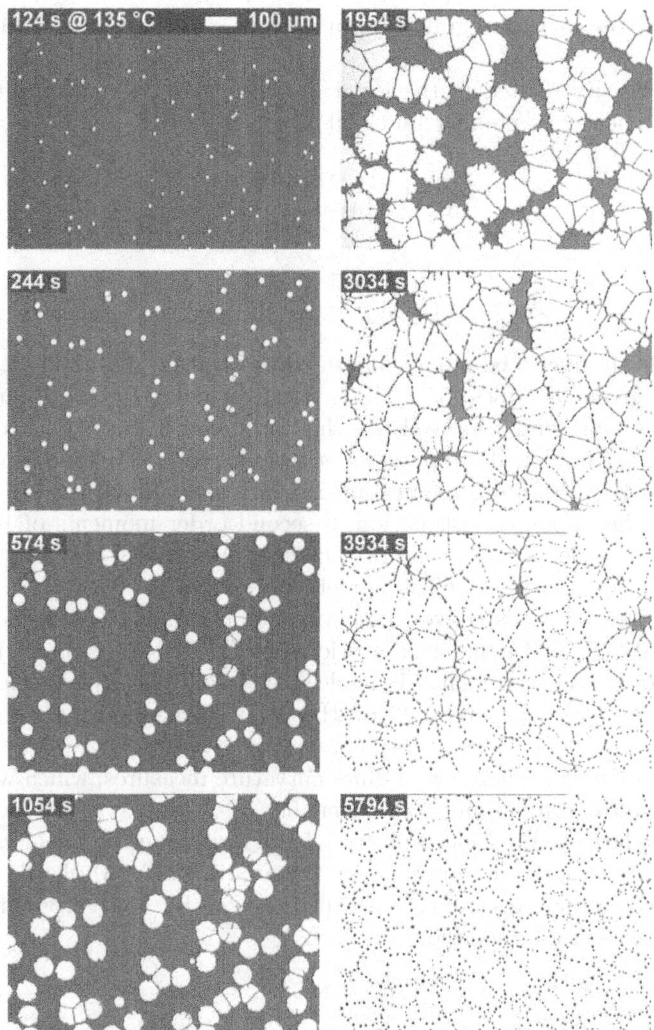

Fig. 1. Series of photographs of a dewetting liquid film viewed through a reflection light microscope: the 80 nm thick polystyrene film (dark) beads off the non-wettable Si substrate (bright) by forming holes at random defects which grow in time given in the left upper corner (see Jacobs et al., 2000).

of the growing germs can be obtained by a dewetting process of thin liquid films. Films rupture at random distributed defects (see Figure 1) and form holes, which grow in time, until the fluid material is pushed completely in thin filaments, which finally break up in droplets. In Figure 2) the time dependence of the experimentally measured Minkowski functionals of the film region is compared with the theoretical expectations of the Minkowski functionals given by Eqs. (2) - (6). The good agreement is a clear indication for an initial Poisson process of holes and a constant speed of growth α. Equation (8) and some generalisations are called Kolmogorov-Johnson-Mehl-Avrami equations. By the way, several authors have studied the empty space regions for very large t, when the volume fraction of the Boolean model is close to one. Hall [22] and Aldous [1] have shown that the distribution of a typical empty region is asymptotically the same as that of a Poisson polygon (or polyhedron), which is the typical cell of a tessellation generated by a system of random lines (or planes) called Poisson line (plane) process. A generalisation of Hall's result can be found in Molchanov [56]. Chiu [14] studied the probability of complete coverage of cubes. A paper by Erhardsson [17] adds to the body of evidence that asymptotically the union of all uncovered regions has a distribution similar to that of a Boolean model with Poisson polygonal grains. This may perhaps explain why such models turned out to be good models for systems of pores, see Serra [61]. Of course, there are also other arguments for the use of this particular Boolean model, namely its polygonal nature and the form of the corresponding covariance.

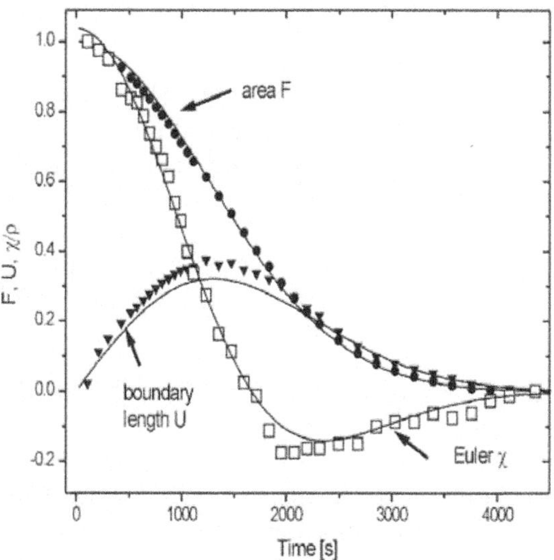

Fig. 2. The time evolution of the Minkowski functionals of the liquid polystyrene film (dark region in Figure 1) can be well captured by the Eqs. (2), (3) and (5), i.e., by the intensities w_k assuming a linear growth behavior of the radii $R = \alpha t$ of Poisson distributed holes (solid lines; see Jacobs et al., 2000).

4 The Non-isotropic and Non-stationary Boolean Model

A fundamental assumption in the use of formula (1) is that the typical grain Ξ_0 is isotropic. In this case Matheron's generalised Steiner formula can be applied; the Boolean model is not only stationary but also isotropic.

If Ξ_0 is not isotropic, then the Boolean model is still stationary, but many calculations become rather difficult. Clearly, volume fraction V_V still is given by

$$V_V = 1 - e^{-\lambda \overline{V}},$$

since V_V is obtained from Eq. (1) by setting $K = \{o\}$. But the determination of the other characteristics discussed in section 2 is difficult, and so it is useful that analytical expressions for particular cases are given by Charollais et al. [13] and Mecke [44, 47].
Weil [71] has thoroughly studied the non-isotropic case and found that it is necessary to use methods of translative integral geometry. Indeed, the following generalisation by Minkowski of the Steiner formula, in which the polynomial form is retained,

$$\nu_d(K \oplus rM) = \sum_{j=0}^{d} r^{d-j} \binom{d}{j} V(K[j], M[d-j])$$

for convex K and M and $r > 0$ includes so-called mixed volumes $V(K[j], M[d-j])$. They are particular cases of so-called mixed functionals $V_{m_1,\ldots,m_k}^{(j)}$, which appear as coefficients in the following iterated translative formula for the intrinsic volume V_j:

$$\int_{\mathbb{R}^d} \cdots \int_{\mathbb{R}^d} V_j(K_1 \cap (K_2 + x_2) \cap \ldots \cap (K_k + x_k))\nu_d(\mathrm{d}x_2) \cdots \nu_d(\mathrm{d}x_k)$$

$$= \sum_{\substack{m_1,\ldots,m_k=j \\ \sum_i m_i=(k-1)d+j}} V_{m_1,\ldots,m_k}^{(j)}(K_1,\ldots,K_k) \quad \text{for} \quad j = 0, 1, \ldots, d.$$

Weil [71] defined constants called 'mixed densities' for these functionals, which appear in the formulas for the intensities of curvature measures of stationary Boolean models. For example, N_A in the planar case satisfies

$$N_A = (\lambda - \overline{A}(X, X^*))e^{-\lambda \overline{A}},$$

where $\overline{A}(X, X^*)$ is a mixed area (instead of volume) density, which can be written as an integral which is determined by the support function and the boundary length measure of the typical grain Ξ_0. In the isotropic case it is $\overline{A}(X, X^*) = \lambda^2 \overline{L}^2 / 4\pi$.

There is another practically interesting case of reduced invariance properties of the Boolean model, which was already studied by Matheron [39] of practical interest: The grains can be i.i.d. disks ($d = 2$) or spheres ($d = 3$), but the Poisson process of germ points is inhomogeneous with intensity function $\lambda(x)$. In this case the Minkowski measures are not stationary, i.e., it does not make sense to speak about intensities. Instead, intensity *functions* are used, i.e. Radon-Nikodym densities of the Minkowski measures with respect to the Lebesgue measure. Examples are the location dependent volume fraction $V_V(x)$ given for $x \in \mathbb{R}^d$ by

$$V_V(x) = P(x \in \Xi) = \lim_{r \downarrow 0} \frac{E\nu_d(\Xi \cap b(x, r))}{b_d r^d}$$

or location dependent specific surface area $S_V(x)$ given analogously by

$$S_V(x) = \lim_{r \downarrow 0} \frac{EH_{d-1}(\Xi \cap b(x, r))}{b_d r^d} \, ,$$

where H_{d-1} is the $(d-1)$-dimensional Hausdorff measure or surface area measure. Several authors were able to give formulas for $V_V(x)$ and $S_V(x)$ for the case discussed here by means of ad hoc methods, see the references in Weil [72] and Mecke [44, 47]. A systematic approach is presented by Weil [72], section 6. He shows that in the planar case the three relevant intensity functions are essentially convolutions of $\lambda(x)$ and functions related to the grain diameter distribution if the grains are i.i.d., see Lantuejoul [37], p. 156. This result can be considered as a particular case of a more general theory for Boolean models with an inhomogeneous Poisson process of germs and location dependent non-isotropic grain distribution, which was created in Mecke [44, 47], Fallert [18] and Weil [72]. Here again mixed functionals play a role.

Thus the situation today is so that theoretically intensity (or intensity function) formulas for rather general Boolean models can be derived and that it is known which price one has to pay for deviations from the usual invariance properties of stationarity and isotropy. The only non-trivial case of anisotropic grains in which explicit formulas can be obtained is that of parallelepipeds, with sides parallel to the coordinate axes as in Mecke [45]. Non-isotropic and non-stationary Boolean models should be exploited further, since many applications are possible on complex fluids, colloidal dispersions and composite and porous media, which show qualitatively rich phase diagrams and spatial structures; see Mecke [47], Brodatzki & Mecke [10] and Groh & Mecke [20]. Other applications are possible in the analysis of inhomogeneous distributions of galaxies or in the estimation of percolation thresholds in inhomogeneous and anisotropic porous rocks, see Arns et al. [3, 4, 5].

5 The Discrete Boolean Model

Since Lenz and Ising introduced the so-called Ising model in the early 1920s, lattice models became the backbone of statistical physics, for instance, for the description of magnets and fluids as well as for morphological image analysis and percolation phenomena. Instead of distributing convex bodies continuously in Euclidean space, these models place cubes, for instance, at discrete spacings. So it is natural to define a discrete Boolean model on a lattice. This discrete model generates spatial sets by unions of i.i.d. unions of polyhedra ('grains') which are centred at lattice points ('germs'); see Figure 3.

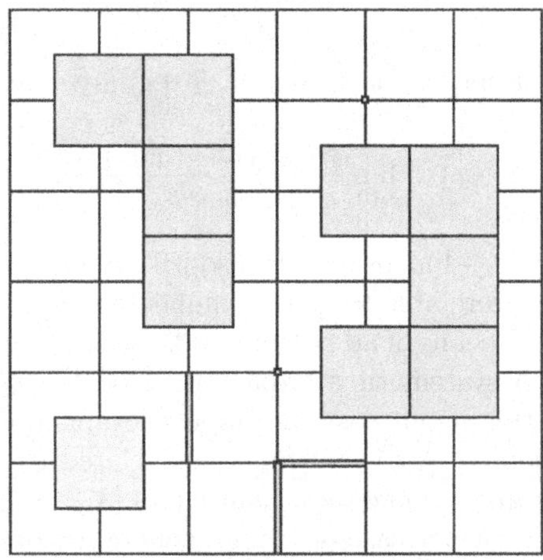

Fig. 3. Realization of a two-dimensional discrete Boolean model (shown in grey) on a 2-dimensional lattice $\Lambda^{(2)}$. In the left lower corner there is a single unit cube $C_i^{(2)}$ (in white) centred at a site x_i.

For theoretical studies of the discrete Boolean model a discrete variant of integral geometry has been developed by Voss [69] and independently by Mecke [44], see also Likos et al. [38] and Mecke [45, 47]. Since discrete objects can be translated and rotated according finite lattice spacings and angles, notions and definitions of integral geometry may be adopted straightforwardly for lattice configurations and lattice groups of motions.

The main idea is as follows. Consider the d-dimensional hypercubic lattice $\Lambda^{(d)} = \mathbb{Z}^d$. At a point $x_i \in \Lambda^{(d)}$ (lattice site) a d-dimensional unit cube $C_i^{(d)}$ can be centred, called a 'cell'. The non-empty intersection of two cells $C_i^{(d)} \cap C_j^{(d)}$ is called an l-cell, if it has dimension $l \leq d$. In the case $d = 3$ the l-cells are corners $C^{(0)}$, edges $C^{(1)}$ and plaquettes $C^{(2)}$ of the cubes $C^{(3)}$. The sets considered are of the form (see Figure 3)

$$A = \bigcup_{(l,m)} C_m^{(l)},$$

i.e. finite unions of cubes $C_m^{(d)}$ and l-cells $C_m^{(l)}$ of any dimensions l. For these discrete objects additive Minkowski functionals W_k can be defined by

$$W_k(A) = \frac{1}{\binom{d}{k}} \sum_{E^{(k)}} \chi(A \cap E^{(k)}) \quad \text{for } k = 0, \ldots, d-1 \tag{9}$$

and

$$W_d(A) = \chi(A),$$

where $E^{(k)}$ denotes a k-dimensional lattice hyperplane, i.e. $E^{(k)} \cap \Lambda^{(d)}$ is congruent to $\Lambda^{(k)}$, and $\chi(B)$ is the Euler-Poincaré characteristic of B. They are related to simple geometrical quantities of the discrete set A, namely the number W_0 of occupied cells $C_i^{(d)}$ of A, the number $2dW_1$ of boundary plaquettes $C_k^{(d-1)}$ etc. In particular, for a l-dimensional cube $C^{(l)}$ it is

$$W_k(C^{(l)}) = \tfrac{k!\, l!}{d!(k+l-d)!} \text{ for } d - l \leq k \leq d \text{ and } W_k(C^{(l)}) = 0 \text{ otherwise.}$$

These functionals W_k differ from the Minkowski functionals defined in continuous integral geometry only by the factor b_k, the volume of the k-dimensional unit sphere. The definition (9) leads to

$$W_k(A) = \sum_{i=d-k}^{d} \frac{(-1)^{d-k+i} k! i!}{d!(k+i-d)!} \#_i(A), \tag{10}$$

where $\#_i(A)$ is the number of i-cells $C^{(i)}$ belonging to A. Analogous to the kinematic formula of continuous integral geometry, for integrals over the lattice group of motions the following kinematic formula holds

$$\int W_k(A \cap A') \mathrm{d}A' = \sum_{i=0}^{k} \sum_{j=0}^{i} \binom{k}{i} \binom{i}{j} W_i(A) W_{k-j}(A') \tag{11}$$

for $k = 0, \ldots, d$ and discrete sets A and A' (shown in Figure 3), see Mecke (1994). The integral $\int dA'$ denotes a sum $\sum_{A'}$ over all discrete rotations and translations of A' by lattice vectors of $\Lambda^{(d)}$. In particular, for $k = 0$ (volume) $W_0(A)W_0(A')$ is obtained and for $k = 1$ (surface area) $W_0(A)W_1(A') + W_1(A)W_0(A') + W_1(A)W_1(A')$. Note the difference to the usual kinematic formula, where the last term $W_1(A)W_1(A')$ does not appear.

Consider now the *discrete Boolean model* where grains Ξ_n with mean Minkowski functionals $\overline{W}_i$ are placed randomly and independently on the lattice $\Lambda^{(d)}$ with intensity λ, i.e. λ is the mean number of germs per lattice point. Applying the kinematic formula (11), the differential equation

$$\frac{\mathrm{d}w_k(\lambda)}{\mathrm{d}\lambda} = \overline{W}_k - \sum_{i=0}^{k} \sum_{j=0}^{i} \binom{k}{i} \binom{i}{j} w_i(\lambda) \overline{W}_{k-j} \tag{12}$$

164   Dietrich Stoyan and Klaus Mecke

can be derived straightforwardly for the densities $w_k(\lambda)$ of the discrete Minkowski functionals for these lattice-homogeneous and isotropic configurations seen as functions of λ, see Mecke [44]. With the initial condition $w_k(0) = 0$ for all k, the differential equation (12) can be solved readily for any dimension d yielding

$$w_0(\lambda) = 1 - e^{-\lambda \overline{V}}$$

and

$$w_k(\lambda) = \sum_{i=0}^{k} (-1)^{i+k+1} \binom{k}{i} e^{-\lambda \sum_{j=0}^{i} \binom{i}{j} \overline{W}_j(A)} \quad \text{for } k = 1, \ldots, d . \tag{13}$$

These densities are of great value for lattice models in statistical physics when studying percolation phenomena [53], [52], fluid phase behaviour ([44] and [38]) or porous media ([2, 3, 4, 5]).

Application of Eq. (10) also yields second-order moments of Minkowski functionals. For example, Mecke (1994) derived a formula for the variance of the Euler-Poincaré characteristic in the case where the typical grain Ξ_0 is the unit-cube of the lattice. Let A be a cubic set of positive volume $V(A)$. Then the normalised variance

$$\sigma^2(\chi) = E(\chi(A \cap \Xi) - E(\chi(A \cap \Xi)))^2 / V(A)$$

satisfies independently on A

$$\sigma^2(\chi) = \sum_{i=0}^{d} \binom{d}{i} f_i^2(q^{2^{d-i}}) \, g_{d-i}(q^{2^{d-i}}) \tag{14}$$

with

$$q = e^{-\lambda}, \quad p = 1 - q, \, f_i(q) = \sum_{j=0}^{i} \binom{i}{j} (-2)^j q^{2^j}$$

and

$$g_i(q) = (-1)^{i+1} + \sum_{j=0}^{i} \binom{i}{j} (-2)^j q^{-2^{-j}} \quad \text{for } i = 0, 1, \ldots, d .$$

Figure 4 shows $\sigma^2(\chi)$ for $d = 1, 2$ and 3 [44]. Second order moments can also be derived for more complex primary discrete grains than a unit cube.

Such normalised variances and also product densities can be calculated also for continuous Boolean models, in particular, with discs and spheres as typical grains Ξ_0 in Euclidean space[48] ; an interesting application in physics is given in Kerscher et al. [33]. Figure 5 shows σ_k^2 (= normalised variances for the W_k) for a planar Boolean model, where Ξ_0 is a disc of constant radius R.

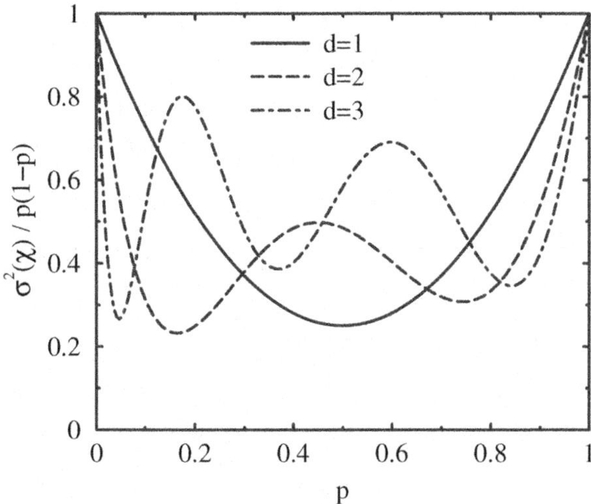

Fig. 4. Normalised variance of the Euler-Poincaré characteristic for the discrete Boolean model described in the text, as function of volume fraction p for $d = 1, 2$ and 3.

Note that the functional form of these characteristics becomes more complex with increasing index k. Figure 6 shows the normalised product density ($=$ pair correlation function) corresponding to W_2 for the same model. Note the pole at $r = 0$ and the discontinuity at $2R$. The calculations leading to these results are rather difficult (for details and explicit expressions see Mecke[48], while the lattice case is a bit simpler and thus more convenient for physical applications. Also for the case of the surface measure corresponding to a Boolean model with identical spheres the product density is known since Doi [16], see also Torquato [66], formula (6.18); the formula can be seen as a generalisation of the well-known relation

$$S_V = (1 - p)H'_s(0).$$

The main advantage of the discrete Boolean model compared to the continuous version is the simplicity of analytic calculations and the feasibility of computer simulations when applied to Gibbs processes as discussed in the following section.

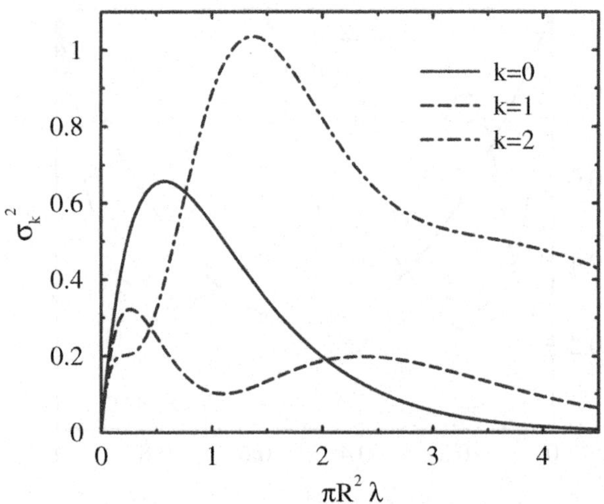

Fig. 5. Normalised variance σ_k^2 of area ($k = 0$), boundary length ($k = 1$) and Euler-Poincaré characteristic ($k = 2$) for a planar continuous Boolean model with identical discs of radius R as function of intensity λ.

6 Beyond the Boolean Model: Morphological Gibbs Processes

For many physical applications the Boolean model of randomly and independently distributed grains is insufficient to describe typical spatial configurations, for instance, of fluids or composite materials. Models with correlations between the grains are necessary in order to capture the correct physical behaviour of measurable quantities such as pressure or specific heat as function of temperature. Therefore, in 1993 a Gibbs process model was introduced based on the Minkowski functionals of overlapping grains Ξ_i in Mecke [44, 46] and Likos et al. [38] for the lattice case, and in Brodatzki & Mecke [10] for the continuous case.

In the canonical case of n bodies the construction is as follows: Each configuration Ξ is assumed to be the union of mutually penetrable convex bodies ('grains') Ξ_i, $\Xi = \bigcup_{i=1}^{n} g_i \Xi_i$, where the g_i are elements of the group $\mathcal{G}$ of motions (translations and rotations) in the $\mathbb{R}^d$. The Boltzmann weights $\exp\{-\beta H\}$ of the Gibbs process are specified by the inverse $\beta = 1/k_B T$ of the temperature T and the Hamiltonian

$$H(\Xi) = \sum_{k=0}^{d} h_k \left(W_k \left(\bigcup_{i=1}^{n} g_i \Xi_i \right) - \sum_{i=1}^{n} W_k(\Xi_i) \right) \tag{15}$$

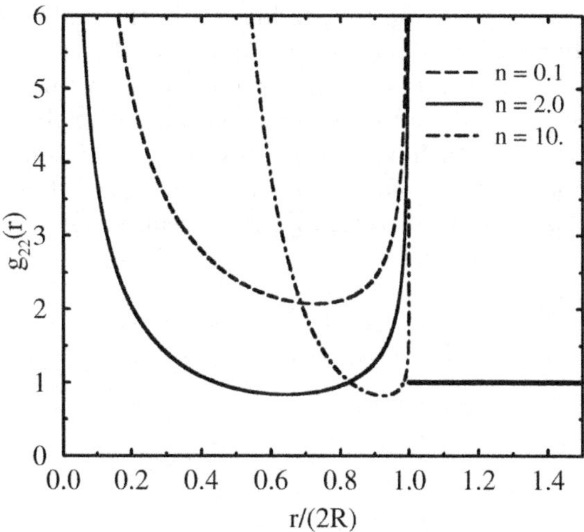

Fig. 6. The normalised pair correlation function $g_{22}(r)$ corresponding to the Euler characteristic W_2 for the Boolean model of overlapping discs of radius R and intensity λ ($n = \pi R^2 \lambda$). Note the pole at $r = 0$ and the discontinuity at $2R$. Integrating $w_k^2(g_{kk}(r) - 1)$ over the distances r gives the variances σ_k^2 shown in Fig. 5.

with suitable weights h_k, which are model parameters. It constitutes a very general model for composite media assuming additivity of the energy H of the mesoscopic components. The configurational partition function (the normalising constant of the Gibbs distribution) is given by

$$Z_n = \frac{1}{n!} \int_{\mathcal{G}} \exp\left\{ - \beta H(\Xi) \right\} \prod_{j=1}^{n} \mathrm{d}g_j \qquad (16)$$

where the integral denotes averages over the motions of the grains with respect to the invariant Haar measure on the group $\mathcal{G}$. Due to the additivity of the Minkowski functionals the partition sum is well defined even for negative values of the Hamiltonian, e.g., for negative Euler-Poincaré characteristics. This model can be generalised to the grand canonical case and the stationary case.

Thermodynamic quantities such as pressure or phase diagrams of this model are then given in terms of additive morphological measures of its constituents. Depending on the relative strength of the energies related to volume, surface area, mean curvature and Euler characteristic of the domains, one finds qualitatively different phase diagrams and spatial structures which resembles the behaviour of microemulsions, for instance ([38, 44, 46]).

In Figure 7, for instance, typical planar stationary (spatially homogeneous) configurations obtained by Monte-Carlo simulations using the Metro- polis algorithm as described in Brodatzki & Mecke [10, 11] are shown for the special case of circular grains with constant radius. Only the midpoints of disks are drawn and solely the Euler-Poincaré characteristic W_d is used as energy H, i.e. $h_k = 0$ for $k < d$ in Eq. (15). Depending on temperature T and intensity λ of the process, one finds not only homogeneous phases but quite complex structured distributions such as solid phases, where the discs are located on regular periodic lattice sites, and even configurations of close-packed strings of discs. This 'glassy' phase shown in Figure 7 reminds of a dense packing of styrene spheres.

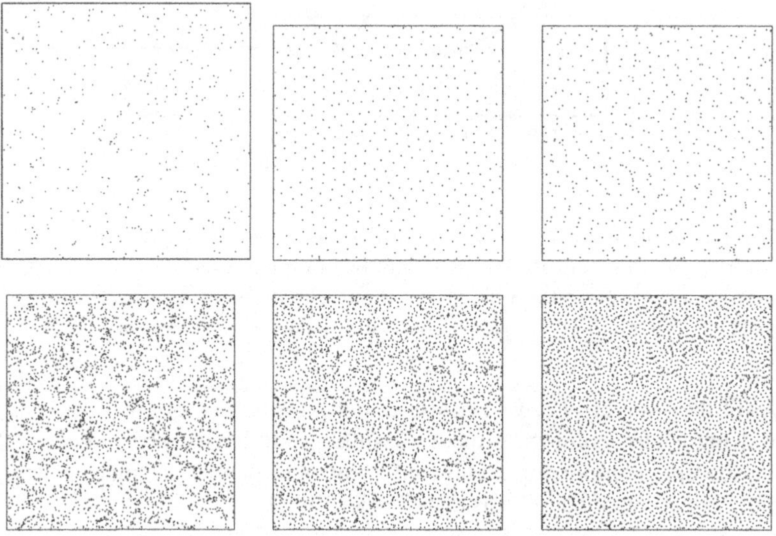

Fig. 7. Typical configurations of the centres of overlapping discs generated by the morphological Gibbs process given by Eqs. (15)-(16) with a Hamiltonian depending only on the Euler-Poincaré characteristic [10]. In the upper (lower) row the temperature T (intensity λ) of the stationary Gibbs process is fixed and the intensity (temperature) increases to the right.

Further Monte Carlo simulations and density functional theory for this morphological interaction model have shown a rich phase behaviour and complex spatial structures, see Brodatzki & Mecke[10] and Groh & Mecke[20]. Although physically inspired (see Widom & Rowlinson[73] and Mecke[44]), the Gibbs process defined above in terms of Minkowski functionals is important also in spatial statistics. The analogous models have appeared under the names 'area interaction' and 'quermass' process, see Baddeley & van Lieshout [6], Baddeley et al. [6], Kendall [31] and Kendall et al. [32]. Statisticians have also studied other forms of interaction, for example nearest neighbour Markov processes, where the interaction is expressed by the corresponding Voronoi tessellation, see Baddeley and Møller [7] and van Lieshout [67]

While in the statistical literature the problem of existence of these models and of model parameter estimation has been studied, standard methods of statistical physics such as expansion of the partition sum (16) of the Hamiltonian (15) in powers of inverse temperature $\beta = \frac{1}{k_B T}$ lead to approximate analytical expressions for the partition sum, i.e. for first- and second-order moments, but also for experimentally measurable correlation functions and phase diagrams for colloidal systems, see Mecke [47] and Groh & Mecke [20].

7 Percolation in the Boolean model

The geometry of Boolean models can be rather complicated: the voids between the grains and the clumps of overlapping grains may have quite interesting shapes and topologies. Its systematic investigation in the mathematical literature began with Kellerer [30] and Hall [22].

One of the most important problems in this context is percolation. Since the introduction of 'percolation processes' to describe gelation and fluids in porous media, percolation models became important for the understanding of many physical properties, see Stauffer & Aharony [62]. Most of the effort is focused on critical exponents of the percolation transition, which show a universal behaviour and can therefore be described by the simplest model exhibiting a percolation threshold. But in designing composite materials it is more important to understand the non-universal behaviour of transport quantities such as electrical and thermal conductivity, diffusion constants or elastic moduli, for instance, in the Boolean model. These non-universal features include the location of the critical threshold and also the dependence of physical quantities on the spatial structure of the component phases away from the critical region. In particular, the prediction of the percolation threshold as a function of volume fraction, shape, orientation, and correlations of the component phases is a key problem in studying random multiphase structures.

In the case of a Boolean model, two features are associated with the word 'percolation':

(i) an 'arbitrary' grain of the Boolean model belongs with positive probability to a clump of infinite order (there are infinitely many $\Xi_i + x_i$ being connected by overlapping);

(ii) the mean number of members of the clump containing an 'arbitrary' grain of the Boolean model is infinite.

Usually, percolation is studied for a fixed typical grain Ξ_0 in dependence on intensity λ. One expects that there is a threshold λ_c such that for $\lambda > \lambda_c$ there is percolation and for $\lambda \le \lambda_c$ not, and many authors believe that the critical value of λ_c is the same for the two cases (i) and (ii) above.

While mathematicians proved until now mathematically strict bounds for λ_c mainly for spherical grains (see Hall[22], Meester & Roy[54] and

Grimmet[19]), physicists derived approximations which are consider to be more useful for applications. The following presents two such approaches.

The first one uses the 'excluded' volume V_{ex} or the volume of the difference body ´, i.e. $V_{\text{ex}} = \nu_d(\Xi_0 \oplus \check{\Xi}_0)$. The idea is that in a percolating Boolean model each particle should have in the average (at least) one neighbor, therefore $\lambda_c V_{\text{ex}} = 1$ is a reasonable percolation criterion.

In the case of isotropic grains, Matheron's generalized Steiner formula yields V_{ex}. It is

$$V_{\text{ex}} = 2\overline{A} + \frac{\overline{U}^2}{2\pi} \quad \text{for } d = 2 \tag{17}$$

and

$$V_{\text{ex}} = 2\overline{V} + \frac{\overline{MS}}{2\pi} \quad \text{for } d = 3 . \tag{18}$$

In the particular case of segments of length l in the plane, Eq. (17) yields the approximation $\lambda_c = \alpha/l^2$ with $\alpha = \pi/2$. Computer simulations showed however over a wide range of lengths l a percolation threshold $\lambda_c \sim 5.7/l^2$, i.e. the predicted qualitative behaviour but a somewhat larger coefficient α. So perhaps the excluded volume approach provides a first insight of the threshold dependence on size, shape and orientation of the grains, but it is far from being satisfactory.

A more interesting approach applies topological arguments based on the observation of Mecke & Wagner [53] that the specific connectivity number N_V of configurations vanishes near the percolation threshold, see also Bretheau & Jeulin [9]. The mean values N_V are polynomials in λ and coincide with the matching polynomials used in percolation theory to calculate exact thresholds for self-matching lattices. Moreover, the connectivity number is positive for isolated cluster configurations but negative for connected sponge-like structures, so that the smallest zero λ_0 of the function $N_V(\lambda)$ given by formula (5) taking λ as variable may provide a good estimation of the percolation threshold λ_c. Often it is more elegant to use n_0 and n_c defined by $n_0 = \lambda_0\overline{A}$ or $\lambda_0\overline{V}$ and $n_c = \lambda_c\overline{A}$ or $\lambda_c\overline{V}$. For $d = 2$ and 3, the zeros are

$$n_0^{(2)} = \frac{4\pi\overline{A}}{\overline{U}^2} \tag{19}$$

and

$$n_0^{(3)} = \frac{48\overline{M}\,\overline{V}}{\pi^2\overline{S}^2}\left(1 - \left(1 - \frac{\pi^3\overline{S}}{6\overline{M}^2}\right)^{1/2}\right) . \tag{20}$$

Table 1 gives some values of λ_c and n_c in comparison to λ_0 and n_0 showing again the good quality of the topological approach. Figure 8 shows for the case of random cylinders of length L and radius r empirical values of λ_c obtained by Monte-Carlo simulations in comparison with the zero λ_0 of the specific connectivity number and percolation estimates obtained by means of V_{ex}.

Table 1. λ_c and n_c compared with λ_0 and n_0 for segments and discs.

$d = 2$		$d = 3$	
discs	segments	balls	discs
$n_c = 1.12$	$l^2\lambda_c = 5.7$	$n_c = 0.34$	$\lambda_c = 0.19$
$n_0 = 1$	$l^2\lambda_0 = \pi$	$n_0 = 0.38$	$\lambda_0 = 0.22$

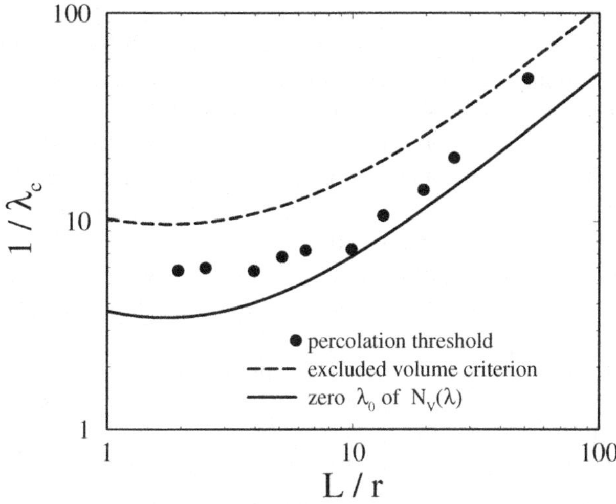

Fig. 8. Percolation thresholds λ_c (dots) for Boolean models with isotropic cylinders of length L and radius r obtained by simulation. The zero λ_0 of $N_V(\lambda)$ given by Eq. (20) (full line) is a good estimate of the percolation threshold. The excluded volume criterion yields in general a less accurate estimate.

Various other grain shapes were tested numerically by Mecke & Wagner [53] and Mecke & Seyfried [52] in order to confirm the assumption that the analytically available zero of $N_V(\lambda)$ is a good estimate of λ_c. Of course, analytic calculations based, for instance, on the cluster expansion of the pair-connectedness function can in principle provide more accurate values (Stauffer & Aharony [62]separately. The advantage of the heuristic zero criterion is the explicit availability of the formula and its dependence on the geometry of the typical grain. Furthermore, one can show that the specific connectivity number has in the case of some particular lattice sets a behaviour already known in site percolation theory, see Mecke & Wagner [53].

The percolation threshold is an extremely important quantity for transport properties of fluids in porous media, because a fluid flow is impossible for porosities below this threshold. Since configurations of Boolean models are often used to mimic porous media, a statistical analysis as presented in the following section is essential for the prediction of percolation thresholds and transport properties based on the measurement of the morphology of a porous media.

8 Statistical Analysis for the Boolean Model with Convex Grains

The aim of statistical analysis for the Boolean model is the estimation of model parameters and testing the distributional hypothesis. In the following we describe methods which solve these problems. They are a small selection from a vast literature, which is better discussed in Molchanov [57] and Stoyan [64]. It is our aim to present methods which are numerically stable, suitable for automatic analysis and conceptionally simple. Methods which work well only in the case of manual measurement or which are otherwise sophisticated are omitted, e.g. methods using curvatures or tangents.

The solution of the test problem goes back to the early 1980s, see the book Serra [61] and its forerunners and the references on p. 85 in Stoyan, Kendall & Mecke [63]. The idea is to use contact distribution functions such as H_l and H_s. For these functions formulas are known in the case of the Boolean model (see section 3), and so it is possible to compare empirical functions $\hat{H}_l$ and $\hat{H}_s$ with their theoretical counterparts. Usually the logarithm of $1 - H_l(r)$ and $1 - H_s(r)$ is taken, which yields polynominals. Division by r yields for H_l a constant value and, in the planar case, for H_s a linear function, see Serra [61], p. 495, and Stoyan, Kendall & Mecke [63], p. 87. For testing the model hypothesis, these functions are plotted and inspected visually. If they are (approximately) constant and linear, respectively, the Boolean model hypothesis is supported. It is obviously very difficult to construct a rigorous significance test for this problem. Probably, simulation tests as described for the case of point process statistics in Stoyan & Stoyan [65] are the appropriate method, see Kerscher et al. [33]. But such tests include simulations of Boolean models and need not only numerical values of λ but also the complete probability distribution of Ξ_0, including size and shape assumptions.

Under such assumptions, the power of the test can be increased by replacing $H_s(r)$ by the volume fraction of $\Xi \oplus b(o, r)$ and $\Xi \ominus b(o, r)$, so obtaining the function $w_0(r)$ for arbitrary real r, i.e. volume fraction as a function of r (see Mecke et. al. [51], Jacobs et al. [26, 27], Arns et al. [3, 4, 5]). This function can be determined analytically for a Boolean model Ξ for $r > 0$ since $\Xi \oplus b(o, r)$ is a Boolean model with grains $\Xi_0 \oplus b(o, r)$, which are convex if Ξ_0 is convex; it is not quite simple to determine the curvature intensities of $\Xi \ominus b(o, r)$, which in general requires numerical algorithms. In order to demonstrate the application of this test, it is applied to the termite nest shown in Figure 9. Its morphology seems to be reminiscent of the structure of level sets of Gaussian random fields, which is therefore a natural model choice for describing the morphology of this system, see Arns et al. [2], [3, 4, 5]. Consequently, there are

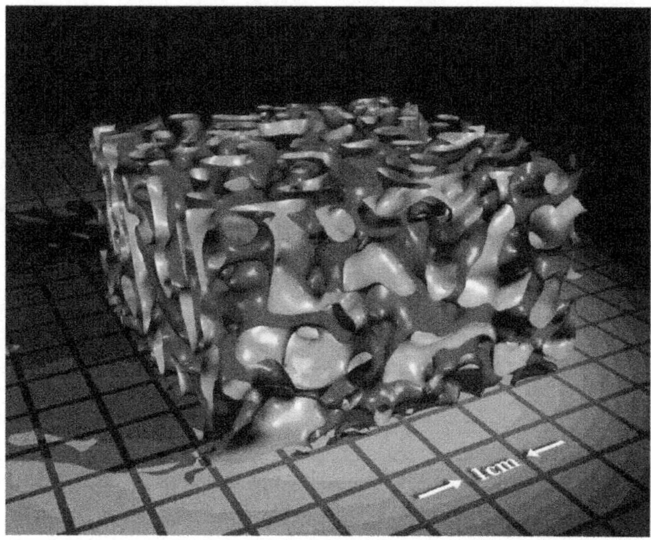

Fig. 9. Tomographic image of a termite nest. Courtesy of Tim Senden.

two alternative classes of models for the structure, namely level sets of Gaussian random fields and Boolean models. The simplest case of a Boolean model is that with $\Xi_0 = b(o, r_0)$, called IOS ("identical overlapping spheres"). For the termite nest, the sphere radius r_0 was estimated by means of the methods described below as $r_0 = 3.3$ mm, while the parameters for the random field models were estimated using the covariance function. Figure 10 shows the empirical covariance for the termite nest data and theoretical covariances for the identical overlapping spheres model (IOS) and two different level sets of Gaussian random fields. Obviously, this second order characteristic is not very helpful in deciding about a suitable model.

In Figure 11 the function $w_0(r)$ is shown as function of the erosion/dilation radius r. Although the volume fraction seems to be matched well by all models, one may observe differences for negative values of r. Thus, the identical overlapping spheres model (IOS) may be ruled out as a model for the termite nest structure. Note that the values of $w_0(r)$ for $r < 0$ were obtained by simulating identical overlapping spheres and erosion; the eroded Boolean model is not a Boolean model – in contrast to a dilated. In order to allow a comparision with voxelized experimental datasets, the discrete Boolean model (section 5) is used here where spheres are approximated by discrete lattice grains. Therefore, one observes in the intensities shown in Figs 11- 13 deviations from the equations given in section 3. The shape of $-\log(1 - H_s(r))/r$ for the data presented in Figure 12 are consistent with may indicates that perhaps another typical grain Ξ_0 (e.g. a sphere with random radius) yields a better fit; this problem is not further considered here. Still the two Gaussian field models are

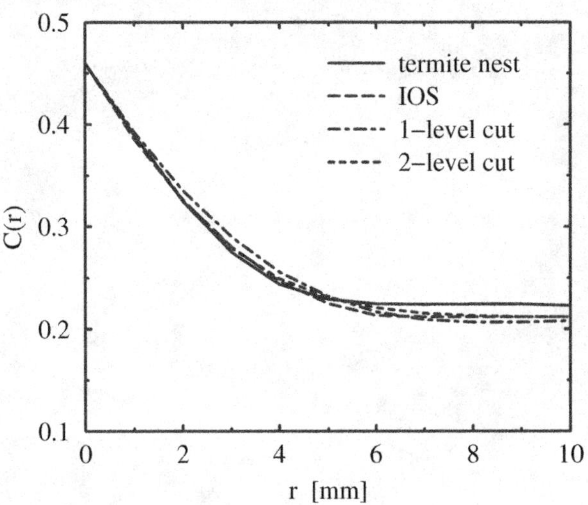

Fig. 10. Empirical and theoretical covariances for the termite nest data. The IOS model is a Boolean model with identical spherical grains. A 1-level-Cut is a thresholded image of a Gaussian random field. In contrast a 2-level-cut denotes the black/white image obtained from a Gaussian random field by thresholding at two different values and viewing only spatial regions with field values in between.

in accordance with the termite nest data. But the model test can be continued by considering also the other curvature intensities, i.e., the functions $w_k(r)$. The corresponding empirical functions are shown in normalized form in Figure 13 both for the termine nest data and for the three models. The conclusion is that none of the models is completely appropriate for the termite nest.

For parameter estimation we recommend the method of intensities, which is a variant of the moment method of statistics and was succesfully applied in many cases. It is described here for the planar case; the case of three-dimensional porous media is considered in Arns et al. [3, 4, 5] and Mecke & Arns [50]. It is based on the following three formulas:

$$A_A = 1 - \exp(-\lambda\overline{A}) , \tag{21}$$
$$L_A = \lambda(1 - A_A)\overline{U} , \tag{22}$$

and

$$N_A = \lambda(1 - A_A)\left(1 - \frac{\lambda\overline{U}^2}{4\pi}\right) . \tag{23}$$

Here A_A is area fraction, i.e. the same as p in section 3. L_A denotes the specific boundary length, i.e. the mean boundary length of the Boolean model per unit area and N_A is the specific connectivity number. All these intensities can be

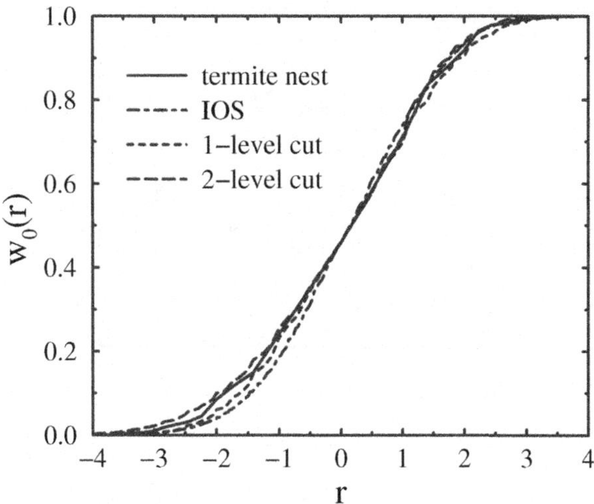

Fig. 11. Volume fraction $w_0(r)$ for the termite nest, the Boolean model of identical overlapping spheres (IOS), and two Gaussian random field models in dependence of r. Positive values of r are related to dilation, negative to erosion.

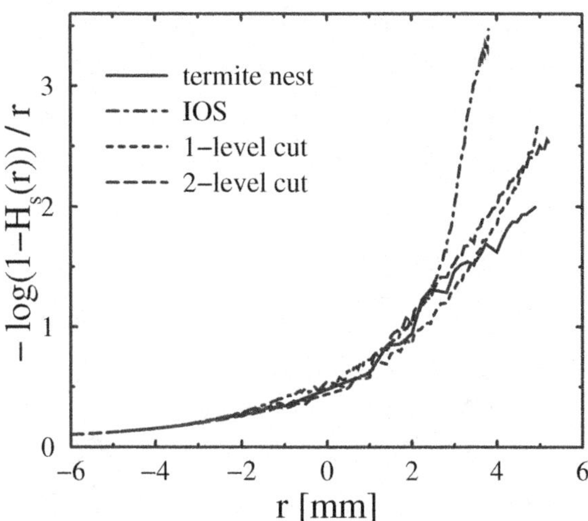

Fig. 12. The function $-\log(1-H_s(r))/r$ for the termite nest and the various models rules out the IOS model.

estimated by means of image analysis. The method yields estimators of λ and of $\overline{A}$ and $\overline{U}$, mean area and perimeter of the typical grain Ξ_0. Thus in the case of spherical grains the first two moments of the diameter distribution can

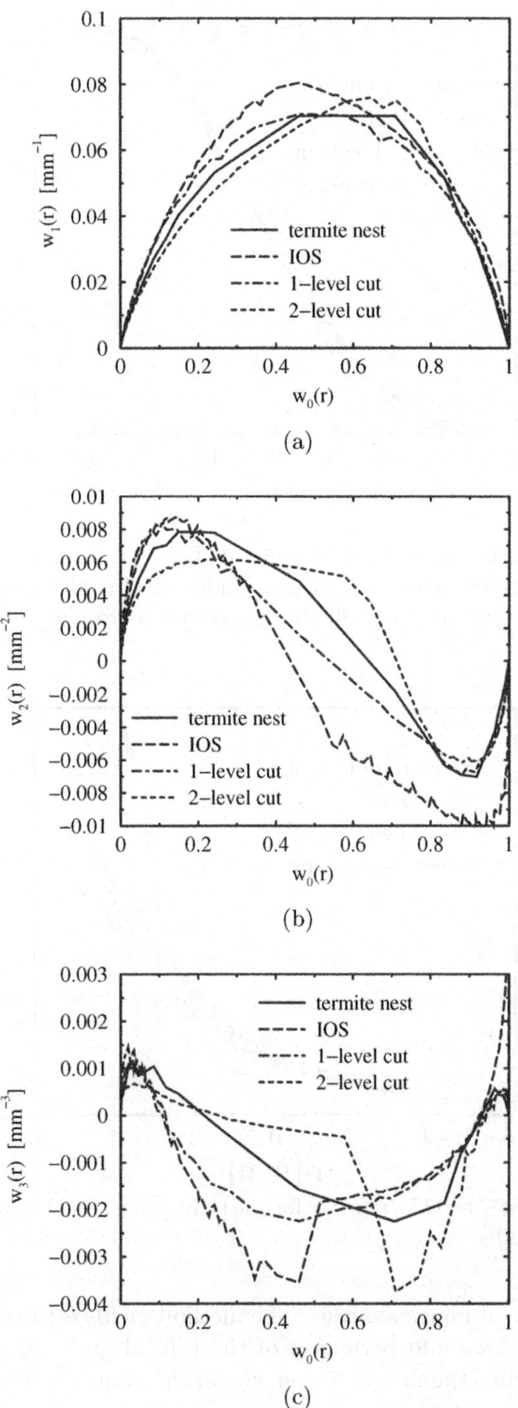

Fig. 13. The functions (a) $w_1(r)$, (b) $w_2(r)$ and (c) $w_3(r)$ for the termite nest (Arns et al., 2004). These functions show very clearly the morphological differences between the nest structure and the various models, clearer than $C(r)$ or $w_0(r)$.

be obtained. Various simulation experiments have shown that the intensity method yields reliable results, even at high volume fractions; the newest are reported in Arns et al. (2002, 2003, 2004) and Mecke & Arns (2004).

The method of intensities has a long history. For the isotropic case it first appeared in Santaló (1976), p. 284, and then in Kellerer (1983). A paper where the specific convexity number N_V^+ was used, is Bindrich & Stoyan (1990), while the stationary (non-isotropic) case was considered in Weil (1988).

An alternative, which can yield more model parameters, is the minimum contrast method. The idea of that method is to minimize a contrast functional, for example

$$\Delta(\theta) = \int\limits_a^b (\hat{f}(r) - f(r,\theta))^2 \mathrm{d}r \, ,$$

so using least squares fitting.

Here $f(\cdot, \theta)$ is a function describing a characteristic of the Boolean model, which depends on some parameters θ; $\hat{f}$ is the empirical counterpart of this function estimated from the data. The idea is to minimize $\Delta(\theta)$ over the set of parameters leading to an optimal choice $\hat{\theta}$, which is called the minimum contrast estimator for θ. A possible choice of f, which we recommend, is $f(r) = \sum_{i=0}^d a_i w_i(r)$ with suitable coefficients a_i. As above, $w_i(r)$ is the intensity of the i-th curvature measure depending on radius r. (We admit that until now we cannot report on good choices of the a_i; perhaps we would start with $a_i = 1$ or $a_i = 1/(1 + i)$.) The particular case of $f(r) = H_s(r)$ was used in Serra [61]. Heinrich [24] gave a mathematical foundation of this estimation method applied to random sets and investigated the asymptotic behaviour of minimum contrast estimators when the sampling window becomes large. If only $C(r)$ can be measured (perhaps by scattering methods), then the choice $f(r) = C(r)$ is natural, see Stoyan, Kendall & Mecke [63], p. 93. A classical paper is Diggle [15], who used this method for ecological data. Particular methods have been developed for the case of spherical grains, where the diameter distribution function has to be determined, see Heinrich & Werner [25].

Examples of statistical analyses with ellipsoidal grains are Charollais et al. [13] and Arns et al. [3, 4, 5]. In the latter papers it could be shown that a Boolean model with two types of ellipsoids yields a good fit of a sample of homogeneous Fontainebleau sandstone. These papers are also methodologically interesting because of the consequent use of lattice methods, based on discrete Boolean models; see also Mecke [44], Likos et al. [38], Arns et al. [2], and Mecke & Arns [50].

9 Conclusion

The Boolean model has become one of the most important stochastic models of stochastic geometry and spatial statistics and probably the most important random set model. It is Matheron's merit that he understood very early its potential and that he developed a general theory, which extends the case considered before, namely that of identical spherical grains. In particular, it has found widespread application in physics, because it allows to generate very diverse spatial patterns, allows elegant calculation of many quantities and structure characteristics, and can explain a wealth of features observed in real physical systems such as porous media (which was the case of Matheron's original interest), composite materials, complex fluids, foams, and even galaxy distributions in the universe. Still there are many problems open for the Boolean model, in particular in the context of its geometry.

Acknowledgement. We thank C. Arns, M. Knackstedt and H. Wagner for a fruitful collaboration and T. Senden for giving us access to the tomographic termite nest dataset. C. Lantuejoul's and D. Jeulin's comments on an earlier version of this paper improved the exposition.

KRM acknowledges support from Deutsche Forschungsgemeinschaft (DFG Grant No. Me1361/6 and Me1361/7).

References

1. Aldous, D., *Probability Approximations via the Poisson Clumping Heuristic.* Springer-Verlag, New York 1989.
2. Arns, C. H., Knackstedt M. A., Pinczewski W. V., and Mecke K. R., Euler-Poincaré characteristics of classes of disordered media, *Phys. Rev. E* **63**, (2001) 31112:1–13.
3. Arns, C. H., Knackstedt M. A., and Mecke K. R., *Characterising the morphology of disordered materials*, p. 40–78 in K. R. Mecke and D. Stoyan (eds.), *Morphology of Condensed Matter - Physics and Geometry of Spatially Complex Systems*, Lecture Notes in Physics, Vol. 600. Springer-Verlag, Heidelberg 2002.
4. C. H. Arns, M. A. Knackstedt, and K. Mecke, Reconstructing complex materials via effective grain shapes, *Phys. Rev. Lett.* **91**, (2003) 21550:6-10.
5. C. H. Arns, M. A. Knackstedt, and K. Mecke, Characterisation of irregular spatial structures by parallel sets and integral geometric measures, *Colloids and Surfaces A* **241**, (2004) 351-372.
6. Baddeley, A. & van Lieshout, M. N. M., Area-interaction point processes. *Ann. Inst. Statist. Math.* **47**, (1995) 601–619.
7. Baddeley, A. J. and Møller, J., Nearest-neighbour Markov point processes and random sets. *Int. Statist. Rev.* **57**, (1989) 89–121.
8. Bindrich, U., & Stoyan, D., Stereology for pores in wheat bread : statistical analyses for the Boolean model by serial sections. *J. Microsc.* **162**, (1990) 231–239.
9. Bretheau, T. and Jeulin, D., Caractéristiques morphologiques des constituants et comportement à la limite élastique d'un matériau biphasé Fe/Ag, *Revue Phys. Appl.* **24**, (1989) 861–869.

10. Brodatzki, U. and Mecke, K. R., *Morphological Model for Colloidal Suspensions*, (2001) cond-mat/0112009 at http://arxiv.org.
11. U. Brodatzki and K. Mecke, Simulating stochastic geometries: morphology of overlapping grains, *Computer Physics Communications* **147**, (2002) 218–221.
12. Capasso, V., Micheletti, A., De Giosa, M., & Mininni, R., Stochastic modelling and statistics of polymer crystallization processes. *Surv. Math. Industry* **6**, (1996) 109–132.
13. Charollais, F., Bauer, M., Coster, M., Jeulin, D. and Trotabas, M. Modelling the structure of a nuclear ceramic obtained by solid phase sintering, *Acta Stereologica* **16**, (1997) 315–321.
14. Chiu, S. N., Limit theorem for the time of completion of Johnson-Mehl tessellations. *Adv. Appl. Prob.* **27**, (1995) 889–910.
15. Diggle, P. J., Binary mosaics and the spatial pattern of heather. *Biometrika* **64**, (1981) 91–95.
16. Doi, M., A new variational approach to the diffusion and the flow problem in porous media. *J. Phys. Soc. Japan* **40**, (1976) 567–572.
17. Erhardson, T., Refined distributional approximations for the uncovered set in the Johnson-Mehl model. *Stoch. Proc. Appl.* **96**, (2001) 243–259.
18. Fallert, H., Quermaßdichten für Punktprozesse konvexer Körper und Boolesche Modelle. *Math. Nachr.* **181**, (1996) 165–184.
19. Grimmet, G., *Percolation.* (2nd ed.) Springer-Verlag, Berlin, Heidelberg, New York 1999.
20. Groh, B. and Mecke, K. R., *Density functional theory for the morphological model*, MPI-preprint, Stuttgart 2003.
21. Hadwiger, H. and Giger, H., Über Treffzahlwahrscheinlichkeiten im Eikörperfeld. *Z. Wahrscheinlichkeitsth. verw. Geb.* **10**, (1968) 329–334.
22. Hall, P., *Introduction to the Theory of Coverage Processes.* J. Wiley & Sons, New York 1988.
23. Heinrich, L., On existence and mixing properties of germ-grain models. *statistics* **23**, (1992) 271–286.
24. Heinrich, L., Asymptotic properties of minimum contrast estimators for parameters of Boolean models. Metrika **40**, (1993) 67–94.
25. Heinrich, L., & Werner, M., Kernel estimation of the diameter distribution in Boolean models with spherical grains. *Nonparam. Stat.* **12**, (2000) 147–176.
26. K. Jacobs, S. Herminghaus, and K. R. Mecke, Thin Liquid Polymer Films Rupture via Defects, *Langmuir* **14**, (1998) 965-969.
27. K. Jacobs, R. Seemann, and K. R. Mecke, *Dynamics of Dewetting and Structure Formation in Thin Liquid Films*, pp. 72-91 in K. R. Mecke and D. Stoyan (eds.), *Statistical Physics and Spatial Statistics - The Art of Analyzing and Modeling Spatial Structures and Pattern Formation*, Lecture Notes in Physics, Vol. 554. Springer-Verlag, Heidelberg 2000.
28. Jeulin, D., *Spatial statistics and micromechanics of materials*, pp. 3–36. In K.R. Mecke and D. Stoyan (eds.), Morphology of Condensed Matter-Physics and Geometry of Spatially Complex Systems. Lecture Notes in Physics, Vol. 600. Springer-Verlag, Berlin, Heidelberg 2002.
29. Jeulin, D. and Ostoja-Starzewski, M. (eds.), *Mechanics of Random and Multiscale Microstructures.* CISM Lecture Notes No. 430, Springer-Verlag, Berlin, Heidelberg, New York 2001.
30. Kellerer, A. M., On the number of clumps resulting from the overlap of randomly placed figures in the plane. *J. Appl. Prob.* **20**, (1983) 126–135.

31. Kendall, W. S., On some weighted Boolean models. In: *Advances in Theory and Applications of Random Sets*. Proc. Int. Symp. Oct 9–11, 1996, Fontainebleau (ed. D. Jeulin). Singapore: World Scientific Publ. Comp. (1997) pp. 105–120.

32. Kendall, W. S., van Lieshout, M. C., & Baddeley, A., Quermass-interaction processes. Conditions for stability. *Adv. Appl. Probab.* **31**, (1999) 315–342.

33. Kerscher, M., Mecke, K. R., Schmalzing. J., Beisbart, C., Buchert, Th., and Wagner, H., Morphological fluctuations of large–scale structure: the PSCz survey. *Astronomy & Astrophysics* **373**, (2001) 1–11.

34. Klain, D.A. and Rota, G.-C., *Introduction to Geometric Probability*. Cambridge University Press, Cambridge 1997.

35. König, P.-M., Roth, R., and Mecke, K., Morphological thermodynamics of fluids: shape dependence of free energies. *Phys. Rev. Lett.* **93**, (2004) 160601-4.

36. Kolmogoroff, A. N., Statistical theory of crystallization of metals. *Bull. Acad. Sci. USSR, Mat. Ser.* **1**, (1937) 355–359.

37. Lantuejoul, C., *Geostatistical Simulation: Models and Algorithms*. Springer-Verlag, Berlin, Heidelberg, New York 2002.

38. Likos, C. N., Mecke, K. R., and Wagner, H., Statistical morphology of random interfaces in microemulsions. *J. Chem. Phys.* **102**, (1995) 9350–9361.

39. Matheron, G., *Fonction aléatoire de type tout ou rien*. Internal report, N. 53 CMM, Ecole des Mines de Paris.

40. Matheron, G., *Eléments pour une théorie des milieux poreux*. Masson, Paris 1967.

41. Matheron, G., *Théorie des ensembles aléatoires. In Cahiers du Centre de Morph. Math.*, Fasc. 4, Forntainebleau 1969.

42. Matheron, G., Ensembles fermés aléatoires, ensembles semi-Markoviens et polyèdres poissoniens, *Adv. in Appl. Prob.* **4**, (1972) 508–541.

43. Matheron, G., *Random Sets and Integral Geometry*. J. Wiley & Sons, New York 1975.

44. Mecke, K.R., *Integralgeometrie in der Statistischen Physik*. Verlag Harri Deutsch, Frankfurt 1994.

45. Mecke, K. R., Morphological Characterization of Patterns in Reaction-Diffusion Systems. *Phys. Rev. E* **53**, (1996a) 4794–4800.

46. Mecke, K. R., Morphological Model for Complex Fluids. *J. Phys.: Condensed Matter* **8**, (1996b) 9663–9667.

47. Mecke, K. R., *Additivity, Convexity, and Beyond: Applications of Minkowski Functionals in Statistical Physics*, pp. 72–184 in K. R. Mecke and D. Stoyan (eds.), *Statistical Physics and Spatial Statistics - The Art of Analyzing and Modeling Spatial Structures and Pattern Formation*, Lecture Notes in Physics, Vol. 554. Springer-Verlag, Heidelberg 2000.

48. Mecke, K. R., Exact moments of curvature measures in the Boolean model. *J. Stat. Phys.* **102**, (2001) 1343–1381.

49. Mecke, K. R., The shapes of parallel surfaces: porous media, fluctuating interfaces and complex fluids. *Physica A* **314**, (2002) 655–662.

50. K. Mecke and C. Arns, Fluids in porous media, *J. Phys.: Condensed Matter*, **17**, (2004).

51. Mecke, K. R., Buchert, Th., and Wagner, H., Robust morphological measures for large-scale structure in the universe. *Astron. Astrophys.* **288**, (1994) 697–704.

52. Mecke, K. R. and Seyfried, A., Strong dependence of percolation thresholds on polydispersity. *Europhys. Lett.* **58**, (2002) 28–34.

53. Mecke, K. R. and Wagner, H., Euler characteristic and related measures for random geometric sets. *J. Stat. Phys.* **64**, (1991) 843–850.

54. Meester, R. and Roy, R., *Continuum Percolation.* Cambridge University Press, Cambridge 1996.

55. Miles, R.E., Estimating aggregate and overall characteristics from thick sections by transmission microscopy. *J. Microsc.* **107**, (1976) 227–233.

56. Molchanov, I. S., A limit theorem for scaled vacancies in the Boolean model. *Stochastics, Stochastic Reports* **58**, (1996) 45–65.

57. Molchanov, I. S., *Statistics of the Boolean Model for Practioners and Mathematicians.* J. Wiley & Sons, Chichester 1997.

58. Santaló, L., *Integral Geometry and Geometric Probability.* Addison-Wesley, Reading MA 1976.

59. Schneider, R., Parallelmengen mit Vielfachheit und Steiner-Formeln. *Geom. Ded.* **9**, (1980) 111–127.

60. Schneider, R., *Convex Bodies: The Brunn-Minkowski Theory.* Cambridge University Press, Cambridge 1993.

61. Serra, J., *Image Analysis and Mathematical Morphology.* Academic Press, London 1982.

62. Stauffer, D. and Aharony, A., *Introduction to Percolation Theory.* Taylor& Francis, London 1992.

63. Stoyan, D., Kendall, W. S., & Mecke, J., *Stochastic Geometry and its Applications.* J. Wiley & Sons, Chichester 1995.

64. Stoyan, D., Random sets: models and statistics. *Int. Statist. Rev.* **66**, (1998) 1–27.

65. Stoyan, D. & Stoyan, H. *Fractals, Random Shapes and Point Fields.* J. Wiley & Sons, Chichester 1994.

66. Torquato, S., *Random Heterogeneous Materials. Microstructure and Macroscopic Properties.* Springer-Verlag, New York, Berlin, Heidelberg 2002.

67. van Lieshout, M. N. M., *Markov Point Processes and their Applications.* Imperial College Press, London 2000.

68. Vogel, H.-J., *Topological Characterization of Porous Media* , p. 75–92 in K. R. Mecke and D. Stoyan (eds.), *Morphology of Condensed Matter - Physics and Geometry of Spatially Complex Systems*, Lecture Notes in Physics, Vol. 600. Springer-Verlag, Heidelberg 2002.

69. Voss, K., *Discrete Images, Objects and Functions in Z^n.* Springer-Verlag, Berlin, Heidelberg, New York 1993.

70. Weil, W., Expectation formulas and isoperimetric properties for non-isotropic Boolean models. *J. Microsc.* **151**, (1988) 235–245.

71. Weil, W., Iterations of translative integral formulae and non-isotropic Poisson processes of particles. *Math. Z.* **205**, (1990) 531–549.

72. Weil, W., Densities of mixed volumes for Boolean models. *Adv. Appl. Prob.* **33**, (2001) 39–60.

73. Widom, J. S., & Rowlinson, B., New model for the study of liquid-vapour phase transition. *J. Chem. Phys.* **52**, (1970) 1670–1684.

Random Structures in Physics

Dominique Jeulin

Centre de Morphologie Mathématique, Ecole des Mines de Paris

1 Introduction

One of the major contributions of Georges Matheron to the Sciences of the 20th century concerns the field of Physics of random media. This important aspect of the work of Georges Matheron is not so well known by the communities of Geostatistics and Mathematical Morphology, but it has a large impact in many domains of engineering sciences. In parallel to his work on flows in porous media, Georges Matheron developed in collaboration with Jean Serra the basis of mathematical morphology (like operations of erosion, dilation, opening,), having in mind the geometrical characterization of complex porous media [48].

He first developed a general methodology for the composition of permeability at different scales through homogenization [48, 49]. This approach, based on perturbation techniques, can be applied to any physical process involving a conservation law and a linear constitutive behavior, such as thermal, electrostatic, or elastic properties. A theoretical study of the genesis of permeability of porous media was made, proving the existence and unicity of the solution of the linearized version of the Navier-Stokes equation (namely the Stokes equation) for random closed sets, and providing useful upper bounds for some random media [53]. Important results were obtained at that time about the dispersion of flows in random media. Later, he studied the properties of geodesics in media with a random refraction index [55], and he derived in a very elegant way bounds of the effective permeability modeled by random functions [57].

In this paper is given a short survey on recent work made on the prediction of physical properties of random media, modelled by random sets or random functions. In the first part, the methodology is recalled and the main results obtained in the field of homogenization of random media. The second part is concerned by the study of fluctuations of effective properties and its important consequences for numerical simulations, in terms of "Representative Volume

Element". The last part introduces some probabilistic models of fracture, which involve the use of specific classes of models of random functions.

2 Homogenization of random media

2.1 Introduction

From a macroscopic point of view, the behavior of a physical system can be considered as the responses of a medium to solicitations. For current engineering practice, standard physical variables are well defined (for instance stresses and strains in mechanics) and are related by constitutive equations expressing physical laws.

The meaning of the used physical variables, and the validity of the constitutive equations always depend on the scale of observation of physical phenomena: for instance in fluid mechanics it is possible to work on a "microscopic scale", at which the physical system is made of a population of moving particles in interaction. At this scale it is possible to speak about velocity momentum, but not about pressure or temperature. These last two "macroscopic" variables have a physical meaning for a volume of matter containing a high number of particles, where only the collective behavior of a population is kept. By a change of scale, we change the physical model, replacing a discrete system by a continuum, for which are defined new variables such as pressure, temperature, concentration...

From a general point of view, we mean by change of scale the problem of the prevision of the macroscopic behavior of a physical system from its microscopic behavior.

This problem is very wide, and is of interest for a large area of applied physics. In this part we limit our purpose to the case of the dielectric properties of composite random media.

What can be expected from a model of change of scale? In addition to the change of physical status and to the emergence of new variables or of macroscopic behavior laws, as mentioned about fluid mechanics, a model of change of scale should give answers to the following problems.

- Does a macroscopic behavior law exist for a given heterogeneous medium (problem of emergence)?
- If it exists, what is its expression and what are its coefficients (or the **effective properties** of the medium)?
- What is the variability of effective properties measured on specimens with a finite size, as a function of their microscopic variability (scale effect on the fluctuations of properties)?

In the next subsections, heterogeneous media are first defined and illustrated, based on the example of electrostatics. Then the principle of calculation of effective properties of heterogeneous media is recalled. The next subsection gives a perturbation expansion of the field developed in a heterogeneous medium, from which a formal calculation of the effective properties of random media is worked out. Finally bounds of the effective properties of random media are reviewed.

2.2 Homogeneous medium and heterogeneous medium

In this subsection, we consider a continuum embedded inside a domain (usually bounded) B within the euclidean space R^d. In every point x of B, we can consider a set of physical properties $P(x)$ (for instance a stress or strain tensor, a velocity, a temperature,...). $P(x)$ builds a field defined on the domain B. The field $P(x)$ is the solution of a problem built from the following conditions:

- application of **conservation** principles (for instance momentum, energy), which from a local balance generally involves systems of partial differential equations (equation 6 for electrostatics);
- **choice of boundary conditions** most often given on the boundary ∂B of the domain B;
- use of **constitutive equations** linking several variables. For instance, for **electrostatics of a dielectric medium**, the electric displacement D is proportional to the electric field E, ϵ being the dielectric permittivity tensor, and ϕ the potential:

$$D = \epsilon E = -\epsilon \, grad\phi \tag{1}$$

It is possible to give a short physical explanation of the origin of these variables: consider a capacitor filled with vacuum. The relationship between its electric charge Q and the applied difference of potential V is given by

$$Q = C_0 V \tag{2}$$

where C_0 is its capacity in vacuum. When replacing the vacuum by a homogeneous dielectric (insulating material) with dielectric permittivity ϵ, its capacity increases to become $C = \frac{\epsilon}{\epsilon_0} C_0$, where ϵ_0 is the dielectric permittivity of vacuum. In the capacitor an electric field E is generated by the charge Q. E is proportional to V. The conservation of the charge Q into a closed domain B can be expressed as the conservation of the flux of a vector D (the electric displacement) on the boundary. This conservation law is the Gauss law. As a consequence, when there is no charge in the domain B ($Q = 0$), the vector D satisfies the partial differential equation 6. From equation 2 and from the Gauss law, D is proportional to E (the proportionality constant being equal to ϵ_0 in vacuum and to ϵ in a dielectric material).

The law (1) is defined for each point x, where only the local information takes part to the relations between the variables. For a **homogeneous medium,** the variable ϵ (which is a positive definite tensor) in equation (1) remains constant in space. For the quoted linear constitutive equations, the existence and uniqueness of the solution for a given geometry can be proved. The geometry can be involved in a complex way in the solution; consider for instance a porous medium, for which elastic properties or flow properties are studied (in this last case, usual boundary conditions prescribe a null velocity for the fluid at the grain boundaries).

The medium is **heterogeneous** when the constitutive equations can change in space, as a result of:

- different types of constitutive equations (for instance linear for some places and non linear for other places within the domain B);
- variations in space of the coefficients of the equations. They can be modelled either by random functions [48, 72], or by periodic fields [72].

In the case of heterogeneous media, the **change of scale** problem can be formulated as follows:

- Is it possible to replace an heterogenous medium (with support B) by an **equivalent homogeneous medium** from a macroscopic point of view?
- If yes, what is the macroscopic constitutive equation and what are the values of its macroscopic coefficients (or effective properties)?

It is generally not straightforward to answer to these questions. To be convinced of that point, it is enough to recall the case of the composition of permeability studied by Georges Matheron [48], for which a macroscopic Darcy's law and a macroscopic permeability exist in the case of uniform flows, while for radial flows between two concentric contours, the macroscopic permeability depends on the geometry and remains different from the case of uniform flows. These difficulties are usually forgotten from the abundant literature on effective properties.

2.3 Principle of calculation of effective properties

For simplicity we will illustrate the case of the dielectric permittivity involving the vector fields E and D (it can be replaced by the composition of permeability, coefficient of diffusion, or thermal conductivity). In [3, 4, 20, 21, 23, 36, 38, 39, 45, 58, 72, 83], results are presented on tensor fields that occur in the linear elasticity of random media. This class of problems involves second order elliptic partial differential equations with random coefficients. The average of any field $P(x)$ in B with the volume V is defined as:

$$\langle P \rangle = \frac{1}{V} \int_B P(x)dx \tag{3}$$

For instance the average of the electric field E and of the displacement field D is:

$$\langle E \rangle = \frac{1}{V} \int_B E(x)dx$$

$$\langle D \rangle = \frac{1}{V} \int_B D(x)dx = \frac{1}{V} \int_B \epsilon(x)E(x)dx \qquad (4)$$

The **effective** dielectric tensor ϵ^* of the equivalent homogeneous medium contained in B is defined in such a way that equation (1) is satisfied on a macroscopic scale:

$$\langle D \rangle = \epsilon^* \langle E \rangle \qquad (5)$$

In general, ϵ^* will depend on the applied boundary conditions on ∂B, as mentioned earlier about radial flows [48], on the local permittivity of the components and on the geometry of the medium.

The relations given by equations (4-5) can be applied to any heterogeneous medium in a bounded region B, provided the field $E(x)$ is expressed from the applied boundary conditions and from the partial differential equation expressing the Gauss' law:

$$div(D) = \sum_i \frac{\partial}{\partial x_i}(D_i(x)) = 0 \qquad (6)$$

This procedure can be used for a periodic medium from the knowledge of the cell building the period [38, 72].

When applying periodic boundary conditions on the period, or homogeneous boundary conditions on a finite domain B (namely imposing either a constant electric field E_0, or a constant displacement field D_0), we get the averages [38, 72]:

$$\langle E \rangle = E_0$$

$$\langle D \rangle = D_0$$

Starting from the local energy $U(x) = \frac{1}{2}E(x)\epsilon(x)E(x) = \frac{1}{2}E(x)D(x)$ and averaging over the domain B, the following relation is satisfied, from which the effective dielectric tensor can equivalently be defined:

$$\langle U \rangle = \frac{1}{2} \langle ED \rangle = \frac{1}{2} \langle E \rangle \langle D \rangle = \frac{1}{2} \langle E \rangle \epsilon^* \langle E \rangle \qquad (7)$$

This result [38], which can be interpreted as a null statistical correlation between E and D, is called the Hill lemma in elasticity of heterogeneous media [21, 45, 72]). It holds for any divergence free field $D(x)$ and any gradient of potential $E(x)$, satisfying the homogeneous or periodic boundary conditions. When a constant electric field E_0 is applied on ∂B, equation (7) yields

$$\langle D \rangle = \epsilon^* E_0 \qquad (8)$$

When a constant displacement field D_0 is applied on ∂B, equation (7) yields

$$\langle E \rangle = \left(\epsilon^{-1}\right)^* D_0 \qquad (9)$$

Therefore, in the case of periodic or of homogeneous boundary conditions applied on ∂B, the effective dielectric permittivity ϵ^*, (or its inverse $(\epsilon^{-1})^*$) is obtained from the average $\langle D \rangle$ (or $\langle E \rangle$). In practice it can be implemented from a numerical solution of the electrostatic problem, knowing the geometry of the medium and the boundary conditions. We will come back to this point in section 3.2.

2.4 Perturbative expansion

From a perturbation expansion is derived an approximate solution of equation (6) for the vector fields D and E [4, 11, 22, 45, 48, 49]. This approximation can be used as input in equations (7-9) to estimate the effective ϵ^*. It can also enter into a variational principle as seen in section 2.6. We consider a domain B containing a heterogeneous medium, submitted to a constant macroscopic field $\overline{E}$ (with components $\overline{E}_i$) applied on ∂B.

We note $E'(x) = E(x) - \overline{E}$ and $\epsilon'(x) = \epsilon(x) - \langle \epsilon \rangle$. The field $E'(x)$ is solution of the following equation, derived from equation (6):

$$div(E\{\epsilon\}E') = -div(\epsilon'E) \tag{10}$$

This comes from the fact that

$$div(D) = 0 = div(\epsilon'E) + E\{\epsilon\}div(E) = div(\epsilon'E) + E\{\epsilon\}div(E')$$

When the medium satisfies $\langle \epsilon_{ij} \rangle = \overline{\epsilon}\delta_{ij}$ (for instance if the medium is isotropic, like a polycrystal with a uniform distribution of orientations, or if $\epsilon_{ij}(x) = \epsilon(x)\delta_{ij}$ with any macroscopic anisotropy), equation (10) can be interpreted as a Poisson equation, after introduction of a potential ϕ' with $E' = -grad\phi'$

$$\overline{\epsilon}\Delta\phi' = div(\epsilon'E) \tag{11}$$

A formal solution of equation (11) is given by means of the Green function $G(x, y)$ (or harmonic potential) solution of:

$$\begin{aligned} \Delta_x G(x,y) &= -\delta(x-y) \text{ for } x \in B \\ G(x,y) &= 0 \text{ for } x \in \partial B \end{aligned} \tag{12}$$

The solution of equation (11) can be written

$$\overline{\epsilon}\phi'(x) = -\int_B G(x,y)div(\epsilon'(y)E(y)) \, dy \tag{13}$$

After integration by parts, and accounting for the condition given in equation (12), the potential follows

$$\overline{\epsilon}\phi'(x) = \sum_i \int_B \frac{\partial}{\partial y_i} G(x,y)(\epsilon'(y)E(y))_i \, dy$$

The components of $E' = -grad\phi'$, and therefore the components of E are obtained by partial derivation of ϕ'

$$\bar{\epsilon}E_j(x) = \bar{\epsilon}\overline{E}_j - \sum_{ik}\int_B \frac{\partial^2}{\partial x_j \partial y_i}G(x,y)\epsilon'_{ik}(y)E_k(y)\,dy$$

$$= \bar{\epsilon}\overline{E}_j - (\Gamma\epsilon'E)_j(x) \tag{14}$$

where we introduced the operator Γ. Equation (14) has the form of the Lippman-Schwinger equation of the quantum mechanical scattering theory [45]. For infinite media with homogeneous boundary conditions, or for periodic media, the operator Γ acts by a convolution, and therefore the solution of Equation (14) can be obtained numerically by iterations (starting from an initial field $E_k(y)$ in the integral), using Fourier transforms on images of the field $\epsilon(x)$ [16, 64].

From equation (14) is deduced

$$\bar{\epsilon}\left[I + \Gamma\frac{\epsilon'}{\bar{\epsilon}}\right]E(x) = \bar{\epsilon}\overline{E}$$

I being the identity operator. The following formal expansion gives $E(x)$:

$$E(x) = \overline{E} + \sum_{n=1}^{n=\infty}(-1)^n\left(\Gamma\frac{\epsilon'}{\bar{\epsilon}}\right)^n\overline{E} = \sum_{n=0}^{n=\infty}E^{(n)}(x) \tag{15}$$

It can be shown that the expansion of equation (15) converges if the dielectric permittivity ϵ can be expressed as $\epsilon = \epsilon_0(I+\gamma)$ with $\|\gamma\| < 1$ [56] (with $\epsilon_0 = \bar{\epsilon}$, the present development converges if $\left\|\frac{\epsilon'}{\bar{\epsilon}}\right\| < 1$). This happens if ϵ and ϵ^{-1} remain bounded (and this excludes media where at some places $\epsilon(x) = 0$ or $\epsilon(x) = \infty$). The reason for this criterion of convergence comes from the fact that the operator Γ is a projector (with a norm less than one) on the subspace of gradients, when $E(x)$ is defined in the Hilbert space with the scalar product $E^1.E^2 = \langle E^1\epsilon E^2\rangle$.

2.5 Formal calculation of the effective properties of random media

If the field $\epsilon(x)$ is modelled by a stationary ergodic random multivariate function, it turns out that for domains B converging to R^d the spatial averages involving ϵ' converge towards mathematical expectations. In these conditions the random field $E(x)$ (and then $D(x) = \epsilon(x)E(x)$) given by equation 15 is stationary and ergodic: from equation (15) it is easy to check that $E\{E_j^{(n)}(x)\} = 0$, since the central correlation function of order n, $E\left\{\epsilon'_{i_2k_2}(x^n)...\epsilon'_{i_1k_1}(x^1)\right\}$, does not depend on x and provides a null contribution after integration by parts. Therefore $E\{E(x)\} = \overline{E}$. In addition we have $\langle E \rangle = \overline{E}$. As a consequence, the averages $\langle . \rangle$ in equations (5, 7, 8, 9) become mathematical expectations $E\{.\}$, and we get:

$$E\{D\} = \epsilon^* E\{E\}$$
$$E\{U\} = E\left\{\tfrac{1}{2}E(x)D(x)\right\} = \tfrac{1}{2}E\{E\}E\{D\} = \tfrac{1}{2}E\{E\}\epsilon^* E\{E\} \tag{16}$$

The two definitions of ϵ^* given in equation (16) are equivalent. The factorization of $E\left\{\tfrac{1}{2}E(x)D(x)\right\}$, derived here from equation (7), can also be derived from the stationarity of D and E [48, 49]: $E'(x)$ is a stationary random field E', with a zero average, deriving from a stationary potential ϕ' with $E'(x) = -grad\phi'$; we have:

$$U(x) = \frac{1}{2}E(x)D(x) = \frac{1}{2}\overline{E}D(x) + \frac{1}{2}E'(x)D(x)$$

By expectation,

$$E\{U\} = E\left\{\frac{1}{2}E(x)D(x)\right\} = \frac{1}{2}EE\{D\} + E\left\{\frac{1}{2}E'(x)D(x)\right\}$$

The second term is equal to zero since $E'(x)D(x) = -div(\phi'(x)D(x))$ from application of equations (1,6). Then $E\{E'(x)D(x)\} = -E\{div(\phi'(x)D(x))\} = -div(E\{\phi'(x)D(x)\}) = 0$ from the stationarity of $\phi'(x)D(x)$.

For a random medium, the effective dielectric tensor defined in equation (16) can therefore be obtained from the expectation $E\{D\} = E\{\epsilon E\}$ when a constant macroscopic field $E\{E\}$ is applied on ∂B. If the random medium is statistically isotropic, all the orientations are equivalent and therefore ϵ^* is an isotropic second order tensor that can be summarized by a scalar.

To estimate the effective properties of the random medium, we need to calculate $E\{D(x)\}$ (cf. equation (16)). We have

$$E\{D(x)\} = E\{\epsilon(x)E(x)\} = E\{(\epsilon'(x) + \overline{\epsilon})(E'(x) + \overline{E})\}$$
$$= E\{\epsilon'(x)E'(x)\} + \overline{\epsilon E} = \epsilon^*\overline{E}(x)$$
$$E\{D_i(x)\} = \sum_j \epsilon^*_{ij}\overline{E}_j = \overline{\epsilon}\overline{E}_i + E\{\sum_j \epsilon'_{ij}(x)E'_j(x)\}$$
$$= \overline{\epsilon}\overline{E}_i + E\{\sum_{jn} \epsilon'_{ij}(x)E_j^{(n)}(x)\}$$

By introduction of the perturbation expansion for $E_j^{(n)}(x)$ and by identification and rearrangement, we obtain

$$\epsilon^*_{ik_1} = \overline{\epsilon}\delta_{ik_1} + \sum_{n=1}^{n=\infty} \frac{(-1)^n}{\overline{\epsilon}^n} \sum_{ji_1k_1...i_nk_n} \int_{B^n} \frac{\partial^2}{\partial x_j \partial x_{in}^n} G(x,x^n)...$$

$$\frac{\partial^2}{\partial x_{i_2}^2 \partial x_{i_1}^1} G(x^2,x^1) E\left\{\epsilon'_{ij}(x)\epsilon'_{i_1k_1}(x^1)...\epsilon'_{i_2k_2}(x^n)\right\} dx^1...dx^n \tag{17}$$

The coefficient of order n of the expansion (17) for $\epsilon^*_{ik_1}$ involves central correlation functions of order $n+1$ of the random field $\epsilon_{ij}(x)$. It is expected to introduce more and more information on the random medium by increasing the order of the development. In [49] it is assumed that $\epsilon' = \bar{\epsilon}\alpha\gamma$ with a small scalar α and a tensor γ. The coefficient of α^n in the expansion is called the Schwydler tensor of order n, generalizing the work of this author made at the order 2 [73, 74].

When the random medium is statistically isotropic, and when $B = R^d$, it can be shown that the second term of the perturbation expansion (for $n = 1$) does not depend on the details of the second order correlation function $E\left\{\epsilon'_{ij}(x)\epsilon'_{i_1k_1}(x^1)\right\}$. In that case, it is necessary to introduce at least the term depending on third order correlation functions to get estimates of effective properties depending on the microstructure. This point motivated the work of Georges Matheron [49], when he was criticizing the rule of geometrical averaging which was in his time more or less systematically applied for the composition of permeability.

The expansion given by Equation (15) can be used to estimate the covariance of $E_i(x)$ and of $E_j(x + h)$ [4] and as a particular case the variance of $E_i(x)$ limited to the first term [48]. In [5] bounds of the variance are derived.

In [41, 42], the propagation of elastic or of electromagnetic waves in random media is studied in a similar way (with appropriate Green functions), from a second order perturbation expansion. Electromagnetic wave propagation is studied by related techniques in [17, 69].

2.6 Bounds and "optimal" random sets

In general, it is not possible to know exactly the effective (or macroscopic) permittivity ϵ^*, except for some specific geometries (some examples are given below). In fact, the exact prediction of ϵ^* requires a very large amount of information (the set of all n point correlation functions, which naturally appear in the perturbation expansion (15) of the solution of equation (6) combined to equation (1) for infinite domains B containing realizations of ergodic random media [4, 45, 48, 49]). Using a limited amount of statistical information, bounds (upper bound ϵ_+ and lower bound ϵ_-) are derived from the perturbation expansion of the electric field and a variational principle on the stored electrostatic energy [3, 4]. From this principle, which is equivalent to the conservation law (6), the replacement of the unknown solution $E(x)$ by a suitable approximation (or trial field) $E^*(x)$ provides an estimate $\langle U^* \rangle$ of the energy, with $\langle U^* \rangle \geq \langle U \rangle$, where $\langle U \rangle$ is obtained for the solution (equation 6). The same inequality is obtained by replacing the unknown $D(x)$ by a trial field $D^*(x)$ with $div(D^*) = 0$.

Wiener and Hashin- Shtrikman bounds

A first application of the classical variational principle gives the generalized Wiener bounds [22]. If the trial field is $E(x) = E\{E\} = \overline{E}$, we get

$$2U = E\{\overline{E}\epsilon(x)\overline{E}\} \geq 2U_0 = \overline{E}\epsilon^*\overline{E}$$

and

$$\overline{E}E\{\epsilon\}\overline{E} \geq \overline{E}\epsilon^*\overline{E}$$

For a trial field $D(x) = \overline{D}$,

$$2U = E\{\overline{D}\epsilon^{-1}(x)\overline{D}\} \geq 2U_0 = \overline{D}(\epsilon^*)^{-1}\overline{D}$$

and

$$\overline{D}E\{\epsilon^{-1}\}\overline{D} \geq \overline{D}(\epsilon^*)^{-1}\overline{D}$$

The obtained inequalities show that the tensors $E\{\epsilon\} - \epsilon^*$ and $E\{\epsilon^{-1}\} - (\epsilon^*)^{-1}$ are positively definite. It results:

$$(E\{\epsilon^{-1}\})^{-1} \leq \epsilon^* \leq E\{\epsilon\} \tag{18}$$

For the scalar case (for instance in the case of a statistically isotropic random medium), equation (18) is applied to the scalars ϵ, $\epsilon^{-1} = 1/\epsilon$, ϵ^*. For a locally isotropic medium with $E\{\epsilon(x)\} = \overline{\epsilon}$, the eigenvalues ϵ^{*i} of ϵ^* satisfy the following inequalities, resulting from equation (18):

$$(E\{\epsilon^{-1}\})^{-1} \leq \epsilon^{*i} \leq \overline{\epsilon} \tag{19}$$

This corresponds to the well known arithmetic and harmonic averages.

Consider now the case of an isotropic multiphase material in R^d, where for every component i (with the volume fraction p_i), ϵ_i is a scalar. The Hashin and Shtrikman (H-S) bounds [19, 20] are obtained after introduction of a reference homogeneous medium (with the permittivity ϵ_0 satisfying $\epsilon_m = \inf_i(\epsilon_i) \leq \epsilon_0 \leq \epsilon_M = \sup_i(\epsilon_i)$) from

$$K = \sum_{j=1}^{j=N} \frac{p_j}{(\epsilon_j - \epsilon_0)^{-1} + 1/(\epsilon_0 d)} \tag{20}$$

Choosing $\epsilon_0 = \epsilon_m$ or alternatively $\epsilon_0 = \epsilon_M$,

$$\epsilon^* \geq \epsilon_- = \epsilon_m + \frac{K}{1 - \dfrac{K}{d\epsilon_0}} \tag{21}$$

$$\epsilon^* \leq \epsilon_+ = \epsilon_M + \frac{K}{1 - \dfrac{K}{d\epsilon_0}}$$

The (H-S) bounds are closer than the Wiener bounds (they introduce additional information about the microstructure, namely the assumption of isotropy). Wiener and (H-S) bounds depend only on the volume fractions and on the values of the properties of the components, and therefore will be the same for very different microstructures.

These first and second order bounds were derived in a very elegant way by Georges Matheron [57], without resorting to any variational principle.

Method of computations of order n bounds

By using the following trial field in the classical variational principle [3, 4]

$$E_j^N(x) = \overline{E}_j + \sum_{n=1}^{n=N} \lambda_n E_j^{(n)}(x)$$

where $E_j^{(n)}(x)$ is the term of order n of the perturbative expansion of the electric field (15), the λ_n minimizing the upper bound of the effective permittivity are obtained as the solution of a linear system with coefficients depending on correlation functions up to the order $2N + 1$ (order 3 for $N = 1$). Similarly for a trial field

$$D_j^N(x) = \overline{D}_j + \sum_{n=1}^{n=N} \lambda_n D_j^{(n)}(x)$$

With for $n > 1$

$$D_j^{(n)}(x) = \overline{\epsilon} E_j^{(n)}(x) + \epsilon' E_j^{(n-1)}(x) - E\{\epsilon' E_j^{(n-1)}(x)\}$$

we get the lower bound of order $2N + 1$. By increasing the order N of the expansion, narrower bounds are obtained, since they involve an increasing amount of information about the microstructure. In practice, the expansion is stopped to the first term, to get third order bounds, correlation functions of higher order being usually unknown.

Third order bounds

The general approach to bounds considers multiphase and even continuous models (scalar or multivariate) [26], for which the calculation of third order bounds can be carried out using the general derivation based on the third order correlation function [4, 22, 33, 45, 59]. An extension of bounds to the complex dielectric permittivity was developed by D. Bergman [8] and by G. Milton [61]. It was applied to various types of random textures [32], showing that third order bounds could generate separate domains in the complex plane.

We consider now random composites made of two phases A_1 (with fraction p) and A_2 (with fraction $q = 1 - p$) having a scalar real dielectric permittivity ϵ_1 and ϵ_2 (with $\epsilon_2 > \epsilon_1$) (the case of tensor permittivity is detailed in [22]). The composite is modelled by a stationary and isotropic random set A (with $A = A_1$ and $A^c = A_2$). The third order bounds are expressed in R^d as a function of ϵ_1, ϵ_2 and of the three-point probability $P(h_1, h_2) = P\{x \in A_1, x + h_1 \in A_1, x + h_2 \in A_1\}$. In R^d ($d = 2, 3$), using G. Milton's [61, 62] and S. Torquato's [81, 82] notations,

$$\epsilon_-/\epsilon_1 =$$
$$\frac{1 + ((d-1)(1+q) + \zeta_1 - 1)\,\beta_{21} + (d-1)\,(((d-1)q + \zeta_1 - 1))\,\beta_{21}^2}{1 - (q + 1 - \zeta_1 - (d-1))\,\beta_{21} + ((q - (d-1)p)\,(1 - \zeta_1) - (d-1)q)\,\beta_{21}^2}$$
$$\tag{22}$$

$$\epsilon_+/\epsilon_2 =$$

$$\frac{1 + ((d-1)(p+\zeta_1) - 1)\,\beta_{12} + (d-1)\,(((d-1)p - q)\,\zeta_1 - p)\,\beta_{12}^2}{1 - (1 + p - (d-1)\zeta_1)\,\beta_{12} + (p - (d-1)\zeta_1)\,\beta_{12}^2} \qquad (23)$$

$$\beta_{ij} = \frac{\epsilon_i - \epsilon_j}{\epsilon_i + (d-1)\epsilon_j} \qquad (24)$$

where the function $\zeta_1(p)$ introduced by G. Milton [62] is obtained from the probability $P(h_1, h_2) = P(|h_1|, |h_2|, \theta)$, with $u = \cos\theta$, θ being the angle between the vectors h_1 and h_2. We have:

$$\zeta_1(p) = \frac{9}{4pq} \int_0^{+\infty} \frac{dx}{x} \int_0^{+\infty} \frac{dy}{y} \int_{-1}^{+1} (3u^2 - 1)P(x, y, \theta)du \text{ in } R^3 \qquad (25)$$

$$\zeta_1(p) = \frac{4}{\pi pq} \int_0^{+\infty} \frac{dx}{x} \int_0^{+\infty} \frac{dy}{y} \int_0^{\pi} P(x, y, \alpha) \cos(2\alpha)\, d\alpha \text{ in } R^2 \qquad (26)$$

The integrals in equations (25) and (26) can be evaluated analytically, but most often numerically (after replacement of $P(x, y, \theta)$ by $P(x, y, \theta) - P(x)P(y)/p$, with $P(h) = P\{x \in A_1, x + h \in A_1\}$.

After exchanging the phases A_1 and A_2 we define the function $\zeta_2(q)$ with $\zeta_2(q) = 1 - \zeta_1(p)$. The function $\zeta_1(p)$ satisfies $0 \le \zeta_1(p) \le 1$. When $\zeta_1(p) = 1$ or $\zeta_1(p) = 0$ (and only in these cases), the two bounds ϵ_+ and ϵ_- coincide, and are equal to the upper (H-S) bound (for $\zeta_1(p) = 1$), or to the lower (H-S) bound (for $\zeta_1(p) = 1$). For these two cases, we obtain from the third order bounds an estimation of the effective permittivity. For given p, ϵ_1 and ϵ_2 (with $\epsilon_1 > \epsilon_2$), ϵ_+ and ϵ_- increase with $\zeta_1(p)$, so that higher values of the effective properties are expected. This gives a guideline to compare the properties of random media from the comparison of their corresponding geometrical function $\zeta_1(p)$. If the two phases A_1 and A_2 are symmetric, the case of $p = 0.5$ produces an autodual random set (the two phases having the same probabilistic properties, as for the mosaic model introduced below), for which the third order central correlation function is equal to zero. Therefore in two and three dimensions, the third order bounds of a symmetric medium present a fixed point at $p = 0.5$. In addition, it is known [43, 48, 49] that for an autodual random set in two dimensions the effective permittivity is exactly equal to the geometrical average of the two permittivities ($\epsilon^* = \sqrt{\epsilon_1\epsilon_2}$). More generally, it was proved by Georges Matheron that this result holds for a log-symmetrical random function, as for instance a permittivity field modelled by a lognormal random function [48, 49].

Recent developments on composites with a non linear behavior propose bounds using the function $\zeta_1(p)$, after introduction of comparison media with a linear behavior [67, 84].

Third order bounds of the Boolean model

We give now examples of application of third order bounds in the case of the Boolean model introduced by Georges Matheron [48, 51], and studied by many authors [25, 26, 75, 76, 77]. It is obtained by implantation of random primary grains A' (with possible overlaps) on Poisson points x_k with the intensity θ: $A = \cup_{x_k} A'_{x_k}$. Any shape (convex or non convex, and even non connected) can be used for the grain A'.

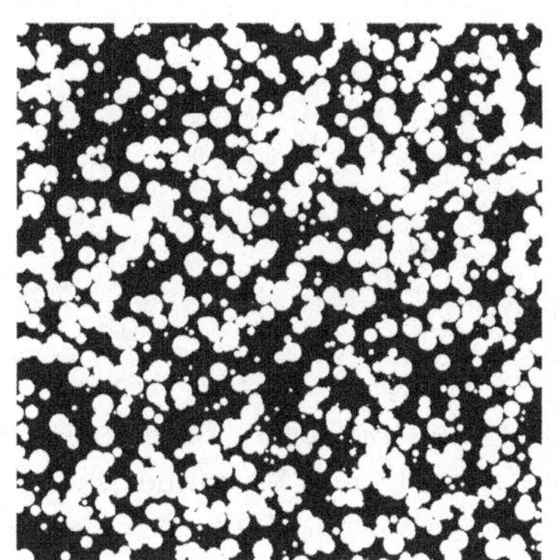

Fig. 1. Realization of a planar section of a Boolean model of spheres ($V_V = 0.5$).

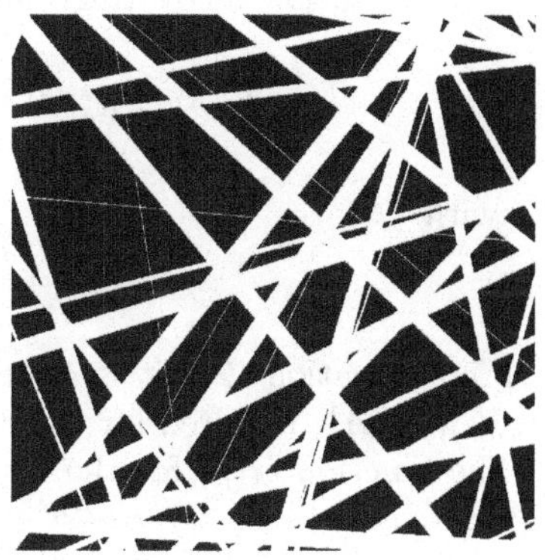

Fig. 2. Section of a Boolean variety model (thick flats) from Poisson flats in 3D.

An example is shown in fig 1 in the case of spheres. In [68], Poisson polyhedra are used for the grains of WC-Co composites, while in [35], parallelepiped random grains are used for the microstructure of gypsum. Replacing the Poisson points by Poisson varieties [51], enables us to generate random sets models with fiber or strata textures [26] (fig. 2). This version of the Boolean model is used in subsection 2.6. Anisotropic versions of this model can be easily proposed, from anisotropic primary grains, as for instance ellipsoids in R^3.

The Choquet capacity $T(K)$ of this model can be deduced from the fact that the number of primary grains hit by K follows a Poisson distribution with average $\theta\overline{\mu}_d(A' \oplus \check{K})$, where $\overline{\mu}_d$ is the average (over the realizations of A') of the Lebesgue measure in R^d and $\check{K} = \{-x, x \in K\}$; denoting $q = P\{x \in A^c\}$, we have:

$$T(K) = P\{K \cap A \neq \varnothing\} = 1 - Q(K)$$

$$= 1 - \exp\left(-\theta\overline{\mu}_d(A' \oplus \check{K})\right) = 1 - q^{\dfrac{\overline{\mu}_d(A' \oplus \check{K})}{\overline{\mu}_d(A')}}$$

$$(27)$$

The three point probability for the random set A^c is given by (A'_h being the set A' after translation by vector $\mathbf{h}$)

$$Q(h_1, h_2) = P\{x \in A^c, x + h_1 \in A^c, x + h_2 \in A^c\} \tag{28}$$
$$= \exp\left(-\theta\overline{\mu}_d(A' \cup A'_{-h_1} \cup A'_{-h_2})\right)$$
$$= q^{3 - r(h_1) - r(h_2) - r(h_2 - h_1) + s(h_1, h_2)}$$

with the geometrical covariogram

$$K(h) = \overline{\mu}_d(A' \cap A'_{-h}) \text{ and } r(h) = \frac{K(h)}{K(0)} \tag{29}$$

and with

$$s(h_1, h_2) = \frac{\overline{\mu}_d(A' \cap A'_{-h_1} \cap A'_{-h_2})}{K(0)} \tag{30}$$

From equation (28) it is easy to derive $P(h_1, h_2) = P\{x \in A, x + h_1 \in A, x + h_2 \in A\}$, required for the calculation of the third order bounds.

The function $\zeta_1(p)$ for the Boolean model is obtained by numerical integration of equations (25) or (26,) using equations (28) and (30). With a good approximation, we get a linear expression with coefficients α and β:

$$\zeta_1(p) \simeq \alpha p + \beta$$

$\zeta_1(0)$ is obtained when $p \to 0$. We have $\zeta_1(0) = a$ with a given by equation (33) below for the random grain A' (or for any size distribution of A'):

- in R^3 $\zeta_1(0) = 0$ for spheres, and $\zeta_1(0) = \frac{1}{4}$ for Poisson polyhedra
- in R^2, $\zeta_1(0) = 0$ for disks and $\zeta_1(0) = 3 - 4\ln 2 \simeq 0.2274$ for Poisson polygons.

The complementary function $\zeta_2(p) = 1 - \zeta_1(1 - p)$ is obtained by exchange of the two phases:

$$\begin{aligned}
\zeta_1(p) &\simeq 0.5615p \text{ for spheres } [9, 79, 80, 81] \\
\zeta_2(p) &\simeq 0.5615p + 0.4385 \text{ for (spheres)}^c \\
\zeta_1(p) &\simeq \tfrac{2}{3}p \text{ for discs in } R^2 [30, 46, 81] \\
\zeta_2(p) &\simeq \tfrac{2}{3}p + \tfrac{1}{3} \text{ for (discs)}^c \text{ in } R^2 \\
\zeta_1(p) &\simeq 0.5057p + 0.2274 \text{ for Poisson polygons in } R^2 [46] \\
\zeta_2(p) &\simeq 0.5057p + 0.2669 \text{ for (Poisson polygons in)}^c R^2
\end{aligned} \tag{31}$$

For the Boolean model, different sets of bounds are obtained when exchanging the properties of A and A^c (which is not the case for the second order H-S bounds), as seen in equation (31). This is illustrated for the case of the thermal conductivity λ of ceramic materials [66], modelled by two Boolean models of spheres with a constant radius (in fig. 3 3 (a) and 3 (b), the morphology of the low conducting bright phase is exchanged): here, the third order bounds increase when $\lambda(x) = \lambda_1 > \lambda_2$ for $x \in A^c$ (fig. 4. This is due to the fact that it is easier for the "matrix" phase A^c to percolate than for the overlapping inclusions building A. We see here that third order bounds

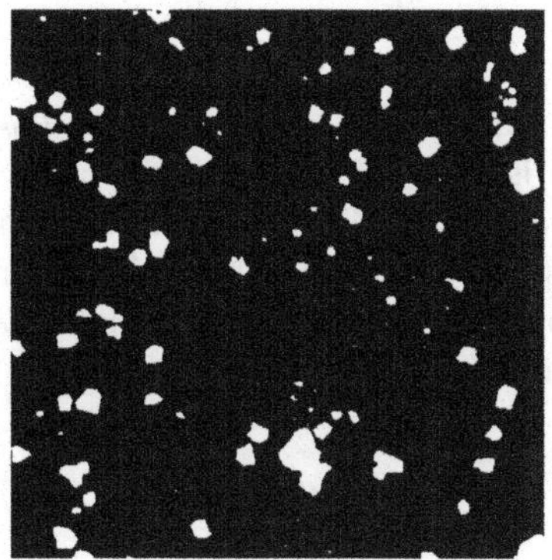 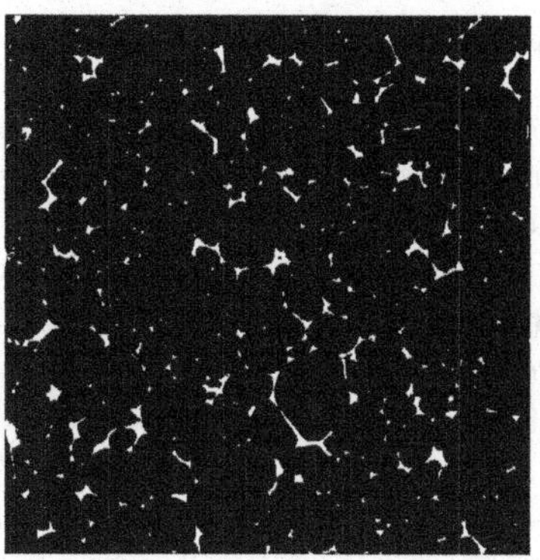

Fig. 3. AlN textures (AlN in black; Y rich phase in white).

are sensitive to the morphology in a more accurate way than the H-S bounds [19].

In [30, 46], various morphologies are studied in two and three dimensions, including the two-phase dead leaves model [31]. Two-scale hierarchical models accounting for local fluctuations of the volume fraction of a Boolean model can be optimized with respect to the third order bounds, as seen below in subsection 2.6.

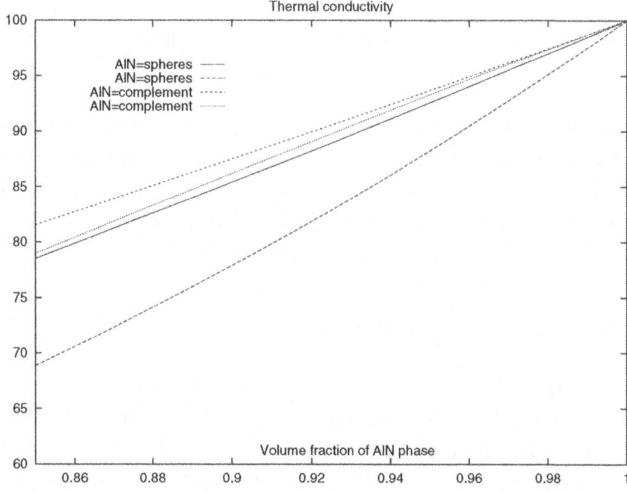

Fig. 4. Bounds of the thermal conductivity (in W/(m.K) of AlN materials ($\lambda_1 = 100$ for AlN $\lambda_2 = 10$ for the second phase). 'AlN = spheres': Boolean model of spheres for the AlN (black phase in fig. 3); 'AlN = complement': Boolean model of spheres for the Y rich phase (in white in fig. 3.

The mosaic model

The mosaic (or cell) model [26, 49, 59] is obtained from a random tessellation of space into classes (each class or grain is denoted by A'). Any class is allocated to the random set A with the probability p and to the set A^c with the probability $q = 1 - p$. The medium is symmetrical in A and A^c (after permutation of p and q). The third order bounds depend on a parameter G introduced by Miller [59, 60], obtained from the function $s(|h_1|, |h_2|, \theta)$:

$$G = \frac{1}{9} + \frac{1}{2} \int_0^{+\infty} \frac{dx}{x} \int_0^{+\infty} \frac{dy}{y} \int_{-1}^{+1} (3u^2 - 1)s(x, y, \theta)du \text{ in } R^3 \qquad (32)$$

we have $\frac{1}{9} \leq G \leq \frac{1}{3}$ in R^3. From the equation (25) we obtain [59, 60, 62]:

$$\zeta_1(p) = p + a(q - p) \text{ with } a = \frac{9G - 1}{2}$$
$$\zeta_2(p) = \zeta_1(p) \qquad (33)$$

and $\zeta_1(p)$ is a linear function of p, with the slope $1 - 2a$ ($0 \leq a \leq 1$).

The coefficient G was obtained for several types of tessellation with spheroidal grains, and are given in the next table, where needles are cylinders with infinite length, and plates are domains between two parallel planes (it is rather difficult to imagine an isotropic random tessellation of space from these two types of grains).

grain	spheres	needles	plates
G	1/9	1/6	1/3

The parameter G is not sensitive to the scale of the grain A', and is therefore invariant for a population of similar grains. Using equation (33), we get $\zeta_1(p) = p$ for spheres, and $\zeta_1(p) = 1 - p$ for plates. In [46, 49] are given the functions $\zeta_1(p)$ for the Poisson mosaic, built from a tessellation of space by Poisson lines in R^2 and by Poisson planes in R^3.

Combination of basic random sets

More complex models can be built from elementary ones. We consider here multi-scale models. An interesting construction is obtained for the union or the intersection of independent random sets [30, 32]. For $A = A_1 \cap A_2$ with $p_1 = P\{x \in A_1\}$, $p_2 = P\{x \in A_2\}$, and three points probability $P_1(h_1, h_2)$ and $P_2(h_1, h_2)$, we have:

$$P(h_1, h_2) = P_1(h_1, h_2)P_2(h_1, h_2)$$

for widely separate scales (in particular for A_2 with a lower scale), we get the following approximate relationship between $\zeta_{H1}(p)$ (corresponding to $A_1 \cap A_2$), $\zeta_{11}(p_1)$ (corresponding to A_1), and $\zeta_{12}(p_2)$ (corresponding to A_2) [30]:

$$pq\zeta_{H1}(p) \simeq p_1 p_2 q_2 \zeta_{12}(p_2) + p_2^3 p_1 q_1 \zeta_{11}(p_1)$$

If the two structures are built from the same random set with $p_1 = p_2$, we obtain after n iterations of the process involving intersections:

$$\zeta_{H1}^{(n)}(p) = \frac{1+p}{1+p^{1/n}}\zeta_1(p^{1/n})$$

for $n \to \infty$ we get asymptotically $\zeta_{H_1}^{(n)}(p) \to \dfrac{1+p}{2}\zeta_1(1)$. Starting from a mosaic model with $a = 1$ (plates in R^3), or starting from a Boolean variety of dilated flats [26] (as in fig. 2, we have $\zeta_1(1) = 0$ and therefore $\zeta_{H_1}^{(\infty)}(p) = 0$. The complementary set is a union of mosaics (or a union of dilated flats), for which $\zeta_{H_1}^{(\infty)}(p) = 1$. The obtained limit structure admits two equal third order bounds [36] for the effective permittivity ϵ^*. If $\epsilon_2 > \epsilon_1$ is attributed to the union of Boolean variety models, we get

$$\epsilon^* = \epsilon_- = \epsilon_+ = \epsilon^{HS+}$$

and if $\epsilon_2 > \epsilon_1$ is attributed to the intersection of Boolean models, we get

$$\epsilon^* = \epsilon_- = \epsilon_+ = \epsilon^{HS-}$$

We obtain the construction of an optimal structure corresponding to these bounds, based on the mosaic model, or on the Boolean variety model. This result is valid for a porous medium (with $\epsilon_1 = 0$), where the upper (H-S) bound becomes the effective permittivity. We have similar results when considering the case of the linear elasticity [36]. These third order bounds were used to predict the elastic properties of gypsum in [35]. The geometrical construction of these multiscale structures is unknown for the mosaic, but is easily obtained and simulated for the Boolean layered random model (fig. 2). This geometry differs from two other kinds of microstructures showing the same optimal properties: the Hashin sphere assemblage [18], made of coated spheres with the two phases and different radii in order to cover the 3D space, and the layered materials proposed by G. Milton [63].

3 Fluctuations of properties, and the Representative Volume Element

3.1 Homogenization of random media from numerical simulations

An alternative way to solve the problem of homogenization is based on the use of numerical solutions of the corresponding partial differential solutions, before estimating the effective properties by spatial averaging of the solution. The input image can be a 3D image of the studied medium obtained by various techniques, like confocal microscopy or microtomography. When such a 3D reconstruction is not available, use can be made of simulations of realizations of an appropriate model of random structure, after identification from 2D images obtained on sections. Numerical simulations open the way to so-called "Digital materials", for which can be predicted a complex non linear behavior.

One of the techniques used to solve the homogenization problem is based on the finite element method, as illustrated in [2]. A convenient way to solve numerically the homogenization problem is to use periodic boundary conditions applied on a periodic cell. For this purpose, we developed simulations of periodic random media, like the Voronoi 3D polycrystal in fig. 5 [14]. Beside the finite element method, efficient iterative techniques operating by Fast Fourier Transform on periodic media were recently developed for micromechanics problems [15, 16, 37]. This numerical approach is based on the convolution form of the Lippman-Schwinger equation (14).

Let us mention that in a pioneering work, Georges Matheron proposed forty years ago a way to estimate the effective permeability by generating appropriate random walks from the permeability field [47].

Fig. 5. Simulation of a 3D Voronoi mosaic, with periodic boundary conditions [14].

3.2 Fluctuations of effective properties

When using numerical simulations as in 3.1, a natural question arises: what is the representativeness of the effective property estimated on a bounded domain of a microstructure? In other words, what is the size of a so-called "Representative Volume Element" RVE [39]? A similar question appears for engineering purpose, when working on parts with dimension of the same order as the microstructure: this occurs for instance for some devices like microbeams in micro-electronics, where a few grains only are seen across the beam [1]. In that case the effective properties of different parts (like the elastic moduli of the beam) fluctuate, so that a specific procedure may be required for the quality control in production, with a more or less important amount of rejected parts.

To answer to these questions, higher order statistical information than average values is required, but unfortunately is not available.

The second order moment of the field over an infinite domain can be worked out when the effective property is known [8, 10, 44, 49, 57, 78], but it does not provide any information on the fluctuations of the average made over a finite domain. This moment of the field is obtained by partial derivation of the effective permittivity ϵ^* with respect to the permittivity of phases, starting from equation (16).

A first order formal estimation of the variance of the average flow rate in porous medias over a finite domain is given by Georges Matheron [48]. It is derived from the first order term of the perturbation expansion of the flow rate. Bounds of the variance are worked out in the same reference, for radial flows and specific covariances of the permeability random function.

Recent developments use a geostatistical approach based on the experimental determination of the integral range [50, 52] from numerical simulations [12, 39, 40].

The integral range

We consider fluctuations of average values over different realizations of a random medium inside the domain B with the volume V. In geostatistics, it is well known that for an ergodic stationary random function $Z(x)$, one can compute the variance $D_Z^2(V)$ of its average value $\bar{Z}(V)$ over the volume V as a function of the central covariance function $Q(h)$ of $Z(x)$ [50] by :

$$D_Z^2(V) = \frac{1}{V^2} \int_V \int_V Q(x - y)\, dx dy \tag{34}$$

For a large specimen (with $V \gg A_3$), equation (34) can be expressed to the first order in $1/V$ as a function of the integral range in the space R^3, A_3, by

$$D_Z^2(V) = D_Z^2 \frac{A_3}{V} \tag{35}$$

$$\text{with } A_3 = \frac{1}{D_Z^2} \int_{R^3} Q(h)\, dh \tag{36}$$

where D_Z^2 is the point variance of $Z(x)$ and A_3 is the integral range of the random function $Z(x)$, defined when the integral in equations (34) and (36) is finite. The asymptotic scaling law (35) is valid for an additive variable Z over the region of interest V.

In the case of a two-phase material with a scalar dielectric permittivity Z_1 for phase 1 (with the volume fraction p), and Z_2 for phase 2, the point variance D_Z^2 of the random variable Z is given by :

$$D_Z^2 = p(1 - p)(Z_1 - Z_2)^2 \tag{37}$$

To estimate the effective dielectric permittivity ϵ^* from equations (8) and (9), we have to compute the averages $\langle D \rangle$ or $\langle E \rangle$. For an appropriate choice of the constant electric field E_0 applied on ∂B (namely of the constant displacement field D_0) and in the isotropic case, ϵ^* is obtained from the estimations of a single component of $\langle D \rangle$ or $\langle E \rangle$. Therefore the variance of the effective property ϵ^* follows the equation (35) when the integral range A_3 of the relevant field is known, which requires the stationarity of the fields D and E, as obtained when using homogeneous boundary conditions (see subsection 2.5).

Since the theoretical covariance of the field (E or D) is not available, the integral range can be estimated according to the procedure proposed by Georges Matheron for any random function [52, 54]: working with realizations of $Z(x)$ on domains B with an increasing volume V, we can estimate the parameter A_3 by fitting the obtained variance according to the expression (35).

Practical determination of the size of the RVE

When considering a material as a realization of a random set or of a random function RF, the idea that there exists one single possible minimal RVE size must be left out. Instead, the size of a RVE can be defined for a physical property Z, a contrast and, above all, a given precision in the estimation of the effective properties depending on the number of realizations that one is ready to generate. By means of a standard statistical approach, the absolute error ϵ_{abs} and the relative error ϵ_{rela} on the mean value obtained with n independent realizations of volume V are deduced from the interval of confidence by:

$$\epsilon_{abs} = \frac{2D_Z(V)}{\sqrt{n}}; \ \epsilon_{rela} = \frac{\epsilon_{abs}}{Z} = \frac{2D_Z(V)}{Z\sqrt{n}} \tag{38}$$

The size of the RVE can now be defined as the volume for which for instance $n = 1$ realization is necessary to estimate the mean property Z with a relative error $\epsilon_{rela} = 1\%$, provided we know the function $D_Z(V)$. Alternatively, we can decide to operate on smaller volumes (provided no bias is introduced), and consider n realizations to obtain the same relative error. In the case of effective elastic moduli, the exact mean value and variance for a given domain size are *a priori* unknown. Using the equation (38), the absolute error on the mean value can be evaluated. This methodology was applied to the case of the dielectric permittivity of various random media [15], and to the elastic properties and thermal conductivity of a Voronoï mosaic [39], and of real microstructures [40].

From various simulations [15, 39, 40], it was shown that the dispersion of the results decreases when the size V of the domain increases for all boundary conditions. The obtained mean values depend on the volume size, but also on the type of boundary conditions. For any property (dielectric permittivity, elastic moduli, or thermal conductivity), the average values converge towards the same limit for large volumes V, which is the wanted effective property. We noticed that the mean value given by the periodic boundary conditions varies slightly as a function of the size of the domain, as compared to the other boundary conditions. Finally, an important bias, due to the effect of the boundaries ∂B, is found in the mean value given by all boundary conditions for small volume sizes, the value being different from the effective one obtained for large specimens, so that the law (35) must be applied with some caution. For small volumes, the average moduli obtained by simulations depend on the boundary conditions: for instance, in the case of the dielectric permittivity, imposing E_0 on ∂B produces results close to the upper Wiener bound, while imposing D_0 on ∂B gives results close to the lower Wiener bound. This bias is well–known [23, 65, 71], and corresponds to the case of the two homogeneous trial fields from which these bounds are derived. It must be taken into account for the definition of the RVE. The result is that the mean value computed on small specimens cannot represent the effective response for the composite material even using the periodic boundary conditions and a sufficient number of realizations. From simulations, it appears also that the mean value obtained with the periodic boundary conditions is unbiased for smaller sizes of B than for the other boundary conditions, but it leads in general to higher variances than for the two other conditions. This requires a larger number of simulations to get a given precision for the effective property.

To illustrate this point, we report in 1 the obtained integral ranges from the estimation of the dielectric permittivity of a 2D random set (a two phase symmetrical dead leaves model [31] with discs with radius 5) [15]. The electrostatic problem was solved for periodic boundary conditions. From the scheme given by equation (14), we obtained the field $E(x)$ on realizations $\epsilon(x)$ by iterations of Fast Fourier transforms on images, starting with an initial approximation of the solution given by the average $\overline{E}$ in the second member of the equation. The effective permittivity ϵ^* was estimated on images of $\epsilon(x)$ with increasing sizes and for an increasing contrast of properties. Since in this case the effective permittivity is obtained by a geometrical average, it is easy to check for which minimal size the simulations give an unbiased estimate of the effective permittivity. For this model, the integral range of ϵ^* is lower than the integral range of the volume fraction (46), and seems to decrease slightly with the contrast (however only 7 realizations were used for the 512^2 simulations). In another situation (3D Voronoï mosaic in [39]), the integral range of effective properties (elastic moduli and thermal conductivity) was of the order of twice the integral range of the volume fraction.

Table 1. Variation of the empirical integral range of the dielectric permittivity with the contrast of a two-phase autodual 2D dead leaves model.

Contrast	10	100	1000	10000
Theoretical ϵ^*	3.162	10	31.622	100
Minimal size	64^2	128^2	256^2	512^2
Integral range of ϵ^*	41	31.36	25.7	14.36

4 Fracture statistics models

Usually, a wide dispersion of strength is observed on brittle materials like ceramics or like steels at low temperature. In addition, size effects are observed in experiments: there is most often a decrease of the average strength with the size of specimens. This is of practical importance for engineers, who have to design parts with a given reliability. Therefore, models based on random structures and accounting for the dispersion of strength and for scale effects are required [26, 27, 28]. The main purpose is to estimate the probability of fracture of a specimen under given loading conditions, as a function of the probabilistic properties of a population of defects. Differently from the case of effective properties involving space averaging, there is in fracture a great sensitivity of the macroscopic fracture behavior to local defects, and mainly to the tail of their probabilistic distribution. One of the main points of fracture statistics approaches deals with scale effects. This point is of major importance when we need to predict the strength of large parts (like in aeronautics, dams or buildings in engineering...) from data available at a laboratory scale; the fact that the average strength may decrease (or sometimes increase) with the size of parts must be known and evaluated in practice. This can be accounted for by models based on the reproduction, at a point scale, of the variations of a fracture criterion (critical stress σ_c, or critical stress intensity factor K_{Ic}, local fracture energy γ, for brittle materials). These models are based on different macroscopic fracture criteria: the weakest link assumption is suitable for the sudden fracture of brittle materials; the models with a damage threshold allow several sites with a crack initiation before fracture; models based on the fracture energy (Griffith's criterion) account for the propagation and arrest of cracks in a random medium; finally, models of random damage mimic the generation of damage zones in quasi-brittle materials. Formally, the different assumptions correspond to various changes types of change of support of the random point-scale criterion:

- Change of support of $\sigma_c(x)$ by the infimum operator $\wedge$ for the weakest link model;

- Change of support of $\sigma_c(x)$ by convolution (moving average) for the damage threshold;
- Change of support of $\gamma(x)$ by the supremum operator $\vee$ along the crack path for the crack arrest criterion.

The strategy followed to solve this problem is to work out efficient models for which these changes of support are available. For every family of models, the probability of fracture is computed as a function of the loading conditions and of the parameters of the selected random structure models. Some interesting aspects for applications, such as the prediction of expected scale effects, are derived. The proposed models can be tested at different scales (including the microscopic scale, by use of image analysis). The diversity of the obtained theoretical distributions for fracture statistics and for scale effects offers new possibilities for the microstructure based interpretation and modelling of mechanical data obtained on materials. A more detailed presentation is given in [36].

4.1 Weakest link model and Boolean random functions

Consider first the case of the weakest link model, where there is a sudden propagation of a crack after its initiation. In this case, the fracture statistics is governed by the most critical defects. The fracture probability is expressed by

$$P\{\text{non fracture}\} = P\{\inf(\sigma_c(x) - \sigma(x)) \geq 0\} \tag{39}$$

where $\sigma_c(x)$ is the random critical stress, $\sigma(x)$ is the applied stress field, and inf is the minimal value over the sample.

Explicit expressions of the fracture probability are available for uncoupled critical and applied stress fields and for specific models (like Boolean RF). They are more general than the standard Weibull model. The expected scale effect is a decrease of strength (to a constant for large samples) with the size of specimens.

For illustration, consider a medium where defects are introduced in a matrix with an infinite strength ($\sigma_c = +\infty$). For instance defects with a critical stress σ_c being a random variable σ'_c inside a closed random set A'_0, implanted in space on Poisson points x_k with the intensity $\theta(\sigma_c)$. We take for σ_c the minimum of the values of σ'_c on overlapping grains, building by this process a infimum ($\wedge$) Boolean RF [24, 25, 26, 75], as illustrated in fig. 6.

In the present case, the resulting field $\sigma_c(x)$ is a mosaic with domains where σ_c is constant. In brittle tensile fracture, we can restrict the purpose to the scalar case with $\sigma > 0$ (σ being the maximal main stress). For a uniform applied stress field, the probability distribution of the apparent fracture stress $\sigma_R = \inf_{x \in B}\{\sigma_c(x)\}$ of a specimen B is obtained from equation (40), generalizing equation (27) obtained for the Boolean random set:

$$P\{\sigma_R \geq \sigma\} = P\{\text{non fracture of } B\} = \exp(-\overline{\mu}(A'_0 \oplus \check{B})\Phi(\sigma)) \tag{40}$$

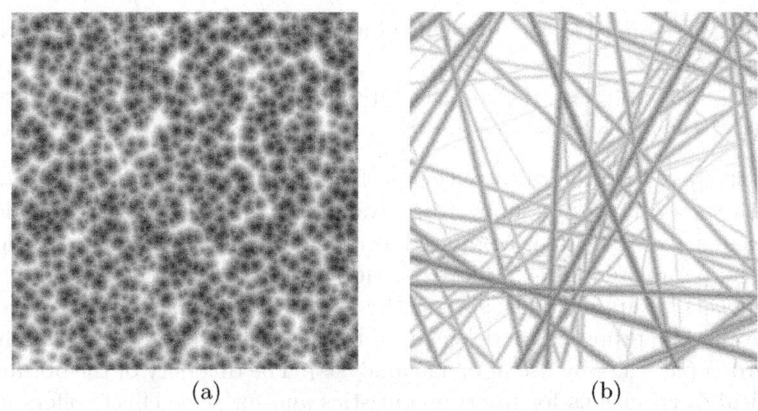

(a)   (b)

Fig. 6. Infimum Boolean RF with cone primary grains (a) and Boolean Variety RF (b). The infimum of observed grey values (dark points corresponding to low values, and bright points to high values) is taken in every point of the image.

where $\bar{\mu}$ is the volume V for grains, $\pi/4S$ (S being the surface area) for fibers, and the integral of mean curvature M for strata, averaged over the realizations of the random set A_0', and $\Phi(\sigma)$ is the intensity of the corresponding Poisson point process (average number of defects per unit volume with strength lower than σ in the case of grain defects):

$$\Phi(\sigma) = \int_0^\sigma \theta(t)dt \tag{41}$$

A model of this kind was used for fracture of ceramics, the function $\Phi(\sigma)$ being estimated by image analysis by some transformation of the distribution function of defects (inclusions and porosities ranging from 20 μm to 70 μm) [6].

When applying a non uniform stress field $\sigma(x)$ and for point defects,

$$P\{\text{non fracture of } B\} = \exp(-\int_B \Phi(\sigma(x))dx) = \exp(-\Phi(\sigma_{eq})) \tag{42}$$

with $\Phi(\sigma_{eq}) = \int_B \Phi(\sigma(x))dx$. This formulation with the equivalent stress σ_{eq} makes possible to put together results of different types of fracture tests, for the identification of a model, as made in the "local approach" [7] for the Weibull statistics, σ_{eq} being the so-called Weibull stress.

For defects with $\theta(\sigma) = \theta_1 m(\sigma - \sigma_0)^{m-1}$ and $m > 1$ and $\sigma > \sigma_0$, we have $\Phi(\sigma) = (\sigma - \sigma_0)^m \theta_1$, and in the case of a homogeneous stress field, σ_R follows a Weibull distribution with $\sigma_u^m = 1/\theta$:

$$P\{\sigma_R \geq \sigma\} = \exp(-\bar{\mu}_d(A_0' \oplus \breve{B})\left(\frac{\sigma - \sigma_0}{\sigma_u}\right)^m) \tag{43}$$

This distribution is well known in the practice of fracture statistics, in the case of point defects, but most users of this model for experimental data are not aware of the underlying Boolean RF model. When $\sigma_0 = 0$, the coefficient of variation of σ_R, $\sqrt{Var[\sigma_R]}/E[\sigma_R]$ does not depend on V, which can be easily checked from data. Scale effects can be seen from the variation of the median strength σ_M, as a function of the specimen volume V:

$$\sigma_M = \sigma_0 + KV^{-1/m} \tag{44}$$

When replacing grains on Poisson points by fibers of strata, the scale effect depend on the surface or on the integral mean curvature of the specimen, which differs from the standard model.

For a mixture of two populations of defects following Weibull distributions with parameters $(\theta_1, \sigma_{01}, m_1)$ and $(\theta_2, \sigma_{02}, m_2)$, resulting into a bimodal Weibull distribution when a uniform stress field is applied:

$$\Phi(\sigma) = \theta_1(\sigma - \sigma_{01})^{m_1} + \theta_2(\sigma - \sigma_{02})^{m_2} \tag{45}$$

Other distributions functions can be derived from other functions $\theta(\sigma)$, like the Pareto distribution, the sigmoid distribution, and so on [26, 28, 36].

4.2 Critical defect fraction

We can relax the weakest link assumption in two different ways, letting the defects (where a potential crack can initiate) reach a critical volume fraction of the domain where $\sigma_c(x) < \sigma(x)$ or a critical density (number of defects per unit volume) [26, 28, 36]. This assumption involves a change of scale by convolution. In both cases, no size effect is expected for the median strength, while the dispersion of the observed strength decreases with the size of the specimen.

4.3 Griffith crack arrest criterion

Fracture statistics models were derived for two dimensional media with a random fracture energy $\Gamma(x)$ [13, 28, 36]. These models can be applied to thin specimens, where the microstructure changes can be neglected across the thickness. The extension to general 3D crack fronts is an open problem, since in that case various fracture criteria based on the fracture energy are generally proposed for homogeneous materials. In the two dimensional case, for a fracture path with length a in a material with fracture energy γ, the energy used in the fracture path is equal to $2\gamma a$. According to the Griffith criterion, a possible fracture path $P(s, d)$ connecting a source s to a destination d must satisfy for every point x of the path:

$$2\Gamma(x) \leq G(x) \tag{46}$$

where the energy release rate $G(x)$ depends on the location of the crack front x, on the loading conditions, and on the crack path connecting s to x. From equation (46) a potential crack path P must satisfy

$$\vee\{2\Gamma(x) - G(x); x \in P\} \leq 0 \tag{47}$$

where $\vee$ is the supremum. For a random medium, the condition (47) is a random event, and

$$P\{\text{fracture}\} = P\{\vee\{2\Gamma(x) - G(x); x \in P\} \leq 0\} \tag{48}$$

where the fracture path $P(s,d)$ separates the specimen in two parts. In general, $P\{\text{fracture}\}$ is difficult to calculate, due to a very complex crack path geometry, related to the microstructure. Usually $G(x)$ is a realization of a random function correlated to $\Gamma(x)$. For simplification, the following situations were considered: straight crack propagation according to the mode I opening of the crack (which propagates in a plane orthogonally to the applied tensile stress) under various loading conditions, resulting in stable or unstable crack propagation. With this assumption, equation (48) implies a change of support by the supremum. The case of a microcrack initiation on defects, followed by an advance or arrest of the crack, was also examined. For a given model of RF concerning the fracture energy $\Gamma(x)$, the following types of probability distributions can be expressed as a function of the model and of the loading conditions: fracture stress σ_R, toughness (standard G_c, and at crack arrest G_a), critical length of defects (unstable microcrack a_c or crack after arrest l_a). Since these probability distributions are related in a coherent way, it is possible to use them in practice to test models from data at different scales:

- mechanical data $(\sigma_R, G_c, \sigma_a, G_a)$,
- size distribution of cracks observed for given loading conditions (a_c, l_a).

Calculations made on a Poisson mosaic, and on a Boolean random mosaic, with the distribution function $F(\gamma) = P\{\Gamma < \gamma\}$ provide various scale effects. An interesting example concerns a model of random fracture energy $\Gamma(x)$ corresponding to the situation of a polycrystal, every crystal showing its own fracture energy. A convenient model for this case is the Poisson mosaic: we start from a Poisson random tessellation with parameter λ delimiting grain boundaries (Poisson lines in the plane in 2D); independent realizations of the fracture energy Γ are attributed to each grain of the tessellation, with the cumulative distribution function

$$F(\gamma) = P\{\Gamma < \gamma\}$$

As a result, it is known that $F(\gamma)$ is the distribution function of the RF $\Gamma(x)$. We consider now the unstable propagation of a surface crack with the initial length $2a$ in a random medium, until it reaches the length $2(a+b)$. For any loading where G increases with x,

$$P_{a+b}\{fracture\} = F(G(a)/2)\exp\left[-\lambda \int_{2a}^{2(a+b)} (1 - F(G(u)/2))du\right] \quad (49)$$

From equation (49), the fracture probability $P_{a+b}\{fracture\}$ increases (and converges to 1)

- with the crack length $2a$, or the applied stress σ
- for lower λ (corresponding to a coarser microstructure): small grains improve the ability to resist the crack growth, as a result of a higher probability to meet grains with a large fracture energy γ, blocking the crack propagation.

The predicted scale effect depends on the distribution $F(\gamma)$ for specimens with a variable size (constant a/b):

- for a finite range of $F(\gamma)$ ($F(\gamma) = 1$ for $\gamma \geq \gamma_c$), $P\{fracture\} \to 1$ when $a \geq a_c$ with $2a_c = 4\gamma_c E/(\pi\sigma^2)$
- for a distribution $F(\gamma)$ with a tail for large fracture energies:

$$1 - F(\gamma) \simeq \gamma^{-\alpha} \text{ for } \gamma \to \infty$$

the scale effects depend on the positive coefficient α

  - if $\alpha = 1$, there is asymptotically no scale effect for $P\{fracture\}$

$$P_{a+b}\{fracture\} = y^{\lambda c} \text{ with } y = a/(a+b) \text{ and } c = E/(2\pi\sigma^2) \quad (50)$$

  - if $\alpha \neq 1$, the large scale behavior of $P\{fracture\}$ is given by

$$P_{a+b}\{fracture\} = \exp\left[-\lambda(2c)^\alpha a^{1-\alpha}(1 - y^{\alpha-1})/(\alpha - 1)\right]$$

Therefore, if $\alpha < 1$, corresponding to a slow growth of $F(\gamma)$ towards 1, the crack is almost surely stopped during its straight propagation by reinforcing obstacles with a high fracture energy. If $\alpha > 1$, $P_{a+b}\{fracture\} \to 1$ for increasing sizes, as in the case of a distribution function of a bounded RV γ.

The scale effect for the median strength σ_M of similar specimens (y remaining constant) depends on α:

- if $\alpha = 1$, there is no scale effect

$$\sigma_M \simeq \sqrt{-\log y}$$

- if $\alpha \neq 1$,

$$\sigma_M \simeq L^{\frac{1-\alpha}{2\alpha}} \left|1 - y^{\alpha-1}\right|^{1/(2\alpha)}$$

where $L = a + b$ is the size of the specimen: σ_M increases with L for $\alpha < 1$ and σ_M decreases with L for $\alpha > 1$.

The probability distributions of the other variables (toughness G_c and G_a, critical length l_c and length at arrest l_a) are found in [28]. Similar conclusions, concerning size effects, were obtained for a model where $\Gamma(x)$ is a Boolean random function [27].

4.4 Models of random damage

The occurrence of damage in a loaded material is followed by a progressive loss of integrity, ending by its ruin. This damage is the result of a local degradation of weaker parts in the material, that can be generated by the presence of defects, which can be accounted for by a probabilistic approach. This is generally made in the case of brittle materials by the weakest link model, as described above. In what follows, we propose probabilistic models of damage based on the presence of random defects. After a presentation of our main assumptions, we consider the case of a random damage generated under a homogeneous loading. More details are given in [34].

We consider a homogeneous elastic material (with the elasticity tensor C) containing point defects. Under the action of a stress field $\sigma(x)$ ($\sigma(x) = \sigma$ in the case of a uniform load), any defect located at x, with a critical stress σ_c lower than $\sigma(x)$, generates a damaged zone with volume v centered in x. The damaged zone can be replaced by a domain v with a constant elasticity tensor C_D. We consider here the extreme case where damage zones behave like pores ($C_D \equiv 0$). This is equivalent to the notion of statistical volume element (containing at least a critical defect) introduced in section 4.5 for the finite elements simulation of damage in composites [29]. One model of this type is obtained as a tessellation of space into statistically independent cells with a volume v, the cell covering x remaining unbroken with probability $p(\sigma(x))$, and generating an evolving two-phase mosaic model. Another model considers point defects distributed according to a Poisson point process, $\Phi(\sigma)$ being the average number of defects with a critical stress lower than σ, per unit volume of material. The function $\Phi(\sigma) = \left(\dfrac{\sigma}{\sigma_u}\right)^m$ gives the Weibull distribution for the "weakest link" model. Here, critical defects at Poisson points x_k generate damaged zones A'_k (namely pores in what follows) with centers in x_k. Therefore the damaged part of a specimen builds a Boolean model with primary grain A'. This primary grain can be random, and can be oriented in a given direction, in the case of anisotropic damage. In examples below, we consider isotropic and uniformly damaged zones, generating a Boolean model with spheres having a constant volume v. If overlaps between damage zones at different points x_k are forbidden, a hard-core process is generated.

The next assumption in the model concerns the redistribution of stress in the non damaged zone. As in the case of the damage of bundles of fibers, we assume a uniform load sharing, the damage generating a uniform amplification of stresses over the non damaged parts. With this simplification, the damage behaves as if it was totally diffuse, no localization being permitted. This is typical of a quasi brittle material, in opposition to a brittle material submitted to the weakest link assumption.

The progression of damage in the material processes as follows: defects are ranked with an increasing severity $(\sigma_c - \sigma(x))$; the most critical defect is converted into a damaged domain; the load is then redistributed over the matrix, and the process goes on for the remaining defects.

We consider now a continuous model in the case of large homogeneous specimens. The average volume fraction of the specimen with a critical stress larger than σ is given by $p(\sigma)$. For the previous cell model, this is the probability for one cell to remain undamaged. For the Poisson point defects, generating a Boolean model, we have:

$$p(\sigma) = \exp\left(-\Phi(\sigma)v\right) \tag{51}$$

Consequently, when the non damaged parts of the material are loaded with the homogeneous stress field σ, the average macroscopic stress σ_M acting on the material is:

$$\sigma_M = \sigma p(\sigma) \tag{52}$$

From equation (52), the average macroscopic stress is expected to reach a maximum σ_{ult}, when it exists, since $p(\sigma)$ decreases with σ. For a specimen of volume V, fluctuations of $p(\sigma)$ (and therefore of $p(\sigma_{ult})$) are expected. Its variance $Var(\sigma, V)$ is given, as a function of the covariance $Q(h)$ of the non damaged material (with $Q(0) = q = p(\sigma)$), by

$$Var(\sigma, V) = \frac{1}{V^2} \int_V \int_V (Q(x - y) - q^2)\, dxdy \tag{53}$$

For a large specimen (with $V \gg A_3$), equation (53) can be expressed as a function of the integral range in the space R^3, A_3, by

$$Var(\sigma, V) = \frac{q(1 - q)A_3}{V} \tag{54}$$

$$\text{with } A_3 = \frac{1}{q(1 - q)} \int_{R^3} (Q(h) - q^2)\, dh$$

For a model with cells having a constant volume v, we get $A_3 = v$. For a Boolean model of spheres, we illustrate this point in 2 by the value of two standard deviations (giving the confidence interval of $q \pm 2\sqrt{Var(\sigma, V)}$ for q expected on realizations) when the specimen is a cube with the volume $V = L^3$ and when the diameter a of spheres (damaged zones) is such that $a/L = 0.1$. The same statistical property follows for σ_{ult}, since from equation (52), the variance of σ_{ult} is obtained from the variance of $p(\sigma_{ult})$, after multiplication by σ_{ult}^2.

Table 2. Variability of the proportion of undamaged material $p(\sigma_{ult})$

$p(\sigma_{ult})$	0.95	0.9	0.8	0.7	0.5
$2\sqrt{Var(\sigma_{ult}, V)}$	0.0054	0.00865	0.0134	0.0167	0.02

The macroscopic average relationship between the stress σ_M and the deformation ε is known, provided the effective elastic tensor $C(\sigma)$ is known as a function of the damage induced in the material at the level of stress σ acting on the matrix. It depends on the geometrical arrangement of the damage (and therefore on its configuration for a given volume V). For simplification, we consider large specimens and assume that $C(\sigma)$ is given by the effective tensor of an infinite porous medium with the solid volume fraction $p(\sigma)$. It is therefore obtained by homogenization of a random medium. We can use as estimate the upper third order bound of the elastic moduli. The average stress-strain relation is deduced in a parametric way (as a function of σ) from

$$\sigma_M = \sigma p(\sigma), \varepsilon = (C(\sigma))^{-1} \sigma p(\sigma) \tag{55}$$

For an isotropic elastic medium in traction, we get the relation between the elongation ε, the traction σ and the Young's modulus $E(\sigma)$: $\varepsilon = \sigma p(\sigma)/E(\sigma)$. To illustrate the approach, we consider specimens in tension, with defects on Poisson points. For defects obeying to the Weibull distribution, the ultimate stress σ_{ult} and the fraction of non damaged material are given by

$$\sigma_{ult} = \sigma_u \left(\frac{1}{mv}\right)^{1/m} \tag{56}$$

$$p(\sigma_{ult}) = \exp(-\frac{1}{m})$$

In 3 is given the variation, as a function of the Weibull modulus m, of σ_{ult}/σ_u (taking $v = 1$), of $p(\sigma_{ult})$ and of the reduced coefficient of variation $c = \sqrt{\frac{1-p(\sigma_{ult})}{p(\sigma_{ult})}}$, from which the ratio $\frac{\text{standard deviation}}{\text{average}}$ of $p(\sigma_{ult})$ and of σ_{ult} is deduced by $c\sqrt{\frac{A_3}{V}}$. The ultimate strength increases with m, while the proportion of damaged material $1 - p(\sigma_{ult})$ and the fluctuations decrease. Increasing m results in a more deterministic behavior. For $m \geq 3$, the volume fraction of the damaged material is lower than the percolation threshold of the phase made of spheres for the Boolean model ($p_c = 0.2895 \pm 0.0005$ [70]), which means that the physical behavior of the model is correct. Concerning the $\sigma_M - \varepsilon$ curve, high values of m generate a more brittle elastic material, while a low value of m results in a ductile type macroscopic behavior, as illustrated on 4.4. The curves own a common point obtained for $\sigma = \sigma_u$. Note that it should be possible to estimate the parameters of the statistical model by identification from an experimental $\sigma_M - \varepsilon$ curve.

Table 3. Effect of m on the ultimate strength, $p(\sigma_{ult})$, and c (Weibull distribution).

m	1	2	3	4	5	10	25	50
σ_{ult}/σ_u	1	0.707	0.693	0.707	0.725	0.794	0.879	0.925
$p(\sigma_{ult})$	0.368	0.606	0.716	0.779	0.819	0.905	0.961	0.980
c	1.311	0.805	0.629	0.533	0.470	0.324	0.202	0.142

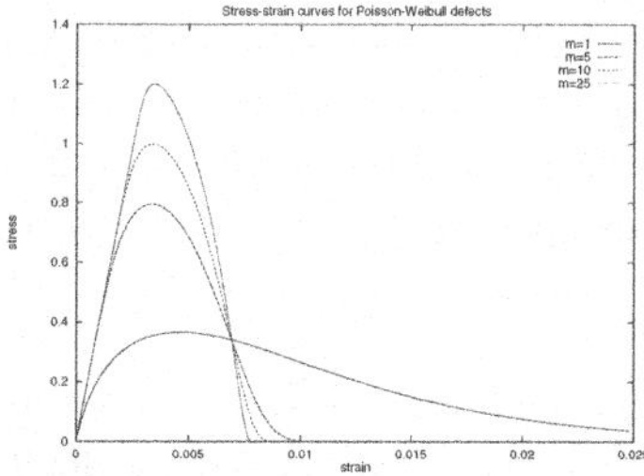

Fig. 7. Average macroscopic stress-strain (σ_M/σ_u) relation for Weibull populations of defects with various m and a constant median $\sigma_u \log 2$.

For defects obeying to a Pareto distribution, we have $\Phi(\sigma) = m \log \dfrac{\sigma}{\sigma_u}$ and $p(\sigma) = \left(\dfrac{\sigma_u}{\sigma}\right)^{mv}$ for $\sigma \geq \sigma_u$. When $mv > 1$, we have $\sigma_{ult} = \sigma_u$ and $p(\sigma_{ult}) = 1$: large specimens of the material are brittle with a constant ultimate stress. When $mv < 1$, we have $\sigma_{ult} = +\infty$ and there is no bounded maximum for $\sigma_M = \sigma p(\sigma)$.

4.5 Fracture statistics models and simulations

When considering damaging materials at different scales, like fiber composites or metals under a ductile fracture, the fracture process must be studied by means of numerical simulations [29], since it is difficult to account for the complex stress field resulting from the interaction between various damage sites. In the case of composite materials, we used realizations of the random fracture criterion, combined with finite elements calculations. The following methodology was developed to study the fracture statistics unidirectional composites for fiber fracture [29], for transverse fracture, and finally for the fracture of laminate composites. The first step is based on the experimental identification of the population of defects (point defects are considered here), by appropriate mechanical tests and by calculation of the local stress field seen by the microstructure; in this experimental part, a statistical volume element (SVE) is determined, as the elementary volume element broken during the progression of damage. Then, the statistical information and the fracture criterion are introduced in a FE calculation: to every SVE is attributed a random strength σ_R corresponding to the population of defects; during the calculation, it is broken if the average stress in the SVE, or more rigorously the equivalent stress σ_{eq} defined by equation (42) is larger than σ_R. Random simulations enable us to study the fracture behavior on a first scale, namely a representative volume element RVE. If necessary, the material can be considered as a set of RVE to study its behavior on the next scale. The main difficulty in running simulations is to determine the appropriate microstructural element converted into a SVE, and to know its fracture statistics. When operating on scales with increasing lengths, the statistical models proposed in the previous sections can be used to generate the necessary random variables. However, the correct corresponding type of assumption (weakest link, critical density,...) has to be introduced on a physical basis corresponding to the behavior of the material, as known from experiments. Similar approaches are developed for ductile fracture statistics in steels (see the chapter by A. Pineau in the book [36]).

From these examples of simulations, general guidelines can be derived. Firstly, a damage parameter (density of point defects, microcrack network parameter, cavity volume fraction, cavity growth rate,...), connected to the microstructure, should be selected at a given scale, the SVE. Secondly, statistical properties of this parameter should be obtained by image analysis (or by micromechanical tests); until now this information was limited to the probability distribution function over domains with a given size, but higher order information can be estimated to recover probabilistic information on the damage parameter, considered as the realization of a random function. Finally, simulations of the damage parameter as initial conditions for the prediction of its evolution by means of finite elements have to be performed; for this step, morphological models of random media, as well as change of scale models to generate correct simulations on different scales, can be useful.

5 Conclusion

From his pioneering work on the composition of permeability in random porous media, Georges Matheron initiated a major contribution for solving change of scale problems in random media. He always considered this field as an important research axis, beside his work on geostatistical estimation, on random sets and on mathematical morphology.

This is still a very active domain, with progress involving more and more simulations and investigations in non linear physics. Efforts are in progress to integrate the models summarized in this paper in numerical simulation techniques and in the design of microstructures. There is no doubt that new developments will be of practical interest in many areas of applications for engineering purpose.

Acknowledgement. A large part of my work on the physics of random media was initiated at the time of my ten-years close cooperation with Georges Matheron, to whom I am very grateful. This time was certainly for me the most exciting period of my work in research. I am glad to dedicate this paper to Georges Matheron.

References

1. Altus E. (2001): 'Statistical modeling of heterogeneous microbeams', *International Journal of solids and structures*, Vol. 38, pp. 5915–5934.

2. Barbe, F., L. Decker., D. Jeulin, G. Cailletaud (2001): 'Intergranular and intragranular behavior of polycrystalline aggregates. Part 1: F.E. model', *Int. J. Plasticity* **17** (4), pp. 513-536.

3. Beran, M.J., J. Molyneux (1966): 'Use of classical variational principles to determine bounds for the effective bulk modulus in heterogeneous media', *Q. Appl. Math.* **24**, pp. 107-118.

4. Beran, M. J. (1968): *Statistical Continuum Theories.* (J. Wiley, New York).

5. Beran M. J. (1980): 'Fields fluctuations in a two phase random medium', *J. Math. Phys.*, vol 21, (10), pp. 2583-2585.

6. Berdin, C., G. Cailletaud, D. Jeulin (1993): 'Micro-Macro Identification of Fracture Probabilistic Models'. In: *Proc. of the International Seminar on Micromechanics of Materials, MECAMAT'93*, Fontainebleau, 6-8 July 1993 (Eyrolles, Paris), pp. 499-510.

7. Beremin, F.M. (1983): 'A local criterion for cleavage fracture of a nuclear pressure vessel steel', *Metall. Trans. A.* **14A**, pp. 2277-2287.

8. Bergman, D. (1978): 'The dielectric constant of a composite material: a problem in classical physics', *Phys. Rep. C* **43**, pp. 377-407.

9. Berryman, G.J. (1985): 'Variational bounds on elastic constants for the penetrable sphere model', *J. Phys. D: Appl. Phys.* **18**, pp. 585-597.

10. Bobeth M., Diener G. (1986): 'Field fluctuations in multicomponent mixtures', *J. Mech. Phys. Solids*, 36, pp. 1-17.

11. Brown W.F. (1955): 'Solid mixture permittivities', *Journal of Chemical Physics*, 23, pp. 1514-1517.

12. Cailletaud, G., D. Jeulin, Ph. Rolland (1994): 'Size effect on elastic properties of random composites', *Eng. Comput.* **11** (2), pp. 99-110.

13. Chudnovsky, A., B. Kunin (1987): 'A probabilistic model of brittle crack formation', *J. Appl. Phys.* **62** (10), pp. 4124, 4129.

14. Decker, L., D. Jeulin (2000): 'Simulation 3D de matériaux aléatoires polycristallins', *Revue de Métallurgie - CIT/Science et Génie des Matériaux*, Feb. 2000, pp. 271-275.

15. Delarue A. (2001): *Prévision du comportement électromagnétique de matériaux composites à partir de leur mode d'élaboration et de leur morphologie*, Thesis, Paris School of Mines.

16. Eyre, D.J., G.W. Milton (1999): 'A fast numerical scheme for computing the response of composites using grid refinement', *Eur. Phys. J. Appl. Phys.* **6**, pp. 41-47.

17. Frisch U. (1968): 'Wave propagation in random media', in: *Probabilistic Methods in Applied Mathematics*, A.T. Bharucha-Reid (ed.), Vol. 1 , pp. 75-198, Academic "Press, New York.

18. Hashin, Z. (1962) 'The elastic moduli of heterogeneous materials', *J. Appl. Mech.* pp. 143-150.

19. Hashin, Z., S. Shtrikman (1962): 'A variational approach to the theory of the effective magnetic permeability of multiphase materials', *J. Appl. Phys.* **33**, pp. 3125-3131.

20. Hashin, Z., S. Shtrikman (1963): 'A variational approach to the theory of the elastic behavior of multiphase materials', *J. Mech. Phys. Solids* **11**, pp. 127-140.

21. Hill R. (1963) 'Elastic properties of reinforced solids: some theoretical principles', J. Mech. Phys. Solids, 11, pp. 357-372.

22. Hori, M. (1973): 'Statistical theory of the effective electrical, thermal, and magnetic properties of random heterogeneous materials. II. Bounds for the effective permittivity of statistically anisotropic materials', *J. Math. Phys.* **14**, pp. 1942-1948.
23. Huet C. (1990): 'Application of variational concepts to size effects in elastic heterogeneous bodies', *J. Mech. Phys. Solids* **38**, pp. 813-841.
24. Jeulin, D., P. Jeulin (1981): 'Synthesis of Rough Surfaces by Random Morphological Models'. In: Proc. 3rd European Symposium of Stereology, *Stereol. Iugosl.* **3**, suppl. 1, pp. 239-246.
25. Jeulin, D. (1987): 'Random structure analysis and modelling by Mathematical Morphology'. In: *Proc. CMDS5*, ed. by A. J. M. Spencer (Balkema, Rotterdam), pp. 217-226.
26. Jeulin, D. (1991): *Modèles morphologiques de structures aléatoires et de changement d'échelle,* Thèse de Doctorat d'Etat, University of Caen.
27. Jeulin, D. (1992): 'Some Crack Propagation Models in Random Media'. In: *Proc. Symposium on the Macroscopic Behavior of the Heterogeneous Materials from the Microstructure,* ASME, Anaheim, Nov 8-13, 1992. AMD Vo. 147, pp. 161-170.
28. Jeulin, D. (1994): 'Random structure models for composite media and fracture statistics'. In: *Advances in Mathematical Modelling of Composite Materials,* ed. by K.Z. Markov (World Scientific Company, Singapore), pp. 239-289.
29. Jeulin, D., C. Baxevanakis, J. Renard (1995): 'Statistical modelling of the fracture of laminate composites'. In: *Applications of Statistics and Probability,* ed. by M. Lemaire, J.L. Favre, A. Mébarki (Balkema, Rotterdam), pp. 203-208.
30. Jeulin, D., A. Le Coënt (1996): 'Morphological modeling of random composites', *Proceedings of the CMDS8 Conference (Varna, 11-16 June 1995),* ed. by K.Z. Markov (World Scientific, Singapore), pp. 199-206.
31. Jeulin, D. (ed) (1997): Proceedings of the Symposium on the *Advances in the Theory and Applications of Random Sets* (Fontainebleau, 9-11 October 1996) (World Scientific, Singapore).
32. Jeulin, D., L. Savary (1997): 'Effective Complex Permittivity of Random Composites', *J. Phys. I/ Condens.Matter* **7**, pp. 1123-1142.
33. Jeulin, D. (1998): 'Bounds of physical properties of some random structure'. In: *Proceedings of the CMDS9 Conference* (Istanbul, Turkey, June 29-July 3, 1998), ed. by E. Inan and K.Z. Markov (World Scientific, Singapore), pp. 147-154.
34. Jeulin, D. (2000): 'Models of random damage'. In: *Proc.Euromat 2000 Conference,* Tours, France, ed. by D. Miannay, P. Costa, D. François, A. Pineau, pp. 771-776.
35. Jeulin, D., P. Monnaie, F. Péronnet (2001): 'Gypsum morphological analysis and modeling', *Cement and Concrete Composites* **23 (2-3)**, pp. 299-311.
36. Jeulin, D. (2001): 'Random Structure Models for Homogenization and Fracture Statistics', In: *Mechanics of Random and Multiscale Microstructures,* ed. by D. Jeulin, M. Ostoja-Starzewski (CISM Lecture Notes N° 430, Springer Verlag).
37. Jeulin D., Delarue A. (2002) Numerical Homogenization of Dielectric Properties of Random Media. Application to Nanocomposites, In: Journée doctorale 2002 Saint-Etienne (20 Novembre 2002), Signaux et milieux complexes, ISBN 2-86272-281-2, Th. Fournel et G. Brun (eds), Presses de l'Université Jean Monnet, pp. 77-87.
38. Jikov V.V., Kozlov S.M., Oleinik O.A. (1994): *Homogenization of Differential Operators and Integral Functionals,* Springer Verlag.

39. Kanit T., Forest S., Galliet I., Mounoury V., Jeulin D. (2003): 'Determination of the size of the representative volume element for random composites: statistical and numerical approach', *International Journal of solids and structures*, Vol. 40, pp. 3647–3679.

40. Kanit T. (2003): *Notion de Volume élémentaire représentatif pour les matériaux hétérogènes: approche statistique et numérique*, Thesis, Paris School of Mines.

41. Karal F. C., Keller J.B. (1964): 'Elastic, electromagnetic and other waves in a random medium', *J. Math. Phys.* 5, pp. 537-547.

42. Keller J.B., Karal F. C. (1966): 'Effective dielectric constant, permeability and conductivity of a random medium and attenuation coefficient of coherent waves', *J. Math. Phys.*7, pp. 661-670.

43. Keller J.B. (1964): 'A theorem on the conductivity of a composite medium', *J. Math. Phys.*5 N° 4, pp. 548-549.

44. Kreher W. (1990): 'Residual stresses and stored elastic energy of composites and polycristals', *J. Mech. Phys. Solids*, 38, pp. 1115-128.

45. Kröner, E. (1971): *Statistical Continuum Mechanics.* (Springer Verlag, Berlin).

46. Le Coënt, A., D. Jeulin (1996): 'Bounds of effective physical properties for random polygons composites', *C.R. Acad. Sci. Paris*, **323**, Série II b, pp. 299-306.

47. Matheron G. (1964): 'Equation de la chaleur, écoulements en milieu poreux et diffusion géochimique', Internal report (Note Géostatistique 55), BRGM.

48. Matheron G. (1967): *Eléments pour une théorie des milieux poreux*, Masson, Paris.

49. Matheron G. (1968): 'Composition des perméabilités en milieu poreux hétérogène: critique de la règle de pondération géomètrique', *Rev. IFP*, vol 23, N° 2, pp. 201-218.

50. Matheron, G. (1971): *The theory of regionalized variables and its applications.* (Paris School of Mines publication).

51. Matheron, G. (1975): *Random sets and integral geometry.* (J. Wiley, New York).

52. Matheron G. (1978): *Estimer et Choisir*, Fascicules du CGMM n°7. (Paris School of Mines publication).

53. Matheron G. (1979): *L'émergence de la loi de Darcy.* (Paris School of Mines publication, N-592 CMM).

54. Matheron G. (1989): *Estimating and Choosing.* (Springer Verlag, Berlin).

55. Matheron G. (1991): 'Géodésiques aléatoires: application à la prospection sismique', *Cahiers de Géostatistique, Fascicule 1, Compte-rendu des Journées de Géostatistique, 6-7 juin 1991*, pp. 1-18.(Paris School of Mines publication).

56. Matheron G. (1992): 'Analyse harmonique et équations de la Physique', Internal Seminar, Centre de Géostatistique, 11 Dec 1992.

57. Matheron G. (1993): 'Quelques inégalités pour la perméabilité effective d'un milieu poreux hétérogène', *Cahiers de Géostatistique, Fascicule 3, Compte-rendu des Journées de Géostatistique, 25-26 Mai 1993*, pp. 1-20.(Paris School of Mines publication).

58. McCoy, J.J. (1970): 'On the displacement field in an elastic medium with random variations of material properties'. In: *Recent Advances in Engineering Sciences*, Vol. 5 (Gordon and Breach, New York), pp. 235-254.

59. Miller, M.N. (1969): 'Bounds for the effective electrical, thermal and magnetic properties of heterogeneous materials', *J. Math. Phys.* **10**, pp. 1988-2004.

60. Miller, M.N. (1969): 'Bounds for effective bulk modulus of heterogeneous materials', *J. Math. Phys.* **10**, pp. 2005-2013.

61. Milton, G. W. (1980): 'Bounds on the complex dielectric constant of a composite material', *Appl. Phys. Lett.* **37**, pp. 300-302.

62. Milton, G. W. (1982): 'Bounds on the elastic and transport properties of two component composites', *J. Mech. Phys. Solids* **30**, pp. 177-191.

63. Milton G. W. (1986): 'Modeling the properties of composites by laminates'. In: *Homogenization and Effective Moduli of Materials and Media*, J.L. Ericksen, D. Kinderlehrer, R. Kohn, J.L. Lions (eds) (Springer Verlag, Berlin), pp. 150-174.

64. Moulinec H., P. Suquet (1994): 'A fast numerical method for computing the linear and nonlinear mechanical properties of composites', *C.R. Acad. Sci. Paris*, **318**, Série II, pp. 1417-1423.

65. Ostoja-Starzewski M. (1998): 'Random field models for heterogeneous materials', *Int. J. Solids Structures*, 35 N° 19, pp. 2429-2455.

66. Pélissonnier-Grosjean, C., D. Jeulin, L. Pottier, D. Fournier, A. Thorel (1997): 'Mesoscopic modeling of the intergranular structure of Y_2O_3 doped aluminium nitride and application to the prediction of the effective thermal conductivity'. In: *Key Engineering Materials*, Volumes 132-136, Part 1 (Transtech Publications, Switzerland), pp. 623-626.

67. Ponte Castaneda, P. (1996): 'Variational methods for estimating the effective behavior of nonlinear composite materials'. In: *Proceedings of the CMDS8 Conference (Varna, 11-16 June 1995)*, ed. by K.Z. Markov (World Scientific, Singapore), pp. 268-279.

68. Quenec'h, J.L., J.L. Chermant, M. Coster, D. Jeulin (1994): 'Liquid phase sintered materials modelling by random closed sets'. In: *Mathematical morphology and its applications to image processing*, ed. by J. Serra, P. Soille (Kluwer, Dordrecht), pp. 225-232.

69. Rytov S.M., Kravtsov Yu. A., Tatarskii V.I. (1989): *Principles of Statistical Radiophysics, vol 3, Elements of Random Fields.* (Springer-Verlag, Berlin).

70. Rintoul, M.D., S. Torquato (1997): 'Precise determination of the critical threshold and exponents in a three-dimensional continuum percolation model', *J. Phys. A: Math. Gen.* **30**, pp. L585-L592.

71. Sab K. (1992): 'On the homogenization and the simulation of random materials', *Eur. J. Mech. Solids*, 11, pp. 585-607.

72. Sanchez Palencia E., Zaoui A. (ed) (1987): *Homogenization Techniques for Composite Media*, Lecture Notes in Physics vol. 272. (Springer Verlag, Berlin).

73. Schwydler M.I. (1962): 'Les courants d'écoulements dans les milieux hétérogènes', *Izv. Akad. Nauk SSSR, Mekh. i; Mas*, N° 3, pp. 185-190; N° 6, pp. 65-7.

74. Schwydler M.I. (1963): 'Sur les caractéristiques moyennes des courants d'écoulements dans les milieux à hétérogènéité aléatoire', *Izv. Akad. Nauk SSSR, Mekh. i; Mas*, N° 4, pp. 127-129; N° 5, pp. 148-150.

75. Serra, J. (1982): *Image analysis and mathematical morphology.* (Academic Press, London).

76. Stoyan, D., W.S. Kendall, J. Mecke (1987): *Stochastic Geometry and its Applications.* (J. Wiley, New York).

77. Stoyan D., Mecke K., this volume.

78. Suquet P., Ponte Castañeda P. (1993): 'Small contrast perturbation expansions for the effective properties of nonlinear composites', *C.R. Acad. Sc. Paris*, 317, Série II, pp. 1515-1522.

79. Torquato, S., G. Stell (1983): 'Microstructure of two-phase random media III. The n-point matrix probability functions for fully penetrable spheres', *J. Chem. Phys.* **79**, pp. 1505-1510.

80. Torquato, S., F. Lado (1986): 'Effective properties of two phase disordered composite media: II Evaluation of bounds on the conductivity and bulk modulus of dispersions of impenetrable spheres', *Phys. Rev. B* **33**, pp. 6428-6434.

81. Torquato, S. (1991): 'Random heterogeneous media: microstructure and improved bounds on effective properties', *Appl. Mech. Rev.* **44**, pp. 37-76.

82. Torquato, S. (2002): *Random heterogeneous materials: microstructure and macroscopic properties.* (Springer Verlag, New York, Berlin).

83. Willis, J.R. (1981): 'Variational and related methods for the overall properties of composites', *Advances in Applied Mechanics*, **21**, pp. 1-78.

84. Willis, J.R. (1991): 'On methods for bounding the overall properties of nonlinear composites', *J. Mech. Phys. Solids* **39**, 1, pp. 73-86.

Mathematical Morphology

Mophological Operators for the Segmentation of Colour Images

Jean Serra

Centre de Morphologie Mathématique, Ecole des Mines de Paris

1 Introduction

During his long and fruitful career, Georges Matheron never tackled the two themes of segmentation and colour processing of images. At the beginning of the nineties, when I was entering the first theme via connected filters, I asked him why he had avoided them. He told me that he could not take up working on all possible problems, and after a silence, added with his marvellous tactful smile: "besides, I guess it is already left to you".

Indeed, G. Matheron was partly right. The work on connected filters contained the seed of a morphological theory of segmentation that arose ten years later [31]. But by this time, my knowledge about colour processing was just a matter of two ideas. First, one represents colour by three grey images, in RGB or HLS modes, second, the compressed bit streams of grey tone images have to be multiplied by two (and not by three) when passing to colour.

I entered the field of colour processing quite accidentally. In 1999, Allan Hanbury, who was one of my Phd students, was developing a methodology for an automatic classification of pieces of woods according to their textures (veins, knots, ...). The directional features were, of course, preponderant and led us to elaborate morphological tools for circular data [11]. Now, in imagery, the two major situations where such data appear, are orientations and hues. Therefore, we decided to try also our algorithms on colour representations involving hue, i.e. on the two HSV and HLS systems. It rapidly led us to a critical analysis of both systems, and to propose more consistent ones [30][15]. In the meantime, A. Hanbury defended his thesis and left the Centre de Morphologie Mathématique (CMM, in brief) in 2002. I carried on the approach with Jesus Angulo, another Phd student at CMM, whose thesis subject did not really involve colour processing, but who was enthusiastic about the question. Together, we started from the new colour systems and discovered that, in the 3-D histograms generated by the L_1 norm representations, the pixels of an arbitrary image used to present specific alignments [4]. What did it mean ?

The new representations had particularly modified the definition of the saturation parameter, allowing us to perceive his meaning more clearly. With J. Angulo, we decided to exploit this parameter to balance luminance versus hue, when segmenting colour images [2].

This thread of ideas is not the only one we developed during these recent years. Since I was thrown into colour problems, it was an excellent opportunity to have a look at RGB representations and to those based upon palettes. I developed the first theme jointly with Marcin Iwanowski [12], and the second with Mariuscz Mlynarczuk [29], both polish Phd students at CMM.

The text which follows surveys these studies on colour image processing. I shall try and show how my way of thinking was pervaded by G. Matheron's one; and since Matheron used to say that the pedagogical order must be the reverse of that of the discovery, we will begin by the most recent developments.

2 The 3-D polar representations of the colour

2.1 Light intensities and gamma correction

Consider a television receiver. It uses three different colour representations. On the one side, the input Hertzian signal is coded as one grey image plus two other ones, associated to green-red and blue-yellow contrasts (i.e. one luminance and two chrominances). On the other side, the image on the monitor is obtained from three electrical signals, which excite three layers of green, red and blue photo-receivers. These two representations are quite different, although technically sound for their respective purposes. However, the manufacturers take none of them for the user's interface, and prefer human adjustments based on light (luminance), contrast (saturation), and, in case of an old receiver, from hue. Hence, this last triplet turns out to be the simplest one for human vision.

What are the relationships between these various representations? Do the technological steps modify the initial light that enters a device? Colour image processing rests on a few basic operations (addition, comparison,..) and properties (increasingness, distances..). Have these tools a physical meaning? In colour imagery the basic notion is the spectral power distribution (SPD) of the light radiating from or incident on a surface. This intensity has the dimension of an energy per unit area, such as watt per m^2. When the light arrives at a photo-receiver, this sensor filters the intensities of each frequency by weighting them according to fixed values. The sum of the resulting intensities generates a signal that exhibits a certain "colour". The CIE (Commission Internationale de l'Eclairage), in its *Rec 709*, has standardized the weights which yield the triplet $R_{709}, G_{709}, B_{709}$ [8]. As energies, the intensities are additive, so that all colours accessible from an RGB basis are obtain by sums of the primary colours R, G, and B and by multiplications by non negative constants.

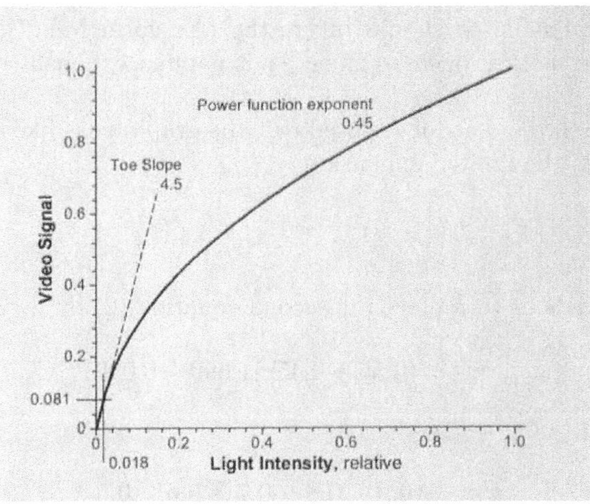

Fig. 1. *Gamma correction function.*

The exploration of the spectrum is lower bounded by $R = G = B = 0$ (zero energy) higher bounded by a maximum red R_0, green G_0 and blue B_0 that are given by the context (illumination, technological limits of the sensors, or of the eye, etc.) in which we work. Generally, each technology fixes the three bounds, which therefore define the reference white, and then introduces the *reduced variables*

$$r = \frac{R}{R_0}, \qquad g = \frac{G}{G_0}, \qquad b = \frac{B}{B_0}. \tag{1}$$

The digital sensitive layers of cameras transform the light intensities into proportional voltages; conversely, the cathodic tubes (CRT) and the flat screens that display images return photons from the electrical current. Now, their response is not linear, but a power function of the input voltage whose exponent γ, (gamma), varies around 2.5 according to the technologies. If we want the light intensities of the CRT to be proportional to those of the scene itself, the gamma effect has to be compensated. In video systems, this *gamma correction* is universally implemented in the camera itself. The *Rec. 709* of CIE proposes the following correction function

$$\begin{aligned} r' &= 4.5r & r &\leq 0.018 \\ r' &= 1.099r^{0.45} - 0.099 & r &> 0.018 \end{aligned} \tag{2}$$

that we write here for the reduced red intensity r, and where $1/\gamma = 0.45$. The same transfer function is applied to both green and blue bands.

Fig. 1, drawn from the excellent book [24] depicts the graph of Rel.(2). The variation domain $[0, 1]$ is the same for the reduced intensities (r) as for the video colours (r'), which implies that the white point $R_0\ G_0\ B_0$ is left invariant. The linear beginning in Rel.(2) minimizes the effect of the sensor noise. An ideal monitor should invert the transform Rel.(2). Indeed, they generally have neither linear segment, nor gamma exponent equal to1/0,45 [24].

Figure (1) shows that for r closed to1, the graph looks like a straight line. More precisely, the limited expansion

$$(1 - u)^{1/\gamma} = 1 - \frac{u}{\gamma} + \epsilon(u) \tag{3}$$

for small u, leads us to replace the second equation (2) by

$$r'^* = (0.55 + 0.45r)1.099 - 0.099 \tag{4}$$

i.e., numerically

r	0.9	0.8	0.7	0.6	0.5
r'	0.949	0.895	0.837	0.774	0.705
r'^*	0.950	0.901	0.851	0.802	0.753
$\frac{r'-r'^*}{r'}$	0.1%	0.6%	1.4%	2.8%	4.8%

In comparison with the noise of the video systems, we can consider the approximation r'^* is perfect for $r \geq 0.8$ and excellent for $0.6 \leq r \leq 0.8$.

2.2 Colour Vector Spaces

Their linearity provide the intensities r, g, b with the structure of a 3 dimensions vector space, or rather of the part E which is limited to the unit cube $[0, 1] \times [0, 1] \times [0, 1]$ of $\mathbb{R}^3$. For colour image processing purposes, it would be wise to go back from the video bands (r', g', b') to the reduced intensities (r, g, b) by the inverse transform of Rel.(2). When starting from the usual 3×8 bits (r', g', b') images, the best should probably be to code in 3×16 bits for computation (or in floating variables). But as a matter of fact, people keeps the (r', g', b') video space, which is implicitly modelled as a part of a vector space, from which one builds arithmetic means, projections, histograms, Fourier transforms, etc.... which often gives significant results.

What are the real consequences of the gamma correction Rel.(2) on the processing of colour data? Formally speaking, one can always consider the unit video cube (r', g', b') as a part, E' say, of a 3-dimensions vector space. This allows us to formulate operations, but their physical interpretations demand we come back to the intensities (r, g, b).

Fig. 2 depicts the unit cube E'. The vector $\mathbf{x}'$, of coordinates (r', g', b') can also be decomposed into two orthogonal vectors $\mathbf{c}'$ and $\mathbf{l}'$ of the chromatic plane and the a-chromatic (or gray) axis respectively. The latter is the main diagonal of the cube going through the origin O and the chromatic plane is perpendicular to the gray axis in O. The two vectors $\mathbf{c}'$ and $\mathbf{l}'$ have the following coordinates

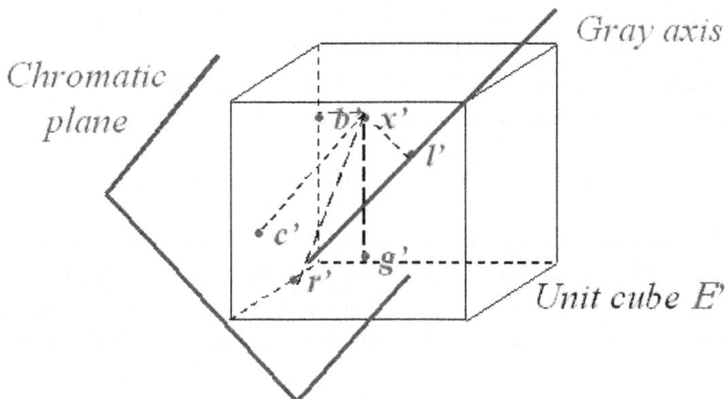

Fig. 2. *Chromatic plane and a-chromatic axis.*

$$3\mathbf{c}' = (2r' - g' - b', 2g' - b' - r', 2b' - r' - g')$$
$$3l' = (r' + g' + b', r' + g' + b', r' + g' + b') \tag{5}$$

Consider the red band $r'(z)$ over a zone Z in a colour image. What meaning can we give to the average red in Z ? As we just saw, the only average that has a physical meaning is the quantity $\bar{r} = \frac{1}{Z} \int (r'(z))^{\gamma} \, dz$, which needs to be corrected into $\bar{r}^{1/\gamma}$ for display purposes (for the moment we neglect the constants 1,099 and 0,099 in Rel.(2)). On the other hand, the usual segmentations aim to split the space into regions Z where the colour is nearly constant. Then at each point $z \in Z$, we can approximate $r(z)$ by the limited expansion

$$r(z) = r'(z)^{\gamma} = \bar{r}'^{\gamma} \left[1 - \frac{\bar{r}' - r'(z)}{\bar{r}'} \right]^{\gamma} = \bar{r}'^{\gamma} \left[1 - \gamma \left(\frac{\bar{r}' - r'(z)}{\bar{r}'} \right) + \varepsilon(r') \right]$$

where $\bar{r}' = \frac{1}{Z} \int_Z r'(z) dz$. Under averaging in Z, the coefficient of the γ term in the right member becomes zero, so that

$$(\bar{r})^{1/\gamma} = \bar{r}' + \bar{\epsilon}(r') \tag{6}$$

Therefore, the arithmetic mean of the video red r' equals, at the second order, the mean of the red intensity r followed by the gamma correction. The result remains true when the coefficients of Rel.(2) are added, when the the average is weighted, and also for the dark zones Z where the first Rel.(2) applies. It extends to the greens and blues. Rel.(6) turns out to be a theoretical justification of the "mosaic" based image segmentations (e.g. waterfall algorithm).

2.3 Brightness

From the point of view of physics, brightness is nothing but the integral of the power spectrum, i.e., here, the sum of the three components r, g, and b, that stand for this spectrum. For colorimetric purposes, this sum has to be weighted relatively to the spectral sensitivity of the eye. The CIE *Rec. 709* defines a white point and three weighting functions of the spectrum which lead to the variables R_{709}, G_{709} and B_{709}, then to the *luminance*

$$Y_{709} = 0.212R_{709} + 0.715G_{709} + 0.072B_{709} \tag{7}$$

and to the luminance Y_W of the associated white point. The three coefficients of Rel.(7) are related to the brightness sensitivity of the human vision and have been estimated by colorimetric measurements on a comprehensive population. The luminance Y_{709}, as a linear function of intensities, is an energy $(watts/m^2)$.

Human vision responds to intensities in a logarithmic way, according to laws of the type $di/i = constant$. Just as we took into account the spectral sensitivity of the eye, we should not ignore its energetic sensitivity. Now, by an amazing coincidence vision response to intensity is closed to the gamma correction of Rel.(2) : for example, when the luminance of a source is reduced to 20%, the eye perceives an intensity reduction of 50% only. Therefore, following many authors, we can consider the transforms

$$r' = r^{1/\gamma} \qquad g' = g^{1/\gamma} \qquad b' = b^{1/\gamma} \tag{8}$$

for $\gamma \simeq 2.2$ as generating *perceptual intensities*. For example, the *Rec. BT 601-E* proposes the *luma* y'_{601} as a perceptual brightness measurement

$$y'_{601} = 0.299r' + 0.587g' + 0.144b'. \tag{9}$$

However, this luma, as established from video values has not an energy dimension, and not any more the deriving additivity properties. The CIE follows the same direction, but defines the *lightness* l^* by taking a slightly different exponent

$$l^* = 116(\frac{Y_{709}}{Y_W})^{1/3} - 16 \qquad Y \geq 0.0089Y_W.$$

As regards the operations of segmentation in image processing, the situation is different. They do not hold on a *perceived* brightness, but on that of the *object under study*. In microscopy, the histological staining usually ranges from blue to violet; the spectrum of a sunset, or that of a human face have nothing to do with the weights given to r, g, and b in Rel.(7) or (9). Thus in the absence of *a priori* informations on the spectra of the objects under study, the purpose of segmentation leads us to take as brightness a *symmetrical function* of primary colours.

As regards the perceived energies now, consider, in the intensity space E, a vector x whose direction is given by $x_o = r_o, g_o, b_o$ but whose intensity varies, i.e.

$$x = (\lambda r_0, \lambda g_0, \lambda b_0) \qquad \lambda \in [0, \lambda_{\max}]$$

The point x describes the segment S which begins in O, goes through (r_o, g_o, b_o) and ends on the edge of cube E. In the video space E' there corresponds to x the point x' :

$$x' = \left((\lambda r_0)^{1/\gamma}, (\lambda g_0)^{1/\gamma}, (\lambda b_0)^{1/\gamma} \right) = \lambda^{1/\gamma} x_0' \tag{10}$$

with $x_0' = r_0^{1/\gamma}$, $g_0^{1/\gamma}$, $b_0^{1/\gamma}$. Similarly, the point x' describes a segment S' in E'. When x varies, if we want its perceptual brightness to seem additive, then Rel.(10) implies that the corresponding brightness of x' is a linear function of the three primary components. Finally, since this "image processing brightness" has to vary from 0 to 1, as r and r' do, the only possibility is to take for it the arithmetic mean m' of the primary colours :

$$m' = \frac{1}{3}(r' + g' + b'). \tag{11}$$

Put $\lambda' = \lambda^{1/\gamma}$. The two expressions

$$|m(x_1) - m(x_2)| = |\lambda_1 - \lambda_2| \, m(x_0)$$
$$|m'(x_1') - m'(x_2')| = \left| \lambda_1^{1/\gamma} - \lambda_2^{1/\gamma} \right| m'(x_0')$$

turn out to be different distances in segments S and S' respectively. The exponent $1/\gamma$ provides the second one with a meaning of *perceptual homogeneity*. But image processing is more demanding, as we must be able to express that a colour point E' (or more generally a set of points) gets closer to another even when these two points are not aligned with the origin. Now, the mean (11) is nothing but the restriction to the cube E' of the L_1 norm, which is defined in the whole space $\mathbb{R}^3$ (i.e. for $r', g', h' \in [-\infty, +\infty]$) by taking $\alpha = 1$ in the relation

$$n(x') = \left(|r'(x)|^\alpha + |g'(x)|^\alpha + |b'(x)|^\alpha \right)^{1/\alpha} \qquad \alpha \geq 1 \tag{12}$$

(This Rel.(12) introduces indeed a family of norms as soon as $\alpha \geq 1$. For $\alpha = 2$, we obtain the Euclidean norm L_2, and for $\alpha = \infty$, the "max" norm). In a vector space V, any norm n generates a distance d_n (see [9], section VII-1-4) by the relation

$$d_n(x_1', x_2') = n(x_1' - x_2') \qquad x_1', x_2' \in V \tag{13}$$

Therefore L_1 is a distance, as well, of course, as its restriction to the unit cube E'.

Thus, for $\alpha = 1$, both brightness $m'(x')$ of Rel. (11) and distance $d(x_1', x_2') = m'(|x_1' - x_2'|)$ in E' derive from a unique concept. This latter relation is important, as in segmentation a number of algorithms which were established for numerical functions extend to vector functions when a distance is provided (e.g. watershed).

2.4 Saturation

The CIE was more interested in various formulations of the brightness (luminance, lightness ...) than in saturation, that it defines as *"the colourfulness of an area judged in proportion to its brightness"*. In other words, it is the concern of the part of uniform spectrum (i.e. of gray) in a colour spectrum, so that any maximal monochromatic colour has a unit saturation and so that any triplet $r = g = b$ has a zero saturation.

Intuitively, what the CIE means here is clear, but its definition of the saturation lends itself to various interpretations. From a given point $x \in E$, one can draw several paths along which the colour density varies in proportion to brightness. For example, in Fig. 2, supposed to represent cube E, we can take the perpendicular xc to the chromatic plane, or the perpendicular xl to the gray axis, or again the axis Ox, etc.. Which path to choose?

Indeed, these ambiguities vanish as soon as we set the context in the chromatic plane. By projecting the cube E onto the chromatic plane, perpendicularly to the a-chromatic axis, we obtain the hexagon H depicted in Fig. 3, which is centered in O.

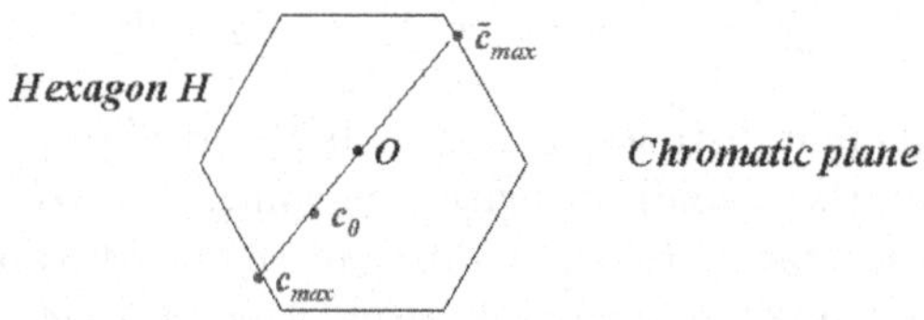

Fig. 3. Chromatic plane and saturation

Consider a point $x_o \in E$, of projection c_0 in H, and such that $c_0 \neq O$. Following the CIE, we define as a saturation any non negative function along the axis Oc_0 that *increases* from O; in O, it equals zero (pure gray) and has its maximum value when the edge of Hexagon H is reached, in $c_{\max}$ say (saturated colour). The hue remains constant along the segment $[0, c_{\max}]$, and the hue of the opposite segment $[0, \bar{c}_{\max}]$ is said to be *complementary* of that of segment $[0, c_{\max}]$. For a point $c \in [0, c_{\max}]$, we have $c = \lambda c_0$, $0 \leq \lambda \leq 1$. Thus, given $c_0 \in H$, the saturation $s(c) = s(\lambda c_0)$ is a function of λ only, and this function is increasing.

We have to go back to the 3-D cube E, as point c_0, projection of x_0, is just an intermediary step (moreover $c_0 \notin E$). The saturation $s(x_0)$ of point $x_0 \in E$ is then defined by

$$s(x_0) = s(c_0)$$

Note that when a point $x \in E$ moves away from the chromatic plane along the perpendicular $c_0 x_0$ to this plane, its gray proportion increases, but its saturation $s(x)$ does not change: it is indeed a matter of *chromatism* and not of *energy* of the light intensity.

As point c describes the radius $[0, \bar{c}_{\max}]$ which is at the opposite of $[0, c_{\max}]$ in the chromatic plane, we have

$$c \in [0, \bar{c}_{\max}] \qquad \Longleftrightarrow \qquad c = \lambda c_0 \qquad \lambda(\bar{c}_{\max}) \leq \lambda \leq 0$$

where λ indicates the proportionality ratio, now negative, between c and $\bar{c}_0$. This purely vector equivalence admits a physical interpretation if we extend the definition of the saturation to all diameters $D(c_0) = [0, c_{\max}] \cup [0, \bar{c}_{\max}]$, $c_0 \in H$, of the hexagon H (saturation was previously introduced for radii only). This can be done by putting $c \in D(c_0)$, $s(c) = s(\lambda c_0) = s(|\lambda| c_0)$. Two opposite points have the same saturation, and more generally if $c_1 \in [0, c_{\max}]$ and $c_2 \in [0, \bar{c}_{\max}]$, then $c_1 + c_2 = (\lambda_1 + \lambda_2) c_0$, with $\lambda_1 \geq 0$ and $\lambda_2 \leq 0$. As s is increasing we have

$$c_1 \in [0, c_{\max}], \qquad c_2 \in [0, \bar{c}_{\max}] \qquad \Longrightarrow \qquad s(c_1 + c_2) \leq s(c_1) + s(c_2). \tag{14}$$

When $c_1 = c_{\max}$ and $c_2 = \bar{c}_{\max}$ we find in particular Newton's disc experiment, reduced to two complementary colours.

When considering the saturation in the video cube E', the conditions of increasingness of s' along the radii (now of H') and of its nullity on the gray axis are still valid. They must be completed by the two constraints of image processing, namely the symmetry w.r.t. r', g', b' and the fact that $s'(x_1' - x_2')$ must be a distance in E'.

We saw that the mean m', in Rel.(11), was the L_1 norm expressed in the unit cube E', and that $3m'(x')$ was both the norm of x' and of its projection l' on the gray axis, i.e.

$$L_1(x') = L_1(l') = 3m'(x')$$

It is tempting to keep the same norm for the hexagon H' of the chromatic plane. By using Rel.(5) we find

$$s'(x') = L_1(c') = \frac{1}{3}\left[|2r' - g' - b'| + |2g' - b' - r'| + |2b' - r' - g'|\right]. \tag{15}$$

By symmetry, $s'(x')$ depends on the three functions $max' = max(r', g', b',)$, $min' = min(r', g', b',)$, and $med' = median(r', g', b',)$ only, which gives

$$s' = \begin{cases} \frac{3}{2}(max' - m') & \text{if} \quad m' \geq med' \\[2mm] \frac{3}{2}(m' - min') & \text{if} \quad m' \leq med' \end{cases} \tag{16}$$

On can find in [30] the derivation yielding s', and that of the following expression h' of the hue (which avoids to bring trigonometric terms into play)

$$h' = \frac{\pi}{3}\left[\lambda + \frac{1}{2} - (-1)^\lambda \frac{max' + min' - 2med'}{2s'}\right] \tag{17}$$

with

$$\lambda = \begin{cases} 0 \text{ if } r > g \geq b \\ 1 \text{ if } g \geq r > b \\ 2 \text{ if } g > b \geq r \\ 3 \text{ if } b \geq g > r \\ 4 \text{ if } b > r \geq g \\ 5 \text{ if } r \geq b > g \end{cases}$$

The hue h', as a coordinate on the unit circle, is defined modulo 2π. The value $h' = 0$ in Eq.(17) corresponds to the red. For $s' = 0$, colour point lies on the gray axis, so that its hue is meaningless. The polar system of the three equations (11), (16) and (17) is called the L_1 *norm representation*. It is inversible. The inverse formulae are given in [15], and the detailed proofs may be found in [30].

The relations (15) and (13) entail that $s\left(c_1' - c_2'\right) = L_1\left(c_1' - c_2'\right)$ is a distance in the chromatic plane, which therefore brings into play both saturation and hue. On the other hand, as L_1 is a norm, Rel.(14) becomes true for all triplets c_1', c_2' and $c_1' + c_2'$ that are on a same diameter of H'. Remark that here the L_1 norm is the concern of the *projections* c', the norm of the vectors x' themselves being their arithmetic mean. Finally, the above comments apply also to the Euclidean norm and to the max-min, which both induce distances in the chromatic hexagon H'.

When passing from the video variables to the intensities, a first result is obtained by observing that the averaging of the saturation s' follows the same law than that of the brightness m', namely Rel.(6), in the zones Z where the colour varies weakly. Moreover, the mapping $x_0' = (r_0', g_0', b_0') \rightarrow x_0 = \left(r_0'^\gamma, g_0'^\gamma, b_0'^\gamma\right)$ shows that $x' = \lambda x_0'$ becomes $x = \lambda^\gamma x_0$, hence

$$s'(x') = L_1\left(c'\right) = \lambda L_1\left(c_0'\right) = \lambda s'(x_0') \Leftrightarrow s(x) = L_1\left(c\right) = \lambda^\gamma L_1\left(c_0\right) = \lambda s(x_0).$$

In other words, the L_1 norm is increasing on the radii, and is zero for the grays, on both chromatic hexagons H of the intensities and H' of the video variables. Thus it represents a saturation *in both spaces*. It seems difficult to go further, as two points $x_0', x_1' \in E'$ whose projections c_0' and c_1' lie on a same radius of H' may have homolog points x_0 and $x_1 \in E$ whose projections are not always aligned with O.

2.5 Two other norms

How to build polar representations which be not contradictory with the previous requirements? Besides the L_1 norm, we can think of two ones. Firstly, the Euclidean norm L_2. In practical image processing, it turns to be less convenient than the L_1 norm, which suits particularly well to linear and morphological operations, and provides nice inverses. In addition, the associated 2-D histograms are rather unclear (see Fig. 5).

Another possibility is to correct the classical HLS system [16], by replacing its saturation by $max(r, g, b) - min(r, g, b)$. In the whole space, the quantity $max - min$ is a semi-norm only: two distinct vectors c and c', whose difference c - c' is a gray have the same $max - min$ [15]. However, in the chromatic plane, $max - min$ becomes a norm. It can be used for the saturation in parallel with m' for the brightness. This is what we will do below each time $max - min$ norm is introduced.

Finally, the norm and distance based approach presents the significant advantage that it *separates the variables* : two points x'_1 and $x'_2 \in E'$ which have the same projection on the chromatic plane (resp. on the gray axis) have the same saturation (resp. the same brightness). However, the last property, on brightness, vanishes when the three bands are given different weights in the means m or m'.

2.6 The classical polar representations

Even though the transformation from RGB to hue, saturation and brightness coordinates is simply a transformation from a rectangular colour coordinate system (RGB) to a three-dimensional polar (cylindrical) coordinate system, one is faced with a bewildering array of such transformations described in the literature (HSI, HSB,HSV, HLS, etc.). Most of them date from the end of the seventies [33], and were conceived neither for processing purposes, nor for the current computing facilities. This results in a confusing choice between models which essentially all offer the same representation. The most popular one is the HLS triplet of System (18), which appears in many software packages. The comments which follow hold on this particular model, but they apply to the other ones. The HLS triplet derives from RGB by the following system

$$\begin{cases} l'_{HLS} = \dfrac{\max(r',g',b') + \min(r',g',b')}{2} \\[4mm] s'_{HLS} = \begin{cases} \dfrac{\max(r',g',b') - \min(r',g',b')}{\max(r',g',b') + \min(r',g',b')} & \text{if} \quad l'_{HLS} \leq 0.5 \\[4mm] \dfrac{\max(r',g',b') - \min(r',g',b')}{2 - \max(r',g',b') - \min(r',g',b')} & \text{if} \quad l'_{HLS} \geq 0.5 \end{cases} \end{cases} \quad (18)$$

One easily checks that the HLS expressions do not preserve the above requirements of linearity (for the brightness), of increasingness (for the saturation) and of variables separation. The HLS luminance both RGB triplets $(1/2, 1/2, 0)$ and $(0, 1/2, 1/2)$ equals $1/4$, whereas that of their mean equals $3/8$, i.e. it is lighter than both terms of the mean. The HLS saturations of the

a) Rubber ring   b) Ana Blanco

Fig. 4. *Two test images.*

RGB triplets $(4/6, 1/6, 1/6)$ and $(2/6, 3/6, 3/6)$ equals $3/5$ and $1/5$ respectively, whereas that of their sum is 1: it is just Newton's experiment denial! Finally the independence property is no more satisfied. Take the two RGB triplets $(1/2, 1/2, 0)$ and $(3/4, 3/4, 1/4)$. One passes from the first to the second by adding the gray $r' = g' = b' = 1/4$. Hence both triplets have the same projection on the chromatic plane. However, the HLS saturation of the first one equals 1 and that of the second $1/2$.

3 2-D histograms and linearly regionalized spectra

In practice, is it really worth deviating from beaten tracks, and lengthening the polar triplets list? What for? We may answer the question by comparing the *luminance/saturation* bi-dimensional histograms for HLS system and for L_1, L_2 and $max - min$ norms. J.Angulo did so on a dozen images [3]. Two of them are depicted below, in Fig.4.

3.1 Bi-dimensional histograms

In the first image, we observe strong reflections on the rubber ring, and various types of shadows. The corresponding histograms are reported in Fig.5, with luminance on the x axis and saturation on y axis. No information can be drawn from HLS histogram, some structures begin to appear in L_2 and $max - min$ norms, but the most visible ones come from L_1 norm.

By coming back to the images, we can localize the pixels which give alignments, as depicted in Fig.6. They correspond to three types of areas :

- shadows with steady hue,
- graduated shading on a plane,

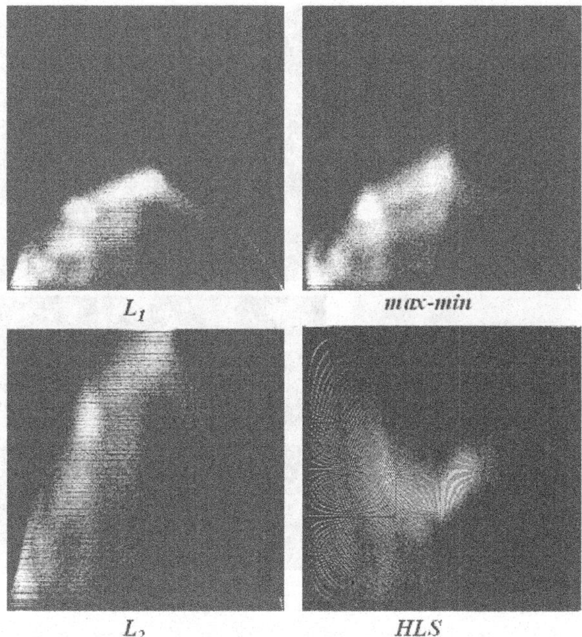

Fig. 5. *Bi-dimensional histograms of the "rubber ring" image. The x-axis corresponds the luminance and the y-axis to the saturation.*

- reflections with a partial saturation.

Consider now the more complex image of "Ana Blanco", in Fig.4b. It includes various sources light (television monitor, alpha-numerical incrustations..), and the light diffused by the background is piecewise uniform over the space. However, there are still alignments, which do not always go through points $(0,0)$, or $(1,0)$, and are sometimes parallel. In the *lum/hue* plane of the L_1 norm representation, several horizontal lines (constant hue) are located at different hue levels, and alternate with elongated clouds of points (Fig.7b).

All in all, we draw from the above histograms four main informations.

1. In the *lum/sat* histogram, there is no accumulation of pixels at point $(1,0)$. It means that the sensors we use are not physically saturated, which make realistic the proposed linear approach;
2. Still in the *lum/sat* histogram, some well drawn alignments can be extrapolated to point $(0,0)$ or point $(1,0)$. The others are parallels to the first ones;
3. However, most of the pixels form clouds in both *lum/sat* and *lum/hue* histograms are not aligned at all, whether the model does not apply, or the homogeneous zones are too small;

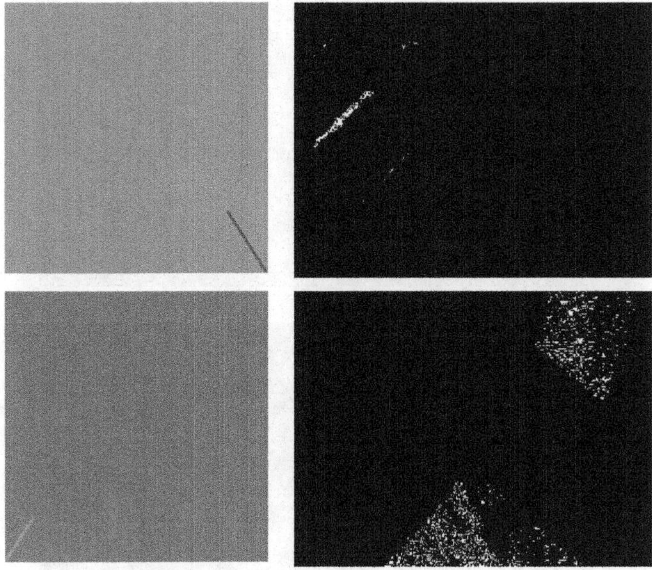

Fig. 6. *Zones of "Rubber ring" associated with alignments. The left two images show the supports of the alignments detected in Fig.5 for the L_1 norm, and the right two images indicate the locations of the aligned pixels in the space of the initial picture.*

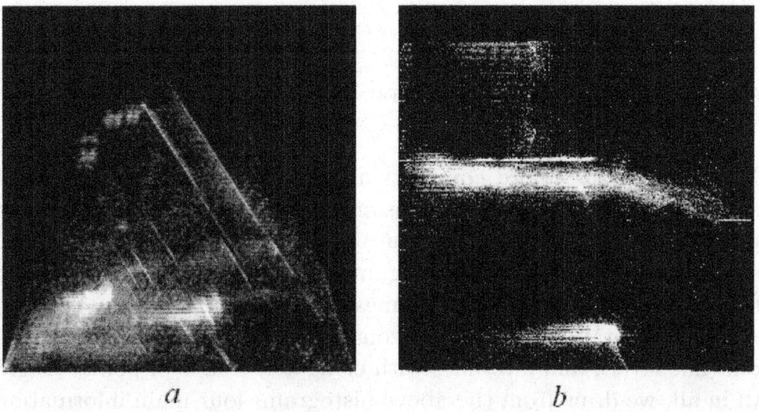

Fig. 7.   *(a) and (b) the two histograms of "Ana Blanco", in the luminance/saturation and the luminance/hue plane respectively, both in L_1 norm.*

4. In the *lum/hue* histogram, most often the aligned pixels exhibit a (quasi) constant hue, i.e. draw horizontal lines. But sometimes, these "lines" turn out to be a narrow horizontal stripe.

Such characteristic structures, such distinct lines suggest we seek a physical explanation of the phenomenon. This is what we will do now. But besides any physical model, a first point is worth to be noticed: the only norm that enables us the extraction of reflection areas, of shadows and gradations is L_1. No other polar model results in such an achievement.

3.2 Linearly regionalized spectra (LR model)

If we assume that the alignments are a property of the spectrum, and not an artefact due to some particular representation, we have to express the spectrum in such a way that the sequence

$$(spectrum) \rightarrow (r'g'b') \rightarrow (m's'h') \rightarrow (m' = \alpha s' + \beta)$$

be true (in the alignments) whatever the weights generating r, g and b are, and also whatever the spectrum itself is. Consider a zone Z of the space whose all pixels yield an alignment in the L_1 histogram. Denote by $sp\,(\nu; z)$ the spectrum of the light intensity at point $z \in Z$. We will say that this spectrum is *linearly regionalized* in Z when for each point $z \in Z$ one can decompose $sp\,(\nu; z)$ into the sum of a first spectrum $sp_0\,(\nu)$, independent of point z, and of a second one, $\overline{\omega}(z)sp_1(\nu)$, which proportionally varies in Z from one point to another. For all $z \in Z$, we have

$$sp\,(\nu; z) = sp_0\,(\nu) + \overline{\omega}(z)sp_1(\nu) \tag{19}$$

where $\overline{\omega}\,(z)$ is a numerical function which depends on z only, and where sp_0 and sp_1 are two fixed spectra.

In the spectrum $sp\,(\nu; z)$, though sp_0 usually corresponds to diffuse light and sp_1 to specular one, we do not need to distinguish between the emitted and reflected components of the light. It can be the concern of the light transmitted through a net curtain, for example, or of that of a TV monitor; but it can also come from passive reflectance, such as those described by Shafer's dichromatic model [32], or by Obein's model of glossiness [22]. But unlike these two models, the term $\overline{\omega}(z)sp_1$ may also represent an absorption, when it is negative. Similarly, we do not need to distinguish between diffuse and specular lights. The term sp_0 may describe a diffuse source over the zone Z, as well as a constant specular reflection stemming from the same zone. But above all, the emphasis is put here on the *space variation* of the spectrum. It is introduced by the weight $\overline{\omega}(z)$, that depends on point z, but not on spectrum sp_1. This weight may bring into play cosines, when the angle of the incident beam varies, or the normal to a glossy surface, etc...

The three spectra sp, sp_0 and sp_1 are known only through the weight functions that generate a (R, G, B) triplet. We use here the notation (R, G, B) in a canonical manner, i.e. it may designate the (X, Y, Z) coordinates of the CIE, or the perceptual system (L, M, S) [35], as well as the (Y, U, V) and (Y, I, Q) TV standards. In all cases it is a matter of *scalar products* of the spectra by such or such frequency weight In particular, the white colour given by $r = g = b = 1$ can be obtained from a spectrum which is far from being uniform. We write

$$r\left(z\right) = \int sp\left(\nu; z\right)\xi(\nu)d\nu = \int \left[sp_0\left(\nu\right) + \overline{\omega}(z)sp_1\left(\nu\right)\right]\xi\left(\nu\right)d\nu = r_0 + r_1\overline{\omega}\left(z\right)$$

(20)

$$g\left(z\right) = \int sp\left(\nu; z\right)\chi\left(\nu\right)d\nu = g_0 + g_1\overline{\omega}\left(z\right) \tag{21}$$

and

$$b\left(z\right) = \int s\left(\nu; z\right)\psi\left(\nu\right)d\nu = b_0 + b_1\overline{\omega}\left(z\right) \tag{22}$$

where ξ, χ and ψ are the three weighting functions that generate the primary colours r, g and b.

As sp_0 and sp_1 are power spectra, they induce *intensities* r, g, and b. Now, in the above histograms, the L_1 norm applies to the *video variables* $r' = r^{1/\gamma}$, $g' = g^{1/\gamma}$, and $b' = b^{1/\gamma}$ (if we neglect the behaviour near the origin). Then we draw from Rel.(20)

$$r'\left(z\right) = \left[r\left(z\right)\right]^{1/\gamma} = \left[r_0 + \overline{\omega}\left(z\right)r_1\right]^{1/\gamma},$$

with similar derivations for the video green and blue bands.

Is the linearly regionalized model able to explain the alignments in video histograms, despite the gamma correction? For the sake of simplicity, we will tackle this question by fixing the order of the video bands as $r' \geq g' \geq b'$, and $m' \geq g'$. Then we have

$$3m'(z) = r'(z) + g'(z) + b'(z)$$
$$2s'(z) = 2r'(z) - g'(z) - b'(z)$$

Alignments with the dark point

In the *luminance/saturation* histograms in L_1 norm, several alignments are in the prolongation of the point $(0, 0)$, of zero luminance and saturation. The shadow regions of the "rubber ring" image illustrate this situation.

Suppose that, in the relation (19) which defines the LR spectrum, the term $sp_0(\nu; z)$ is identically zero. Then $r(z)$ reduces to $\overline{\omega}(z)r_1$, which gives

$$r'(z) = r^{1/\gamma} = \overline{\omega}^{1/\gamma}r_1^{1/\gamma} = \overline{\omega}^{1/\gamma}(z)r_1',$$

with similar derivations for two other bands. Therefore we have

$$3m'(z) = \overline{\omega}^{1/\gamma}(z)\left[r_1^{1/\gamma} + g_1^{1/\gamma} + b_1^{1/\gamma}\right] = 3\overline{\omega}^{1/\gamma}(z)m_1'$$

and

$$2s'(z) = 2r'(z) - g'(z) - b'(z) = \overline{\omega}^{1/\gamma}(z)\left[2r_1^{1/\gamma} - g_1^{1/\gamma} - b_1^{1/\gamma}\right]$$

hence $m'(z)s_1' = m_1's'(z)$. In the space E of the intensities, we find in the same way that $m(z)s_1 = m_1 s(z)$. Therefore the nullity of the constant spectrum $sp_0(\nu)$ entails that both m' and s' on the one hand, and m and s on the other one, are proportional. Each *video* alignment indicates a zone where the *intensities* spectrum varies proportionally from one point to another.

Alignments with the white point

The "rubber ring" image generates also an alignment along a line going through the point $(1,0)$, i.e. the point with maximum luminance and zero saturation. That suggests to suppose the spectrum $sp_0(\nu; z)$ constant and equal to 1, and in addition that the three colors r_1, g_1, b_1 are not identical (if not, the saturation s' should be zero). We have

$$r(z) = 1 + \overline{\omega}(z)r_1 \tag{23}$$

and the two sister relations for $g(z)$ and $b(z)$. Under gamma correction, $r(z)$ becomes

$$r'(z) = (1 + \overline{\omega}(z)r_1)^{1/\gamma}.$$

Now, to say that the alignment is closed to a point of maximum luminance comes down to saying that $r_1, g_1,$ and b_1 are small with respect to 1, or again that

$$r'(z) = 1 + \frac{\overline{\omega}(z)}{\gamma}r_1 + \varepsilon(r_1), \tag{24}$$

hence $m'(z) = 1 + \frac{\overline{\omega}(z)}{\gamma}m_1$ and $s'(z) = \frac{\overline{\omega}(z)}{\gamma}[2r_1 - g_1 - b_1]$. We observe that the two conditions $r_1 \geq 0$ and $r_1' \leq 1$, jointly with Rel.(24) imply that the coefficient $\overline{\omega}(z)$ is negative. Moreover, as the three colours r_1, g_1, b_1 are distinct, the condition $s'(z) \geq 0$ implies in turn that the quantity $2r_1 - g_1 - b_1$ is strictly negative. By putting $\sigma_1 = -(2r_1 - g_1 - b_1) > 0$ (σ_1 is not the saturation at point z_1), we obtain the following linear relation with positive coefficients

$$m'(z) = 1 - \frac{m_1}{\sigma_1}s'(z). \tag{25}$$

As in the previous case, but without approximations, the mean $m(z)$ and the saturation $s(z)$ of the intensities are linked by the same equation (25): it is a direct consequence of Eq.(23). Again, both video and intensity histograms carry the same information, and indicate the zones of almost white reflections.

Alignments with a gray point

There appears in some images, as "AnaBlanco", series of parallel alignments. Their supports go through points of (quasi) zero saturation but their luminance is strictly comprised between 0 and 1. The interpretation we just gave for the case of reflections extends to such a situation. It is still assumed that $r_0 = g_0 = b_0$, but with $0 < r_0 \leq 1$, and that the terms $\overline{\omega}(z)r_1$, $\overline{\omega}(z)g_1$, and $\overline{\omega}(z)b_1$ are small with respect to r_0. Then we have,

$$r'(z) = (r_0 + \overline{\omega}(z)r_1)^{1/\gamma} = r_0^{1/\gamma} + r_0^{1/\gamma - 1}\frac{\overline{\omega}(z)}{\gamma}r_1,$$

and the two sister relations for g' and b'. Hence

$$m'(z) = r_0^{1/\gamma} + r_0^{1/\gamma - 1}\frac{\overline{\omega}(z)}{\gamma}m_1,$$

$$s'(z) = -r_0^{1/\gamma - 1}\frac{\overline{\omega}(z)}{\gamma}\sigma_1,$$

so that, finally

$$m'(z) = r_0^{1/\gamma} - \frac{m_1}{\sigma_1}s'(z).$$

When the colour component (r_1, g_1, b_1) remains unchanged, but that the gray component (r_0, g_0, b_0) takes successively various values, then each of them induces an alignment of the same slope $\frac{m_1}{s_1}$. The property extends to the histograms of the intensities themselves. Finally, we derive from Eq.(17) that, in the three cases, the hue remains *constant* in each of these zones.

4 Saturation weighted segmentations

The most radical change between the classical HLS system and those based on norms holds on the saturation equation. In system (18), when $\min(r, g, b) = 0$, (with $l \leq 0.5$) or when $\max(r, g, b) = 1$ (with $l \geq 0.5$), then the saturation equals 1. Now for human vision, the most significant parameter is the hue in high saturated areas, and it turns to luminance when saturation decreases. Any person whose reaction to colours is normal can easily check it. In the darkness, or, at the opposite, in white scenes (e.g. a landscape of snowy mountains), the eye grasps the contours by scrutinizing all small grey variations, whereas when the scene juxtaposes spots of saturated colours, then the eye localizes the frontiers at the changes of the hue. But how to transcribe quantitatively such a remark by a saturation function that takes its maxima precisely when the colours loose their saturation, as the classical HLS system does?

The norms based representations correct this drawback, so that their saturations may serve to split the space into hue-dominant versus grey-dominant

Fig. 8. *Representation of the"Parrots" image 14 in L_1 norm : a) luminance, b) saturation, c) hue.*

regions. This very convenient key to entering the segmentation of colour images was initially proposed by C.Demarty and S.Beucher [10]. They introduce the function $max - min$ on the image under study, and threshold it at a level s_0 that depends on the context. Then they adopt the HSV representation, but they replace its saturation by 1 in the regions above s_0 and by 0 elsewhere. Their downstream segmentations become easier and more robust.

However, they did not take the plunge of a new representation, and they worked at the pixel level, which is not the most informative. In order to go further in the same way of thinking, J.Angulo and J.Serra propose, in [2], the following two steps segmentation procedure:

1. to *separately* segment the luminance, the saturation and the hue in a correct Newtonian representation;
2. to combine the obtained partitions of the luminance and of the hue by means of that of the saturation: the later is taken as a *criterion* for choosing at each place either the luminance class, or the hue one.

The three bands of the "parrots" image of Fig.14, in L_1 representation, are depicted in Fig.8(a-c). Each band is segmented by the *jump connection* algorithm [31] (one groups in same classes all points x where $f(x)$ differs by less than k of an extremum in the same connected component, these classes are then withdrawn from the image, and one iterates). The method depends only on the jump positive value k.

As the parameter k increases, the over-segmentations reduce, but in compensation heterogeneous regions appear. A satisfactory balance seems to be reached for $k = 20$ (for 8-bits images), up to the filtering of a few very small regions. We obtain the two segmentations depicted in Fig.9.

4.1 Synthetic partition

How to combine the two partitions of images 9a and 9b? The idea consists in splitting the saturation image into two sets X_s and X_s^c of high and low

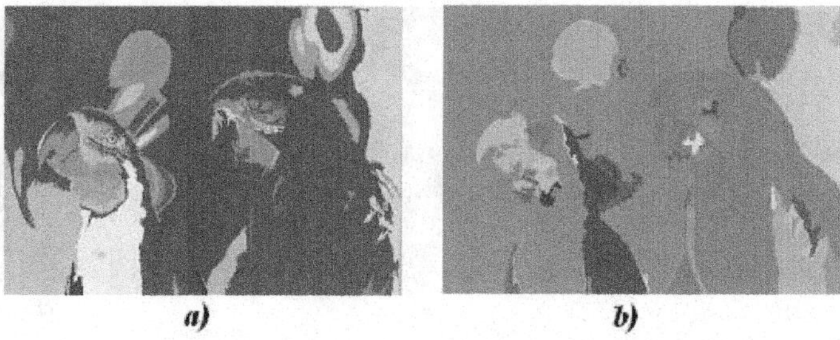

Fig. 9. *Grey segmentations of the luminance (a) and the hue (b). Both are depicted in false colour.*

Fig. 10. *a) Segmentation of the saturation (presented in grey tones); b)optimal threshold of a); c) final synthetic partition, superimposed to the initial image.*

saturations respectively, and in assigning the hue partition to the first set, and the luminance one to the second. A class of the synthetic partition is either the intersection of a luminance class with the low saturation zone X_s^c, or the intersection of a hue class with the high saturation zone X_s. If the classes of the luminance, the hue, and the synthetic partition at point x are denoted by $A_m(x)$, $A_h(x)$, and $A(x)$ respectively, we have

$$A(x) = A_m(x) \cap X_s^c \quad \text{when} \quad x \in X_s^c \qquad (26)$$
$$A(x) = A_h(x) \cap X_s \quad \text{when} \quad x \in X_s.$$

The simplest way to generate the set X_s consists, of course, in thresholding the saturation image. But this risks to result in an irregular set X_s, with holes, small particles, etc. Preferably, one can start from the mosaic image of the saturation provided by the same segmentation algorithm as for the the hue and the luminance (Fig.10a). An optimal threshold on the saturation histogram determines the value for the a-chromatic/chromatic separation (Fig.10b). By applying Rel.(26) we finally obtain the composite partition depicted in Fig.10c, which is excellent.

5 The unit circle and the hue

The unit circle, like the round table of King Arthur's knights, has no order of importance, and no dominant position. In mathematical terms, this signifies that we cannot construct a lattice on the unit circle, unless assigning it an arbitrary origin. This is a severe verdict against morphological treatments (i.e. operators relying on lattices) when we use them on the unit circle.

However, is it really impossible to bypass this interdiction? If we consider the standard morphological operators, three paths at least seem possible, that A. Hanbury and J. Serra investigated in [11]. Only he first path is presented here, because it was designed it by transposing Matheron's idea of working on increments, that underlies the whole linear geostatistics. He did it because the range of fluctuations of the grades, in some mineral deposits, seems practically infinite, although the increments at finite distances remain finite. In the case of the hue, we are not disturbed by the infinity, but by the choice of an origin, which arbitrarily forces the reds to be the smallest values and the purples largest ones (see for example the histogram of Fig.12). Can we transfer to the circular case the class of operators which bring into play differences only, such as gradients, top-hats, medians, etc.?

5.1 Circular centered operators

We fix an origin a_0 on the unit circle C with centre o by, for example, choosing the topmost point, and indicate the points a_i on the circle by their curvilinear coordinate in the trigonometric sense between 0 and 2π from a_0. Given two points a and a', we use the notation $a \div a'$ to indicate the value of the acute angle aoa', i.e.

$$a \div a' = \begin{cases} |a - a'| & \text{if } |a - a'| \leq \pi \\ 2\pi - |a - a'| & \text{if } |a - a'| \geq \pi \end{cases} \tag{27}$$

If the a_i are digital values between 0 and 255 (for example), the expression "$\leq \pi$" becomes "≤ 127", and "2π" becomes "255". However, we continue using the notation in terms of π, as it is more enlightening. Rel.(27) appears in [23] applied to the treatment of the hue band of colour images.

5.2 Circular hue gradient

We know that in the Euclidean space $\mathbb{R}^n$, to determine the modulus of the gradient, at point x, of a numerical differentiable function f, one considers a small sphere $S(x, r)$ centered on x with radius r. Then one takes the supremumq of the increments $|f(x) - f(y)|$, where y describes the small sphere $S(x, r)$, i.e.

$$g(x, r) = [\vee \{|f(x) - f(y)|, y \in S(x, r)\}] / r \tag{28}$$

Finally, one determines the limit of the function $g(x, r)$ as r tends to zero. This limit exists as the function f is differentiable in x. In the two-dimensional

Fig. 11. *a) Pere Serra's Painting of the Virgen (detail, St Cugat Monastery, Barcelona); b) corresponding image of the hue in HLS system (image size 352 × 334 pixels).*

digital case, it is sufficient to apply Rel.(28), taking for $S\,(x,r)$ the unit circle centered on x (square or hexagon). This is the classic Beucher algorithm for the gradient.

Consider now an image of hues or of directions, i.e. a function $h : E \to C$, where E is an Euclidean or digital space, and C is the unit circle. As the previous development only involves increments, we can transpose Rel.(28) to the circular function h by replacing all the $|h\,(x) - h\,(y)|$ by $|h\,(x) \div h\,(y)|$. This transposition then defines the modulus of the gradient of the circular distribution. For example, in $\mathbb{Z}^n$, $K\,(x)$ indicates the set of neighbours at distance one from point x, hence

$$|\mathrm{grad}\ \mathrm{h}|\,(x) = [\vee\,\{|h\,(x) \div h\,(y)|\,, y \in K\,(x)\}] \qquad (29)$$

As an illustration, consider the hue component of Fig.11a, shown in Fig.11b. This image was chosen as it is mostly red in colour, and in the angular hue encoding, red usually has hue values around 0°. This means that pixels which appear red could have low hue values (for example, 0° to 30°) and high hue values (330° to 360°). A large discontinuity is therefore visible in the hue image, with red pixels appearing at the extremities of the histogram (Fig.12). A classical gradient on this hue band produces a large number of spurious high-valued pixels, as shown in Fig.13a.

These high-values are present even though the neighbouring pixels appear very similar in colour, and are due to the discontinuity in the hue encoding. A good illustration of this is the outer part of the halo, which appears smooth in Fig.11a, but results in very high gradients in Fig.13a. The gradient calculated using Eq.(29), shown in Fig.13b, overcomes this problem. As its range is rela-

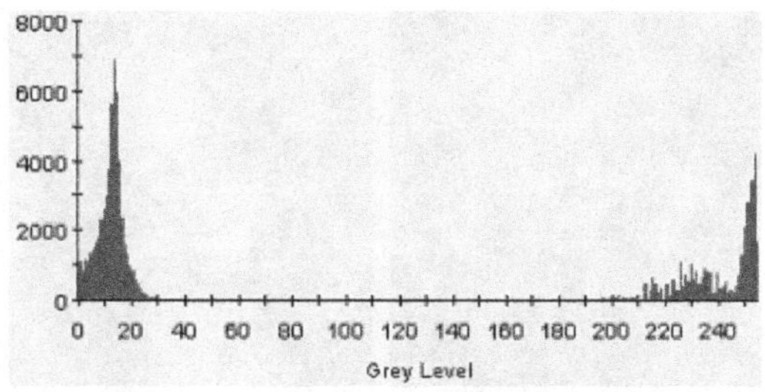

Fig. 12. *Histogram of the hue band.*

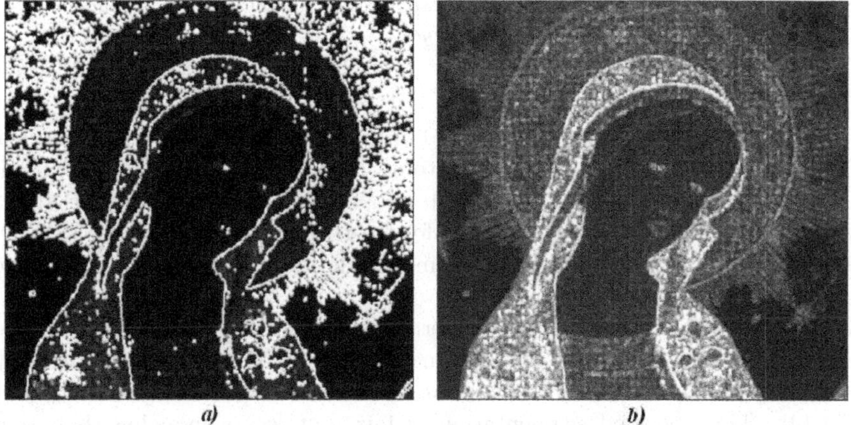

Fig. 13. *a) Classical morphological gradient on the hue band; b) Circular centered gradient on the hue band. The gradients were calculated using a* 3×3 *structuring element*

tively small, we reinforced its contrast for the display, and the texture of the underlying canvas becomes visible. Note that if we rotate the hue band pixel values by π, the classical gradient will be the same as the angular gradient. The angular gradient is, however, invariant to rotations of the pixel values.

5.3 A use of the circular hue gradient

Consider a colour function f, and its L_1 norm representation. The circular gradient of the hue of f can be inserted in a global gradient |grad f| which takes the three components of f into account. There are number of such gradients, but the above comments on the saturation of Eq. (16) in L_1 norm representation suggests us to construct a *barycentric* gradient, where the L_1 saturation balances the effects of luminance and hue gradients, i. e.

$$|\text{grad f}| = s \times |\text{grad h}| + (1 - s) \times |\text{grad l}|. \tag{30}$$

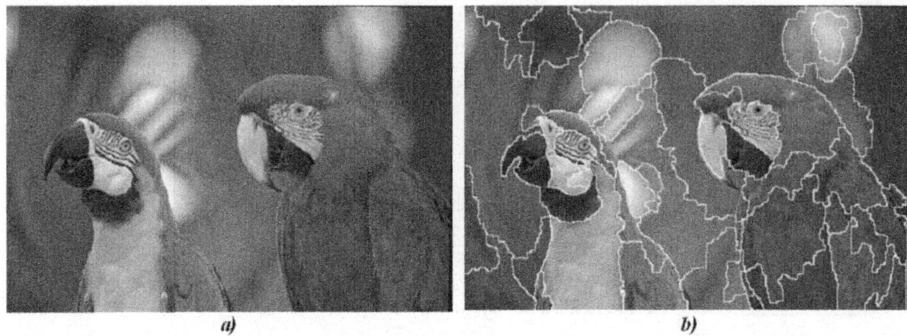

Fig. 14. *a) Initial "Parrots" image; b) watershed partition at level 4 of the waterfall pyramid (superimposed to the initial image).*

It is well known that the watershed of the gradient |grad f| of a numerical function f provides us with the contours of f [19]. This property extends to the vector functions of the space, because then the term |grad f| is still a numerical function, which is sufficient for the theory. We can then use Rel.(30) to segment the parrots image, and compare the result with that we obtained by combined partitions in Fig.10c .

Under iteration, the watershed operator turns out to generate the so-called waterfall pyramid, which is non parametric, and increases the partitions [19]. For the parrots, the best fit occurs at the fourth level of the hierarchy, see Fig.14b. The obtained segmentation is fair, but less convincing than that by combined partitions of Fig.10c : in the watershed process, the reduction (colour)→(numerical) by Rel.(30) arises too early.

6 Colour Interpolation

6.1 Morphological interpolators

During the nineties, the need for image interpolation arose with the development of video coding. Several proposed algorithms were efficient, though no theory justified them; one could not say whether they were optimal in some sense, just as kriging derives from a variance minimization for example. I asked G.Matheron which concept could serve as a substitute for variance in morphological interpolation, and he suggested me to see whether Haudorff metric admitted geodesics. The Hausdorff distance ρ is a metric defined on the class of the non empty compact sets of $\mathbb{R}^n$ by the relation

$$\rho(X, Y) = \inf \{\lambda : X \subseteq \delta_\lambda(Y) \; ; \; Y \subseteq \delta_\lambda(X)\} \tag{31}$$

where δ_λ stands for the Minkowski dilation by the ball of size λ. The value ρ indicates a degree of similarity between the two sets. A geodesic (if it exists) between sets X and Y from distance μ apart is an ordered family $\{Z_\lambda,\ 0 \leq \lambda \leq \mu\}$ of sets. Each Z_λ is at distance λ from X and $\rho - \lambda$ from Y. They are optimal interpolators in the sense that for any λ the sum $\rho(X, Z_\lambda) + \rho(Z_\lambda, Y) = \lambda + (\mu - \lambda)$ is minimum.

Matheron's intuition was right: the Hausdorff distance admits geodesics, and more than one [28]. However:

1. the intermediary sets Z_λ are always inflated, but one geodesic swells less than the others;
2. they are not self dual, as the interpolator of X^c and Y^c differs from the complement of the interpolator between X and Y ;
3. when one of the two sets, X say, is shifted or rotated, the Z_λ's are modified.

Moreover, the duality between dilation and erosion w.r.t. complementation suggests to play with both Hausdorff distances by dilations, on $X \cap Y$, and by erosions, on $X \cup Y$. Then, the less inflating geodesic permits a self dual variant which gives, for $\lambda = 0.5\rho$, the so called *morphological median*

$$M(X,Y) = \bigcup_{\sigma > 0} [(X \cap Y) \oplus \sigma B] \cap [(X \cup Y) \ominus \sigma B] \tag{32}$$

(assuming $X \cap Y \neq \emptyset$). Concerning the third point, one can always find a displacement that minimizes the distance $\rho(X, Y)$, hence optimizes the median [28]. Rel.(32) establishes a common theoretical basis to the previous works of F. Meyer [20], and S. Beucher [5]. Nice developments based on distance function have also been proposed by P. Soille [34], J.R. Casas [7] and by P. Moreau and Ch. Ronse [18].

M. Iwanowski found a method to determine the best displacement, and generated sequences of geodesics by subdividing morphological medians [14] [13]. A binary example is depicted in Fig.15.

6.2 A false colour case

In Rel.(32), the passage to the corresponding numerical version is straightforward. As the operator increases with both arguments, it suffices to replace all $\cup$ and $\cap$ by $\vee$ and $\wedge$ respectively. The extension to colour images is more subtle. For the first time in this paper, we have to produce a *colour* image, and neither a partition, nor a modulus of gradient. By treating separately the three bands, of any representation, we would obtain new colours that risk to parasite the quality of the result. The only way to be sure that every colour vector of the transform is already present in the initial image consists in *totally* ordering the set of the colour vectors. There are many ways to generate such a lattice. For the scenes of the everyday life M. Iwanowski tried several possibilities and finally adopted an ordering with the luminance $l = 0.3r + 0.6g + 0.1b$ first, then g and finally r [12].

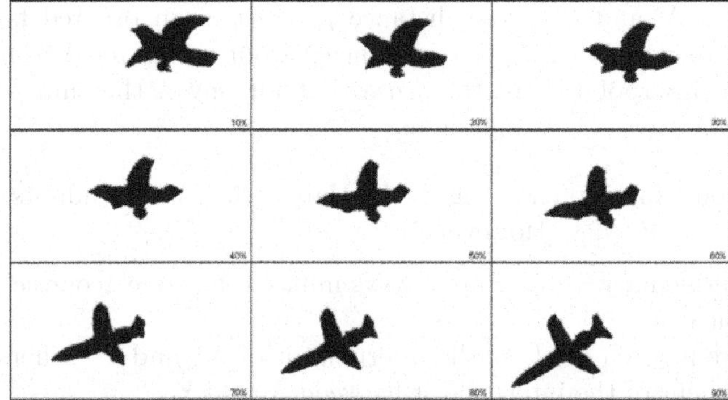

Fig. 15. Morphological interpolations based on the Hausdorff distance.

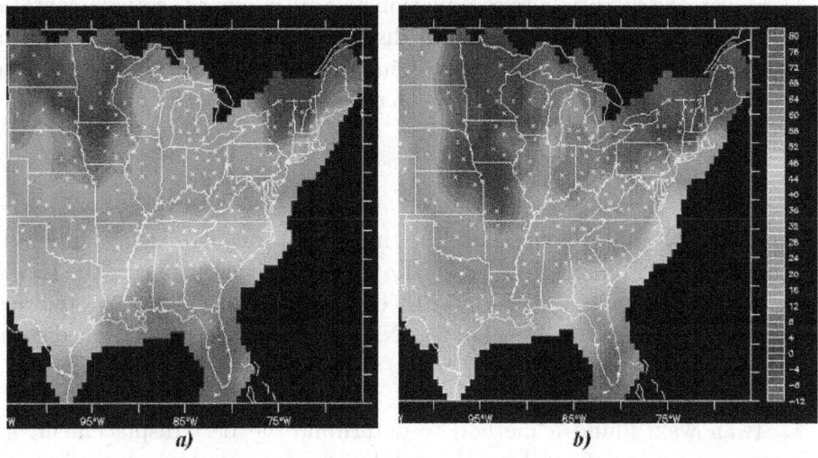

Fig. 16. *Maps of the temperatures at two consecutive days.*

But the reference to G. Matheron leads me propose an example in the themes of geostatistics. The two maps depicted in Fig.16 come from the website of the American Meteorological Institute, and indicate the daily maximum temperature on January the 5th and the 6th, 1996. They are themselves interpolations from point measurements, but we do not mind about it and take them as an input. The complete series comprises seven maps, for one week, and which cover the whole country. The final interpolated sequence contains 102 images. Here we just calculate the morphological median between the two maps of Fig.16.

The gamut of colours, artificial, has been established in such a way that the temperatures linearly decrease with the hue. Moreover, for graphic reasons, the saturation is maximum everywhere. Not only the circularity argument

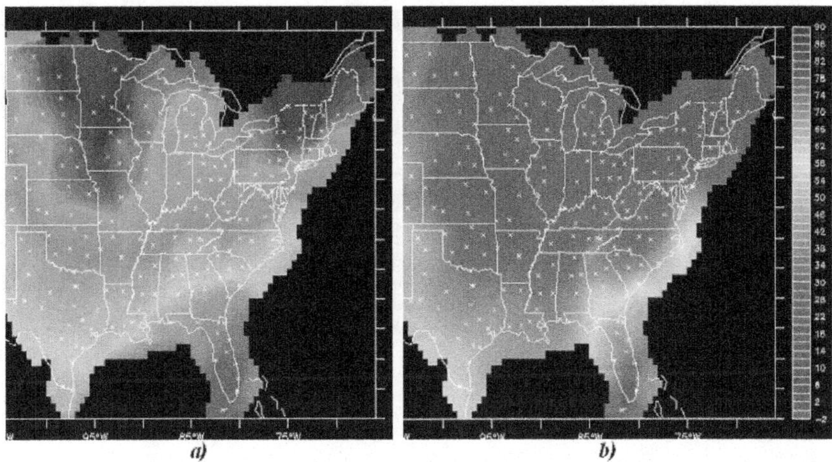

Fig. 17. *a) Morphological median between the two maps of figure 16; b) linear interpolation between the same maps.*

against the hue vanishes, but this parameter is the most representative of the physics of the phenomenon. Therefore the convenient ordering is as follows:

$$f'(h,l,s,) \geq f(h,l,s) \text{ when either } h' < h$$
$$\text{or } h' = h \text{ and } l' > l$$
$$\text{or } h' = h, l' = l \text{ and } s' > s$$

It results in the morphological median depicted in Fig.17a. The arithmetic mean of the same maps is placed at the right (Fig.17b) for the sake of comparison. The morphological operator better follows the fronts because it keeps unchanged the number of gradations of the temperatures scale.

7 A morphological approach to multivariate analysis

This last study on colour imagery commemorates G. Matheron less by the approach than by the theme of the application. Indeed mathematical morphology is born on the occasion petrographical problems [17], and it is my pleasure to conclude the survey by an example which recalls our beginnings.

In digital processing, it occurs that a series of different descriptors be attached to each pixel, and that one wishes to segment the space into the homogenous regions on the basis of this information. The descriptors may combine data that are physically heterogeneous. For example, in geographic information systems (G.I.S.), each pixel is sometimes assigned a slope and a population density. These two quantities, although connected, are physically

Fig. 18. *a) and b) two polarized views of the Lipka sandstone thin section.*

non comparable, and their mean, hence a linear descriptor, is just a physical nonsense. Such an heterogeneity becomes more complicated again when some of the quantities vary over the unit circle (as the colours obtained by various polarization angles), and when the colour information has been coded by palettes. It will be the case here, but our approach, though designed for this current case, applies to any type of heterogeneous multivariate data. The underlying idea is the following. By segmenting separately each of the n variables, we obtain n *partitions*, which are data of the same type, and without the initial heterogeneity. The goal then consists in combining these partitions for producing a synthetic one. One possible solution is developed below: a numerical function is generated by adding indicator functions associated with each partition, then this numerical function is segmented.

7.1 Polarized thin sections

Polarized light is a useful tool for examination of rock thin sections. It assigns a specific hue to each rock component and allows distinguishing and separating the grains according to the crystalline orientations [21]. But it is generally used in a qualitative way.

In this section we purpose to use a partition based approach for reaching the same goal, but in a quantitative manner. In other words, starting from a sequence of polarized images of a same microscopic field, we propose to determine automatically the contours of the grains.

J. Serra and M.Mlynarczuk have treated four rocks of different microscopic appearance, in order to base the proposed method on a significant spectrum of rock structures [29]. They are two sandstones from Tumlin and Lipka, one quartzite from Wiśniówka and one dolomite from Redziny. From each selected rock a thin section was prepared. Then a chosen field was observed in a polarized microscopy, in such a way that the polarization prism was turned 15 times by 12 degrees. The obtained sequence (15 images of 8 bits depth

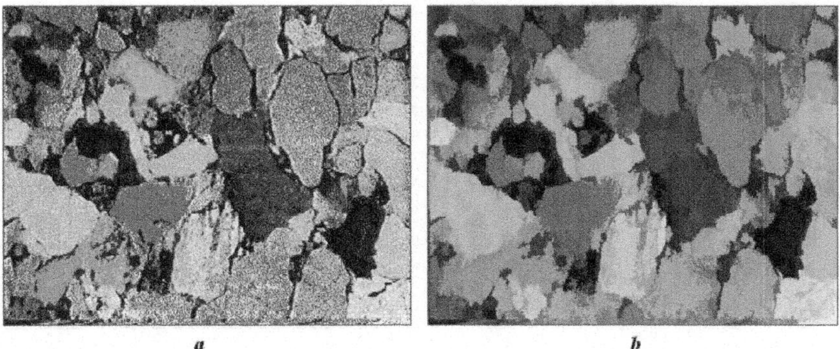

Fig. 19. *a) artificial grey tone image which is associated with the palette of Fig. 18a; b) flat zones filtering of image a), namely image fi.*

and of size 352 x 268), was digitized by means of a CCD camera. The colour of each image has been coded on 256 levels by using a specific palette for each image. Fig.18 gives an outline of the obtained images. By observing the images, one can notice that:

- for the same thin section, the variations of colours between the images observed in different polarizations are considerable;
- for a same grain, and under a given polarization, the hue remains more or less uniform;
- the crystals are placed side by side, without any visible border;
- finally, we can notice that the images are of poor quality.

7.2 Proposed approach

A palette is a look-up table, i.e. a mapping $[0, 255] \times [0, 255] \times [0, 255] \rightarrow [0, 255]$. Therefore, the 256 output values do not constitute a perceptual greytone axis. In a same image the luminance associated with the level 120, say, may be higher or lower than that of level 140. Moreover, if at point x, two different images of a same sequence have the same numerical value, this does not mean that the two images have the same colour or the same luminescence at point x, because the palette changes from one image to another. But, fortunately, in each image, the majority of points of the same crystals admit the same numerical values. This happens in most cases, but not always, because the crystals are not absolutely pure.

Classically, a processing of multidimensional data begins by reducing the number of dimensions by means of linear multivariate analysis techniques, such as principal component analysis or the Karunen-Loève transformation. When there are no more than one or two major variables, different filters and

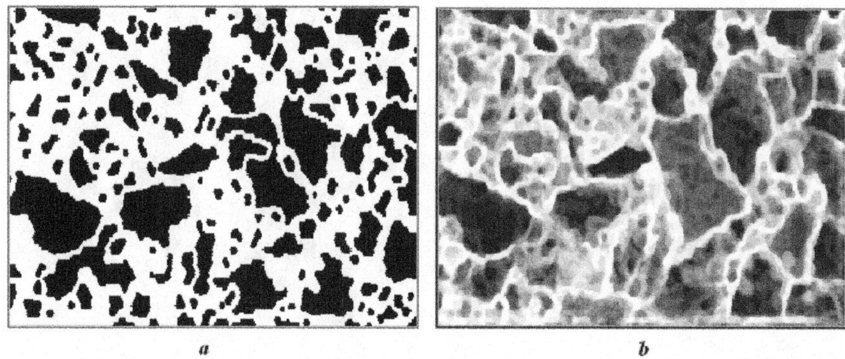

Fig. 20. *a) Black and white image m_i obtained by partition erosion ; b) Sum of all m_i images of the sequence, resulting in a probability map.*

segmentations conclude the processing. Here linearity is inadvisable and we have to invent another approach [29].

Our procedure turns out to be the exact opposite of multivariate analysis. If we want an approach to be valid for dimensionally heterogeneous data, it is recommended not to try and mix them in a first stage, which should be more or less linear. On the contrary, we must:

i)   first associate a partition of the space with each individual image. A partition $\mathcal{D}$ of a space E is a segmentation of E into classes $D(x)$ that do not overlap, and that cover the whole space E. For this stage, we will have to *erode* partitions, in order to enhance boundaries, by applying the following result

Proposition 1. *Given an arbitrary set E, every set erosion $\varepsilon^* : \mathcal{P}(E) \to \mathcal{P}(E)$ for which $\varepsilon^*(\emptyset) = \emptyset$ induces, on the lattice $\mathcal{D}$ of the partitions of E, a unique erosion ε defined by:*

$$(\varepsilon D)(x) = \begin{cases} \varepsilon^*[D(x)] & when\ x \in \varepsilon^*[D(x)] \\ \{x\} & when\ not\ . \end{cases}$$

ii)   and on the second stage only, group the obtained partitions into a synthetic one. To do this, one considers them as different realizations of a random partition, and one estimates the probability for each pixel to be at a given distance d from the border of its class. In this probability map, the closer to the border a pixel is, the higher is its numerical value. Finally, the watershed of this probability map provides the wanted segmentation.

7.3 An example

We will illustrate the algorithm by means of the Lipka sandstone sequence.

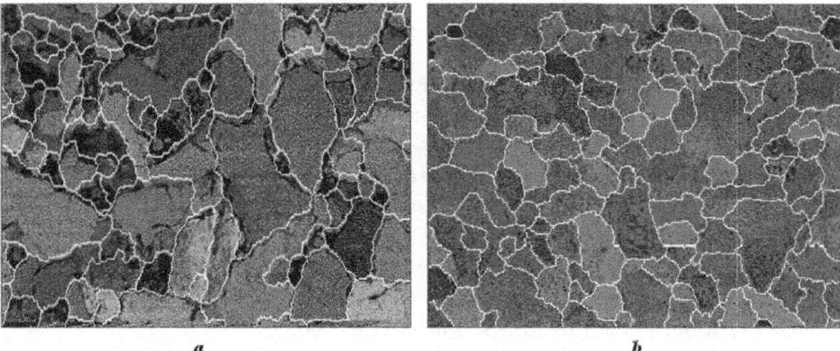

a   b

Fig. 21. *a) Watershed of the probability map Fig. 20 b), in superimposition with one of the polarized views. Same procedure, and same parameters applied to a quartzite from Wisniowka.*

First stage:

Filter each image of the sequence by flat structuring elements. The optimal filter is the smaller that eliminates the various defects, since the defects are always smaller than the grains. The solution adopted here consists an alternating sequential filter by reconstruction of size two [26]. We call f_i the images after filtering, where i indicates the label of the image in its sequence (Fig.19b). The obtained flat zones are then eroded by applying Proposition 1 for an hexagonal erosion of size two. In such a process, each class of the partition, i.e. each flat zone, is narrowed independently of the others, and the rest of the space is occupied by classes reduced to points. In Fig.20a, the non point classes are given value zero, and the others value one. Let m_i denote the obtained image.

Second stage

The border probability map is estimated by the estimate $m = 255 - 15(\sum m_i)$. Fig.20b depicts the image m associated with Lipka sequence. The final detection of the contours is provided by the watersheds of image m. Since we are interested in large regions without internal details, we take as a marker, for watershed, the set of the maxima of function m, after an opening of size 3. On Fig.21a the final watershed lines are superimposed on one of the polarized images of the initial sequence. Fig.21b depicts the result of the same algorithm, with the same parameters, for Wiśniówka quartzite sequence.

Acknowledgement. The author gratefully thanks Prof. F.Meyer, Dr P.Dokladal and Dr J.Angulo for their valuable comments and the improvements they suggested.

References

1. J. Angulo, Colour segmentation using bivariate histograms in 3D-polar colour spaces, *Rapport Technique CMM-Ecole des Mines de Paris, N-03/03/MM*, Jan. 2003.
2. J. Angulo, J. Serra, Color segmentation by ordered mergings, in *Proc. of IEEE International Conference on Image Processing (ICIP'03)*, IEEE, Vol. 2, Barcelona, Spain, Sept. 2003, p. 125–128.
3. J. Angulo, *Morphologie mathématique et indexation d'images couleur. Application à la microscopie en biomédecine.* Thèse doctorale, Centre de Morphologie Mathématique, Ecole des Mines, Paris, Dec. 2003.
4. J. Angulo, J. Serra, Traitements des images de couleur en représentation luminance/saturation/teinte par norme L_1, *to be published in traitement du signal in 2005.*
5. S. Beucher, Interpolation of sets, of partitions and of functions, *Mathematical Morphology and its applications to image and signal processing*, H.Heijmans and J. Roerdink eds Kluwer,1998.
6. T. Carron, *Segmentations d'images couleur dans la base Teinte-Luminance-Saturation: approche numérique et symbolique*,Thèse doctorale, Univ. de Savoie, 1995.
7. J.R. Casas, *Image compression based on perceptual coding techniques*, PhD thesis, UPC, Barcelona, March 1996.
8. Commission Internationale de l'Eclairage (CIE), Colorimetry, Second Edition. *CIE Publication No. 15.2*, Vienna, 1986.
9. G. Choquet, "Topologie", Academic Press, N.Y., 1966.
10. Claire-Hélène Demarty and Serge Beucher, Color segmentation algorithm using an HLS transformation. In *Proceedings of the International Symposium on Mathematical Morphology (ISMM '98)*, p. 231-238, 1998.
11. A. Hanbury, J. Serra, Morphological operators on the unit circle, *IEEE Transactions on Image Processing*, Vol. 10, No. 12, 2001, p.1842–1850.
12. M. Iwanowski and J. Serra, Morphological Interpolation and Colour Images, *Proc. International Conference on Image Processing* ICIAP'99 Venice, Sept.1999.
13. M. Iwanowski and J. Serra, The Morphological-affine object deformation, *Mathematical Morphology and its Applications to Image and Signal Processing*, J. Goutsias, L. Vincent, D.S. Bloomberg (Eds.) Kluwer Ac.Publ. 2000, pp.81-90.
14. M. Iwanowski *Application de la Morphologie Mathématique pour l'interpolation d'images numériques* Phd thesis Ecoles des Mines de Paris- Ecole Polytechique de Varsovie, 15 nov. 2000.
15. A. Hanbury, J.Serra, Colour Image Analysis in 3D-polar coordinates, in *Proc. of DAGM symposium*, Vienna, April 2003.
16. H. Levkowitz, G.T. Herman, GLHS : a generalized lightness, hue and saturation color model, *Graphical Models and Image Processing*, Vol. 55, No. 4, 1993, p.271–285.
17. G. Matheron and J. Serra, The birth of Mathematical Morphology, *in Mathematical Morphology*, H. Talbot, R. Beare, Edts Proc. ISMM 2002, CSIRO Sydney 2002 pp.1-16.
18. Moreau P., and Ronse Ch., Generation of shading-off on images by extrapolations of Lipschitz functions, *Graph. Models and Image Processing*, **58**(6), July 1996, pp. 314-333.

19. F. Meyer, S. Beucher, Morphological Segmentation, *J.of Visual Communication and Image Representation*, Vol. 1, No. 1, 1990, p.21–46.

20. F. Meyer A morphological interpolation method for mosaic images, in *Mathematical Morphology and its applications to image and signal processing*, Maragos P. et al. eds. Kluwer, 1996.

21. M. Młynarczuk, Opis wybranych struktur skalnych przy użyciu metod morfologii matematycznej i analizy obrazów, *PhD thesis*, IMG PAN, (1998)

22. G. Obein, K. Knoblauch, F. Viénot, Perceptual scaling of the gloss of a one-dimensional series of painted black samples, *Perception*, Vol. 31, Suppl., 2002, p.65.

23. R. A. Peters II. Mathematical morphology for angle-valued images, in *Non-Linear Image Processing VIII*. SPIE volume 3026, 1997.

24. Ch. Poynton *A technical Introduction to Digital Video*. New York: Wiley, 1996. Chapter 6, "Gamma" is available online at *http://www.poyton.com/PDFs/TIDV/Gamma.pdf*

25. J. Pokorny, V.C. Smith, S.J. Starr, Variability of color mixture data I. The effect of viewing field size on the unit coordinates, *Vision Research*, Vol. 16, 1976, p.1095–1098.

26. P. Salembier and J. Serra, Flat Zones Filtering, Connected Operators, and Filters by Reconstruction, *IEEE Transactions on Image Processing*. Aug. 1995, vol. 4, n° 8, 1153-1160 .

27. J. Serra Connectivity for sets and functions, *Fundamenta Informaticae, 41 (2000) 147-186*

28. J. Serra, Hausdorff distance and Interpolations, *Mathematical Morphology and its applications to image and signal processing*, H.Heijmans and J. Roerdink eds Kluwer,1998, pp.107-115

29. J. Serra and M.Mlynarczuk, Morphological merging of multidimensional data, *Proc. STERMAT'2000*, Cracow, Sept. 2000, pp.385-390 and 455.

30. J. Serra, Espaces couleur et traitement d'images, *Rapport Technique CMM-Ecole des Mines de Paris, N-34/02/MM*, Oct.2002.

31. J. Serra, A lattice approach to Image segmentation, *Rapport Technique CMM-Ecole des Mines de Paris, N-02/04/MM*, 87 p. Janv. 2004 (to be published by JMIV)

32. S.A. Shafer, Using color to separate reflection components from a color image, *Color Research and Applications*, Vol. 10, No. 4, 1985, p. 210–218.

33. A.R. Smith, Color gammet transform pairs, *Computer Graphics*, Vol. 12, No. 3, 1978, p.12–19.

34. P. Soille, Spatial Distribution from Contour Lines: An efficient Methodology based on Distance Transformations, *J. Vis. Com. and Im. Under.* Vol.2 n°2 , June 1991, pp.128-150.

35. A.Tremeau, Ch. Fernandez-Maloigne, P. Bonton, *Image numérique couleur*, Ed. Dunod, Paris, 2004.

Automatic design of morphological operators

Junior Barrera[1], Gerald J. F. Banon[2], and Edward R. Dougherty[3]

[1] Departamento de Ciência da Computação. Universidade de São Paulo
[2] Divisão de Processamento de Imagens, Instituto Nacional de Pesquisas Espaciais
[3] Department of Electrical Engineering, Texas A & M University

1 Introduction

A central paradigm in mathematical morphology is the decomposition (representation) of complete lattice operators (mappings) in terms of four classes of elementary operators: dilations, erosions, anti-dilations and anti-erosions. The rules for performing these representations can be described as a formal language, the *morphological language* [4]. The vocabulary of this language is composed of the four classes of elementary operators and the lattice operations of intersection and union. A phrase of the morphological language is called a *morphological operator.*

The adequacy of morphological operators to solve image and signal processing problems show that the morphological language is expressive, meaning that many useful operators can be represented by relatively few words. The completeness of the morphological language was first studied by Matheron in 1975, in his classical book, *Random Sets and Integral Geometry* [35], p.219). In the context of translation invariant set operators (operators that commute with translation), Matheron introduced the notion of the operator kernel, which is a family of sets whose images by the operator cover the origin. He proved that any increasing (i.e., inclusion preserving) and translation invariant set operator can be represented by a union of erosions, with structuring elements in the operator kernel. Ten years later, in 1985, Maragos [33], and Dougherty and Giardina [16], independently, proved that Matheron's decomposition result could be simplified in the sense that just a minimal subset of the kernel, called the operator *basis*, is enough to perform the decomposition. Maragos included a topological condition for the existence of the basis, that operator to be decomposed be upper semicontinuous under the hit-miss topology. Maragos extended these results to function operators [34].

Some years later, in 1991, Banon and Barrera proved that any translation invariant set operator could be decomposed as the union of sup-generating operators (i.e., intersections of erosions and anti-dilations), with structuring elements that are extremities of intervals included in the operator kernel [2].

Banon and Barrera also proved the existence of a basis for the family of upper semi-continuous operators under the hit-miss topology. The results of Matheron, Maragos, and Dougherty and Giardina for increasing set operators are particular cases of the general result. From a suggestion of Matheron, Banon and Barrera also applied their decomposition to operators that can be built as the intersection of an increasing and a decreasing operator (i.e., an operator that inverts the inclusion relation).

A couple of years later, in 1993, Banon and Barrera found the notion of morphological connection, which extends the notion of Galois connection. While a Galois connection induces a dual isomorphism between the set of dilations and erosions, the morphological connection induces an isomorphism between the set of morphological sup-generating operators and a set formed by the Cartesian product of dilations and anti-dilations. This property and the generalization of the notion of kernel are the keys to finding a constructive representation for any lattice operator in terms of a union of sup-generating operators. This decomposition also admits a simplified form parameterized by the operator basis. All previous decomposition results become particular cases of this general decomposition of lattice operators [3].

A remarkable property of the lattice-operator decomposition is that it applies for both continuous and discrete lattices. The decomposition of discrete operators that are both translation invariant and locally defined, meaning the value of a transformed signal in a given point depends just of a neighborhood of this point in the input signal, permits the development of a technique for the design of morphological operators based on the observation of examples of the desired transformation, the sequence being of pairs formed from the input and the desired ideal output. This technique consists essentially in the estimation of the basis of an operator that minimizes a given statistical error measure. After about a decade of research Dougherty, Barrera and their collaborators have obtained theoretical and applied results that prove the adequacy of discrete techniques for solving many image and signal processing problems [12], [8], [10], [14]. A salient example is the design of debluring filters that have better performance than the classical optimal Wiener filter [27]. Two points are at issue here. First, the Wiener filter is optimal among linear filters, and linear filters cannot provide sufficient restoration for many types of blur. Second, the morphological deblurring operators are not increasing, which is what gives them the ability to perform optimal nonlinear deblurring.

The techniques developed by Dougherty and Barrera can also be included in the context of automatic programing for image processing. In this case, the input and ideal output image pairs are the program formal specification and the designed program is expressed as a morphological operator. In fact, the intuition for the development of this automatic operator design approach comes from years of experience designing morphological operators by the classical approach developed by Serra and his cooperators. They design image transformations to solve image processing problems from a toolbox of morphological operators, that are integrated by operator composition and lattice operations.

The choice of the appropriate tools and their integration is guided by the experience and intuition of the morphologist. The morphological operator created is tested against a dataset of typical images and the ones that produce smallest error are chosen. In the sense of the morphologist, error is an intuitive visual notion. Dougherty and Barrera modeled the design process considering that image pairs are realizations of join processes and the goal is designing an operator that predicts ideal output images from a transformation of the corresponding input images. A critical issue for choosing good operators was finding a statistical measure that could mimic the morphologist intuition. In general the formal measures do not agree with intuitive human criteria, but at the limit they do. Human intuition and formal measures agree at least when two images are very similar and, happily, this is enough to design useful operators. However, the knowledge of the morphologist remains crucial in automatic design. Complex problems would require prohibitive amount of training data, what is overcome by optimization constraints based on operators properties studied in mathematical morphology.

Following this introduction, Section 2 recalls the canonical decomposition of set W-operators. Section 3 recalls the design of set W-operators. Section 4 recalls the morphological and loose canonical decompositions of lattice operators. Section 5 presents applications of the canonical lattice decompositions to gray-scale image operators design. Finally, Section 6 discusses the impact of this research and shows some future perspectives of the field.

2 Set W-operator decomposition

In this section, we present the family of set W-operators and give their representation, in terms of sup-generating operators, and characterization, in terms of Boolean functions.

2.1 Set W-operator

The set E is assumed to be an Abelian group with respect to a binary operation denoted $+$. The zero element of $(E, +)$ is denoted by o. This zero element is also called the origin of E.

Let X^t denote the *transpose* of a subset X of E, that is, $X^t \triangleq \{y \in E : y = -x, x \in X\}$.

For any $h \in E$ and $X \subseteq E$, the set $X_h \triangleq \{x \in E : x - h \in X\}$ is called the *translation* of X by h. In particular, $X_o = X$.

Let $\mathcal{P}(E)$ denote the power set of E. A *set operator* is any mapping defined from $\mathcal{P}(E)$ into itself. The set $\text{Fun}(\mathcal{P}(E), \mathcal{P}(E))$ of all the operators from $\mathcal{P}(E)$ to $\mathcal{P}(E)$ inherits the complete lattice structure of $(\mathcal{P}(E), \subseteq)$ on setting, $\forall \Psi_1, \Psi_2 \in \text{Fun}(\mathcal{P}(E), \mathcal{P}(E))$, $\Psi_1 \leq \Psi_2 \Leftrightarrow \Psi_1(X) \subseteq \Psi_2(X)$,

$\forall X \in \mathcal{P}(E)$. The supremum and infimum of a subset $\mathcal{T}$ of the complete lattice $(\mathrm{Fun}(\mathcal{P}(E), \mathcal{P}(E)), \leq)$ verify $(\bigvee \mathcal{T})(X) = \bigcup \{\Upsilon(X) : \Upsilon \in \mathcal{T}\}$ and $(\bigwedge \mathcal{T})(X) = \bigcap \{\Upsilon(X) : \Upsilon \in \mathcal{T}\}$, $\forall X \in \mathcal{P}(E)$.

A set operator Ψ is called *translation invariant (t.i.)* if and only if (iff), $\forall h \in E$, $\Psi(X_h) = \Psi(X)_h$, $\forall X \in \mathcal{P}(E)$.

Let W be a finite subset of E. A set operator Ψ is called *locally defined on* W iff, $\forall h \in E$, $h \in \Psi(X) \Leftrightarrow h \in \Psi(X \cap W_h)$, $\forall X \in \mathcal{P}(E)$.

Let $\mathcal{T}_W$ denote the collection of t.i. operators locally defined on W. The elements of $\mathcal{T}_W$ are called *set W-operators* or, simply, *W-operators*. The pair $(\mathcal{T}_W, \leq)$ constitutes a sublattice of the lattice $(\mathrm{Fun}(\mathcal{P}(E), \mathcal{P}(E)), \leq)$.

2.2 Representation

The *kernel* $\mathcal{K}(\Psi)$ of a W-operator Ψ is the subcollection of $\mathcal{P}(W)$ defined by $\mathcal{K}(\Psi) \triangleq \{X \in \mathcal{P}(W) : o \in \Psi(X)\}$.

Proposition 1. *The mapping $\mathcal{K}$ from $\mathcal{T}_W$ to $\mathcal{P}(\mathcal{P}(W))$ defined by, for any $\Psi \in \mathcal{T}_W$,*

$$\mathcal{K}(\Psi) \triangleq \{X \in \mathcal{P}(W) : o \in \Psi(X)\}$$

constitutes a lattice isomorphism between the lattices $(\mathcal{T}_W, \leq)$ and $(\mathcal{P}(\mathcal{P}(W)), \subseteq)$. The inverse of the mapping $\mathcal{K}$ is the mapping $\mathcal{K}^{-1}$ defined by, for any $\mathcal{X} \subseteq \mathcal{P}(W)$ and $X \in \mathcal{P}(E)$,

$$\mathcal{K}^{-1}(\mathcal{X})(X) \triangleq \{x \in E : X_{-x} \cap W \in \mathcal{X}\}.$$

As a consequence of the last proposition, the following equalities hold: for any $\Psi_1, \Psi_2 \in \mathcal{T}_W$, $\mathcal{K}(\Psi_1 \wedge \Psi_2) = \mathcal{K}(\Psi_1) \cap \mathcal{K}(\Psi_2)$ and $\mathcal{K}(\Psi_1 \vee \Psi_2) = \mathcal{K}(\Psi_1) \cup \mathcal{K}(\Psi_2)$.

The set operator Γ, defined by $\Gamma(X) = X^c$, $\forall X \in \mathcal{P}(E)$, is called *negation*. Let $A, B \in \mathcal{P}(E)$. The operations

$$A \oplus B \triangleq \bigcup \{A_b : b \in B\} \text{ and } A \ominus B \triangleq \bigcap \{A_{-b} : b \in B\}$$

are called, respectively, *Minkowski addition and subtraction*. Let $B \in \mathcal{P}(E)$. The t.i. set operators Δ_B and E_B defined by $\Delta_B(X) = X \oplus B$ and $E_B(X) = X \ominus B$, for any $X \in \mathcal{P}(E)$, are called, respectively, dilation and erosion by B. The parameter B that characterizes a dilation or an erosion is called a *structural element* or a *structuring element* .

Let $A, B \in \mathcal{P}(W)$ such that $A \subseteq B$. The t.i. set operator $\Lambda_{[A,B]}$ defined by, for any $X \in \mathcal{P}(E)$,

$$\Lambda_{[A,B]}(X) \triangleq \{x \in E : A \subseteq X_{-x} \cap W \subseteq B\},$$

is called *sup-generating* operator. This operator was first stated by Serra ([42], p.39) in the form $\Lambda_{[A,B^c]}$ and called *hit-miss operator*.

Note that $\Lambda_{[A,B]}$ is locally defined on W and can be, equivalently, represented by, for any $X \in \mathcal{P}(E)$,

$$\Lambda_{[A,B]}(X) = E_B(X) \cap \Gamma \Delta_{B^{tc}}(X),$$

where the complement of B is taken relative to W.

Given $A, B \in \mathcal{P}(W)$, the subcollection $[A, B]$ of $\mathcal{P}(W)$ defined by

$$[A, B] \stackrel{\Delta}{=} \{X \in \mathcal{P}(W) : A \subseteq X \subseteq B\}$$

is called an *interval*.

Proposition 2. *The kernel of a sup-generating operator is an interval, that is,*

$$\mathcal{K}(\Lambda_{[A,B]}) = [A, B].$$

Theorem 1. *If Ψ is a W-operator, then, for any $X \in \mathcal{P}(E)$,*

$$\Psi(X) = \bigcup \left\{ \Lambda_{[A,B]}(X) : [A, B] \subseteq \mathcal{K}(\Psi) \right\}.$$

Proof. $\mathcal{K}(\Psi) = \bigcup \{[A, B] \subseteq \mathcal{P}(W) : [A, B] \subseteq \mathcal{K}(\Psi)\}$, since any subset of a complete lattice can be built by the union of its intervals; thus, $\mathcal{K}(\Psi) = \bigcup\{\mathcal{K}(\Lambda_{[A,B]}) : [A, B] \subseteq \mathcal{K}(\Psi)\}$, by Proposition 2, and

$$\Psi = \bigvee \left\{ \Lambda_{[A,B]} : [A, B] \subseteq \mathcal{K}(\Psi) \right\},$$

since $\mathcal{K}$ is a lattice isomorphism between $(\mathcal{T}_W, \leq)$ and $(\mathcal{P}(\mathcal{P}(W)), \subseteq)$.

The decomposition of the last theorem is called *canonical sup-decomposition*. Though this decomposition is quite general, it may lead to inefficient computational representation for most W-operators, in the sense that a smaller family of sup-generating operators may be sufficient to represent the same operator.

An interval $[A, B]$ is called maximal in a collection of intervals **I** iff, $[A, B] \in$ **I** and, $\forall [A', B'] \in$ **I**, $[A, B] \subseteq [A', B']$ implies that $[A, B] = [A', B']$. The set $\mathbf{B}(\Psi)$ of all maximal intervals contained in $\mathcal{K}(\Psi)$ is called *basis* of Ψ.

Theorem 2. *If Ψ is a W-operator, then, for any $X \in \mathcal{P}(E)$,*

$$\Psi(X) = \bigcup \left\{ \Lambda_{[A,B]}(X) : [A, B] \in \mathbf{B}(\Psi) \right\}.$$

Proof. Once W is finite, we can express $\mathcal{K}(\Psi)$ in terms of its maximal intervals, that is, $\mathcal{K}(\Psi) = \bigcup \{\mathcal{K}(\Lambda_{[A,B]}) : [A, B] \in \mathbf{B}(\Psi)\}$. The result follows by the same arguments used to prove Theorem 3.

We should observe that there are also dual representations in terms of the kernel and basis of a W-operator. These dual representations are called *inf-canonical,* since they are the intersection of *inf-generating operators* (i.e., dual of sup-generating operators).

In particular, when the set W-operator Ψ is increasing $[A, B] \in \mathbf{B}(\Psi)$ implies that $B = W$ and the canonical sup-decomposition reduces to an union of erosions, that is, for any $X \in \mathcal{P}(E)$,

$$\Psi(X) = \bigcup \{E_A(X) : A \in \mathcal{B}(\Psi)\},$$

$\mathcal{B}(\Psi) = \{A : [A, B] \in \mathbf{B}(\Psi)\}$. This last representation is exactly the one proposed by Matheron in ([35], p.219).

2.3 Characterization

We have seen that W-operators can be represented by their kernel or basis. We will study now a third way of representing W-operators: the equivalent Boolean function.

Let T be the mapping defined from $\mathcal{T}_W$ to $\mathrm{Fun}(\mathcal{P}(W), \{0, 1\})$ defined by, for any $X \in \mathcal{P}(E)$,

$$T(\Psi)(X) \triangleq \begin{cases} 1 & \text{if } o \in \Psi(X) \\ 0 & \text{otherwise} \end{cases}$$

The mapping T constitutes a lattice isomorphism between the complete lattices $(\mathcal{T}_W, \leq)$ and $(\mathrm{Fun}(\mathcal{P}(W), \{0, 1\}), \leq)$ and its inverse T^{-1} is defined by, for every $\psi \in \mathrm{Fun}(\mathcal{P}(W), \{0, 1\})$, for any $X \in \mathcal{P}(E)$,

$$T^{-1}(\psi)(X) \triangleq \{x \in E : \psi(X_{-x} \cap W) = 1\}.$$

3 Set W-operator design

The morphological representation theory provides a natural framework for automatic operator design. An operator is desired to optimally estimate an image when it is observed after going through a system.

3.1 Unconstrained design

To frame the problem, binary digital images are modeled as discrete random sets. The task is to design an operator Ψ so that, given an observed random set S, $\Psi(S)$ is probabilistically close to a desired (ideal) random set I. The closeness of the ideal and the estimator $\Psi(S)$ is measured by some probabilistic error measure $Er[I, \Psi(S)]$. Assuming the operator belongs to some family $\mathcal{I}$, an *optimal operator* relative to $\mathcal{I}$ is an operator $\Psi_{opt} \in \mathcal{I}$ for which

$Er[I, \Psi_{opt}(S)] \leq Er[I, \Psi(S)]$, for all $\Psi \in \mathcal{I}$. If every operator $\Psi \in \mathcal{I}$ has a representation, then optimization can be viewed as finding the representation defining an operator possessing minimum error $Er[I, \Psi_{opt}(S)]$.

When $\mathcal{I}$ is the family of set W-operators, estimation of I from S by a W-operator Ψ requires finding a Boolean function ψ to minimize error. Since, Ψ is translation-invariant, we make the modeling assumption that I and S are jointly strict sense stationary. Hence, if $\mathbf{X}$ is the random vector of binary values in the translate of W to z and $Y \triangleq I(z)$, then the joint probability distribution for $\mathbf{X}$ and Y is independent of z, so that estimating Y from $\mathbf{X}$ yields a translation invariant operator. We denote random variables and random vectors by upper-case italic and bold face letters, respectively. Realizations of the random variable Y and the random vector $\mathbf{X}$ will be denoted by y and $\mathbf{x}$, respectively.

For operator optimization we require a loss function $l : \{0,1\}^2 \rightarrow [0, \infty)$, where $l(a, b)$ measures the cost of the difference between a and b, with $l(0,0) = l(1,1) = 0$. Relative to the loss function (and owing to stationarity), filter error, $Er(\Psi)$, is given by the expected loss from estimating $Y \triangleq I(z)$ by $\psi(\mathbf{X}) \triangleq \Psi(S)(z)$, where z is an arbitrary pixel:

$$Er(\Psi) \triangleq E[l(Y, \psi(\mathbf{X})] = \sum_{\{\mathbf{x}:\psi(\mathbf{x})=0\}} l(1,0)P(Y = 1|\mathbf{x})P(\mathbf{x})$$

$$+ \sum_{\{\mathbf{x}:\psi(\mathbf{x})=1\}} l(0,1)P(Y = 0|\mathbf{x})P(\mathbf{x})$$

where $P(\mathbf{x})$ denotes $P(\mathbf{X} = \mathbf{x})$. An optimal operator is one whose Boolean function ψ minimizes $Er(\Psi)$. Although there can be more than one operator achieving minimal error, we shall denote "the" optimal operator and its Boolean function by Ψ_{opt} and ψ_{opt}, respectively, the convention being that, from the optimization view point, operators having minimal error are equivalent.

The *mean-absolute error* (MAE) loss function is defined by $l(y, \psi(\mathbf{x})) = |y - \psi(\mathbf{x})|$. Since y and $\psi(\mathbf{x})$ are binary-valued, the loss function is given by $l(1,0) = l(0,1) = 1$ and $l(0,0) = l(1,1) = 0$. An optimal operator is determined by $\psi_{opt}(\mathbf{x}) = 1$ if $P(Y = 1|\mathbf{x}) > 0.5$ and $\psi_{opt}(\mathbf{x}) = 0$ if $P(Y = 1|\mathbf{x}) \leq 0.5$. The MAE for an optimal operator is given by

$$MAE[\Psi_{opt}] = \sum_{\{\mathbf{x}:\psi(\mathbf{x})=1\}} P(Y = 0|\mathbf{x})P(\mathbf{x}) + \sum_{\{\mathbf{x}:\psi(\mathbf{x})=0\}} P(Y = 1|\mathbf{x})P(\mathbf{x})$$

If a suboptimal operator Ψ is used instead of Ψ_{opt}, then there is an increase in error. To quantify the error increase, if $P(\mathbf{x}) > 0$, define the *advantage* of $\mathbf{x}$ by

$$Ad_l(\mathbf{x}) \triangleq (E[l(Y,0)|\mathbf{x}]) - E[l(Y,1)|\mathbf{x}]))P(\mathbf{x})$$

$Ad_l(\mathbf{x}) > 0$ iff $\mathbf{x} \in \mathcal{K}[\psi_{opt}]$. An increase in error can arise in two ways from using Ψ instead of Ψ_{opt}: $\mathbf{x} \in \mathcal{K}[\psi_{opt}]$ but $\mathbf{x} \notin \mathcal{K}[\psi_{opt}]$, or $\mathbf{x} \notin \mathcal{K}[\psi_{opt}]$ but $\mathbf{x} \in \mathcal{K}[\psi_{opt}]$. The total error increase from using Ψ instead of Ψ_{opt} is

$$\Delta(\Psi, \Psi_{oopt}) \triangleq Er[\Psi] - Er[\Psi_{opt}] = \sum_{\mathbf{x} \in \mathcal{K}[\psi_{opt}] \Delta \mathcal{K}[\psi]} |Ad_l(\mathbf{x})|$$

the last sum being over the symmetric difference between the kernels. For MAE, the absolute advantage for $\mathbf{x}$ is given by $|1 - 2P(Y = 1|\mathbf{x})P(\mathbf{x})|$.

In practice, the optimal operator is statistically estimated from image realizations. Operator design involves using a random sample S_n of pairs $(\mathbf{X}^{(1)}, Y^{(1)})$, $(\mathbf{X}^{(2)}, Y^{(2)})$,..., $(\mathbf{X}^{(n)}, Y^{(n)})$ to form an estimate ψ_n of ψ_{opt}. The error $Er[\psi_n]$ can not be less than the error $Er[\psi_{opt}]$. Letting $\Delta_n \triangleq Er[\psi_{opt}] - Er[\psi_n]$ denote the *design cost* (operator estimation error), the error of the designed operator is decomposed as $Er[\psi_n] = Er[\psi_{opt}] + \Delta_n$. Hence, the expected error of the designed operator is $E[Er[\psi_n]] = Er[\psi_{opt}] + E[\Delta_n]$. The standard design approach is to estimate the conditional probabilities composing the decision criterion ($\psi_{opt}(\mathbf{x}) = 1$ if $P(Y = 1|\mathbf{x}) > 0.5$) and to use these estimates to determine ψ_n. This method yields a consistent estimate of $Er[\psi_n]$, that is, $E[\Delta_n] \to 0$ as $n \to \infty$.

Example 1. This example [10] presents a solution for the classical problem of detection of defect lines on euthetic alloy images. The window considered is the 5×5 square centered at the origin and the training images used are the ones of Figure 1. The training data gives 29040 examples from which 11785 are distinct. The basis of the designed operator has 418 intervals. Some results of application of the designed operator are given in Figure 2.

3.2 Constrained design

Satisfactory image filtering typically requires large windows, and it is often impossible to get large enough samples to sufficiently reduce $E[\Delta_n]$. To ease the design problem, optimization is constrained to some subclass $\mathcal{C}$ of operators. If $\psi_{\mathcal{C}}$ is an optimal operator in $\mathcal{C}$ with error $Er[\psi_{\mathcal{C}}]$ and design error $\Delta_{n,\mathcal{C}}$, then $Er[\psi_{\mathcal{C}}] \geq Er[\psi_{opt}]$ and $E[\Delta_{n,\mathcal{C}}] \leq E[\Delta_n]$. The error of a designed constrained operator, $\psi_{n,\mathcal{C}}$, possesses the decomposition $Er[\psi_{n,\mathcal{C}}] = Er[\psi_{\mathcal{C}}] + \Delta_{n,\mathcal{C}}$. The cost of constraint is given by $\Delta_{\mathcal{C}} = Er[\psi_{\mathcal{C}}] - Er[\psi_{opt}]$. Hence, $Er[\psi_{n,\mathcal{C}}] = Er[\psi_{opt}] + \Delta_{\mathcal{C}} + \Delta_{n,\mathcal{C}}$, and

$$E[Er[\psi_{n,\mathcal{C}}]] = Er[\psi_{opt}] + \Delta_{\mathcal{C}} + E[\Delta_{n,\mathcal{C}}]$$

A constraint is statistically beneficial iff $E[Er[\psi_{n,\mathcal{C}}]] \leq E[Er[\psi_n]]$, which is true iff

$$\Delta_{\mathcal{C}} \leq E[\Delta_n] - E[\Delta_{n,\mathcal{C}}]$$

The saving in design error must exceed the cost of constraint. Since for a consitent estimator $E[\Delta_n] - E[\Delta_{n,\mathcal{C}}] \to 0$ as $n \to \infty$, a constraint can only be beneficial for samples that are not too large.

A fundamental problem is to find constraints for which $\Delta_{\mathcal{C}} \leq E[\Delta_n] - E[\Delta_{n,\mathcal{C}}]$. The benefit of a constraint depends on the class of ideal and observed signals under consideration. For instance, suppose a signal is degraded by an extensive operator (the transformed set always includes the input set). Then, letting $\mathcal{C}$ be the class of anti-extensive operators (the transformed set always is included in the input set) yields $\Delta_{\mathcal{C}} = 0$, so that there is no constraint error. Such situations rarely occur in practice. Here we describe some constraints that have been studied and proven useful. In many cases the error has been analyzed and quantified; however, we leave that to the literature.

The most studied constraint is that the operator be increasing [12]. Several design methods have been employed to design increasing operators. One way is to find the kernel of the optimal operator and then apply a switching algorithm that derives the optimal increasing operator from the optimal operator by switching structuring elements in and out of the kernel of the designed optimal operator to obtain an increasing operator for which the switching error is minimal [36][25][43]. Once the increasing operator is designed, its basis representation in terms of erosions can be obtained by logically reducing the kernel expansion. A second way to proceed is to estimate the erosion expansion directly. This is achieved by first obtaining MAE estimates for single erosion operators and then recursively obtaining MAE estimates for multiple-erosion operators[31]. The MAE of an m-erosion operator Ψ_m can be expressed in terms of a single-erosion operator with structuring element B_m and two $(m-1)$-erosion operators Ψ_{m-1} and Φ_{m-1}:

$$MAE[\Psi_m] = MAE[\Psi_{m-1}] - MAE[\Phi_{m-1}] + MAE[B_m],$$

where the bases are given by $\mathcal{B}(\Psi_{m-1}) = \{B_1, B_2, ..., B_{m-1}\}$, $\mathcal{B}(\Psi_m) = \{B_1, B_2, ..., B_m\}$ and $\mathcal{B}(\Phi_{m-1}) = \{B_1 \cup B_m, B_2 \cup B_m, ..., B_{m-1} \cup B_m\}$. Further constraint can be imposed by limiting the basis size or constraining the search to a subclass of potential structuring elements [30]. Two adaptive procedures have been proposed to design increasing operators. One is based on gradient-type structuring-element adaptation akin to the classical LMS algorithm for linear filters [39]. Another utilizes genetic algorithms to adjust the structuring elements [22]. Comparison of design error for increasing and non-increasing operators has been investigated [17].

One way to classify constraints is to verify whether they are independent or dependent[14]. A constraint is *independent* if the decision whether to place a vector $\mathbf{x}$ in the kernel is constrained by a condition involving only $\mathbf{x}$ itself, and no other vectors. A *dependent* constraint is one that cannot be applied to each vector independently. This means there are required relations among vectors that do not reduce to independent constraints. Increasingness is a dependent constraint.

Envelope constraint is an independent constraint that involves two humanly designed operators α and β, such that $\alpha \leq \beta$. $\mathcal{C}$ is the class of all operators ψ such that $\alpha \leq \psi \leq \beta$. Envelope design is a form of human-machine operator design [6]. If the human designed envelope contains the optimal operator, then $\Delta_{\mathcal{C}} = 0$; if not, then $\Delta_{\mathcal{C}} > 0$ and the constraint is beneficial iff $\Delta_{\mathcal{C}} \leq E[\Delta_n] - E[\Delta_{n,\mathcal{C}}]$. The key to design is having a sufficiently tight envelope so that design error is reduced with overly increasing constraint error.

In image processing, it is often the case that the variables near the center of the window contribute most to the operator, whereas those on the window periphery contribute less, while enormously increasing the demand for data. In this case, one can apply secondary constraint [41]. The variables near the window center are used in an unconstrained manner, whereas those at the periphery are constrained in how they contribute to the operator. The situation here is that $\Delta_{\mathcal{C}}$ is not too large owing to the lesser importance of peripheral variables in determining Y.

Iterative design involves an operator-decomposition constraint. A large window W is decomposed into a Minkowski sum of windows, $W = W_1 \oplus W_2 \oplus \ldots \oplus W_q$, and $\mathcal{C}$ constrains operators of the form $\Psi = \Psi_q \Psi_{q-1} \ldots \Psi_1$, where ψ_k is defined on W_k [40], [26]. Not only do iterative operators require less sample data, they can also possess implementation advantages. $\Delta_{\mathcal{C}}$ depends on the degree to which the optimal operator can be approximated by an iterative operator, relative not only to algebraic decomposition, but also relative to the action of the operator on the random signal process. Iteration can be utilized in conjunction with increasing constraint [20].

Rather than operating on an image at a given resolution, a mapping can be applied to reduce the resolution [15]. While information is lost with resolution reduction, thereby introducing a constraint error, the number of pixels in the window is reduced and this results in smaller design error. One way to take advantage of this is to try to find a resolution at which the optimal operator has minimum error. A better way is, for each observation, decide whether it has been observed sufficiently in training to be confident in the optimal operator at its resolution. If it has been, then apply the operator designed at that resolution; if not, then take lower resolution versions of the vector until it has been observed sufficiently, and then apply an operator designed at an appropriate lower resolution.

Constrained design represents a use of prior information: based on an image degradation model or experience, the cost of constraint is known to be small. Another way to utilize prior knowledge is to begin with a heuristically chosen operator and redefine the kernel if there is sufficient data to indicate a change [7]. Specifically, if ψ_{pri} is the prior operator, define $\psi_{opt}(\mathbf{x}) = \psi_{pri}(\mathbf{x})$ unless the estimate of $P(Y = 1|\mathbf{x})$ is sufficiently precise to change the definition. The method has been used successfully in digital document processing with the prior operator being the identity, in which case it is known as the method of differencing filters [18].

A more sophisticated form of prior knowledge assumes the conditional probability $P(Y = 1|\mathbf{x})$ posses a prior distribution and the data is used to obtain the Bayes estimate of $P(Y = 1|\mathbf{x})$ relative to the prior distribution [13]. The method is difficult to use owing to the large number of prior distributions required. To help alleviate this problem, it has been used in conjunction with multi-resolution analysis, where its potential effectiveness has been demonstrated [28].

4 Decomposition of lattice operators

In the lattice theory, a subset $\mathcal{X}$ of a lattice $(\mathcal{L}, \leq)$ is a *sup-generating family for* $(\mathcal{L}, \leq)$ if any element of $\mathcal{L}$ is the supremum of elements of $\mathcal{X}$. Dually, $\mathcal{X}$ is an *inf-generating family for* $(\mathcal{L}, \leq)$ if any element of $\mathcal{L}$ is the infimum of elements of $\mathcal{X}$ ([23], p.28).

The concepts of sup- and inf-generating family apply directly to the problem of decomposition of a lattice element in terms of some predefined lattice elements.

We denote by $\bigvee \mathcal{X}$ (resp. $\bigwedge \mathcal{X}$) the supremum (resp. infimum) of a subset $\mathcal{X}$ of a lattice $(\mathcal{L}, \leq)$.

One can prove the following general result.

Proposition 3. *(decomposition in terms of a sup-generating family) - The subset $\mathcal{X}$ of $\mathcal{L}$ is a sup-generating family for the lattice $(\mathcal{L}, \leq)$ iff, for any Y in $\mathcal{L}$,*

$$Y = \bigvee \{X \in \mathcal{X} : X \leq Y\}.$$

Dually, we have the decomposition in terms of an inf-generating family.

A very nice and now historical example of sup-generating family for a special class of image operators is the Matheron's decomposition ([35], p.219) recalled in Section 2.

We know that the set of translation invariant increasing operators on binary images is a complete sub-lattice of the binary image operator lattice ([24], Proposition 3.1).

Hence, Matheron's decomposition can be stated by saying that the binary image erosions form a sup-generating family for the lattice of translation invariant increasing operators on binary images.

Banon and Barrera's decomposition [2] which is an extension of Matheron's decomposition to non necessarily increasing operators, also recalled in Section 2, provides another example of sup-generating family.

They have proved that the set of Hit-or-Miss operators form a sup-generating family for the lattice of the translation invariant operators on binary images.

In this section, we show that the decomposition of binary image operators, presented in Section 2, is a special case of **two** types of decomposition of lattice operators (i.e., mappings between complete lattices).

In the next two subsections, we introduce the so-called loose sup-generating operators and then the so-called morphological sup-generating operators.

4.1 Loose sup-generating operators

Let $(\mathcal{L}_1, \leq)$ and $(\mathcal{L}_2, \leq)$ be two complete lattices. We denote by O and I, respectively, the least and greatest elements of these lattices. From now on, we will consider the set of operators from $\mathcal{L}_1$ to $\mathcal{L}_2$ equipped with the point-wise ordering. This structure is again a complete lattice by inheritance of the complete lattice structure of $(\mathcal{L}_2, \leq)$. It is denoted by $(\mathrm{Fun}(\mathcal{L}_1, \mathcal{L}_2), \leq)$ and called power lattice.

Let α and β be two mappings from $\mathcal{L}_2$ to $\mathcal{L}_1$, we denote by $\overline{\alpha\beta}$ the expression defined by, for any X in $\mathcal{L}_1$:

$$\overline{\alpha\beta}(X) \stackrel{\triangle}{=} \bigvee \{Y \in \mathcal{L}_2 : \alpha(Y) \leq X \leq \beta(Y)\}.$$

Definition 1. *(loose sup-generating operator) - Let $(\mathcal{L}_1, \leq)$ and $(\mathcal{L}_2, \leq)$ be two complete lattices. An operator ϕ is a* loose sup-generating operator *from $(\mathcal{L}_1, \leq)$ to $(\mathcal{L}_2, \leq)$ if there exist two mappings α and β from $\mathcal{L}_2$ to $\mathcal{L}_1$, such that $\phi = \overline{\alpha\beta}$.*

The set of loose sup-generating operators is denoted by Φ.

Let us give examples of loose sup-generating operators. Let U and V be two elements, respectively, in $\mathcal{L}_1$ and $\mathcal{L}_2$, and let $\psi_{(U,V)}$ be the operator from $\mathcal{L}_1$ to $\mathcal{L}_2$ defined by, for any X in $\mathcal{L}_1$,

$$\psi_{(U,V)}(X) \stackrel{\triangle}{=} \begin{cases} V & \text{if } X = U \\ O & \text{otherwise} \end{cases}.$$

Proposition 4. *(examples of loose sup-generating operators) - For any U and V, respectively, in $\mathcal{L}_1$ and $\mathcal{L}_2$, the operator $\psi_{(U,V)}$ is a loose sup-generating operator, that is $\psi_{(U,V)} \in \Phi$.*

Proof. Let U and V be two elements, respectively, in $\mathcal{L}_1$ and $\mathcal{L}_2$, and let $\alpha_{(U,V)}$ and $\beta_{(U,V)}$ be two mappings from $\mathcal{L}_2$ to $\mathcal{L}_1$ defined by, for any Y in $\mathcal{L}_2$,

$$\alpha_{(U,V)}(Y) \stackrel{\triangle}{=} \begin{cases} U & \text{if } Y = V \\ I & \text{otherwise} \end{cases} \quad \text{and} \quad \beta_{(U,V)}(Y) \stackrel{\triangle}{=} \begin{cases} U & \text{if } Y = V \\ O & \text{otherwise} \end{cases}.$$

For any U and V, respectively, in $\mathcal{L}_1$ and $\mathcal{L}_2$, and any X in $\mathcal{L}_1$,

$$\bigvee \{Y \in \mathcal{L}_2 : \alpha_{(U,V)}(Y) \leq X \leq \beta_{(U,V)}(Y)\} = \bigvee \begin{cases} \{V\} \text{ if } X = U \\ \emptyset \quad \text{ otherwise} \end{cases}$$

$$\text{(definition of } \alpha_{(U,V)} \text{ and } \beta_{(U,V)}\text{)}$$

$$= \begin{cases} V \text{ if } X = U \\ O \text{ otherwise} \end{cases}$$

$$\text{(supremum definition)}$$

$$= \psi_{(U,V)}(X)$$

$$\text{(definition of } \psi_{(U,V)}\text{)}.$$

That is, there exist α and β, namely $\alpha \stackrel{\Delta}{=} \alpha_{(U,V)}$ and $\beta \stackrel{\Delta}{=} \beta_{(U,V)}$ such that $\psi_{(U,V)} = \overline{\alpha\beta}$, in other words $\psi_{(U,V)} \in \Phi$.

We observe that more than one pair of mappings α and β may lead to the same loose sup-generating operator. For example, if $\alpha_1(Y) = O$ and $\beta_1(Y) = I$ and if the pair of mappings α_2 and β_2 differs from the pair α_1 and β_1 just at $Z \leq Y$, $Z \neq Y$, then we still have $\overline{\alpha_1\beta_1} = \overline{\alpha_2\beta_2}$.

The loose sup-generating operators have the following nice property.

Proposition 5. *(decomposition in terms of loose sup-generating operators) - The set Φ of loose sup-generating operators from $\mathcal{L}_1$ to $\mathcal{L}_2$ is a sup-generating family for the power lattice* $(\mathrm{Fun}(\mathcal{L}_1, \mathcal{L}_2), \leq)$*. Equivalently, for any ψ in* $\mathrm{Fun}(\mathcal{L}_1, \mathcal{L}_2)$,

$$\psi = \bigvee \{\phi \in \Phi : \phi \leq \psi\}.$$

Proof. From Proposition 3 it is sufficient to prove the second assertion. Let us divide the proof in two parts.

i. For any ψ in $\mathrm{Fun}(\mathcal{L}_1, \mathcal{L}_2)$, $\{\phi \in \Phi : \phi \leq \psi\}$ has ψ as an upper bound and $\bigvee \{\phi \in \Phi : \phi \leq \psi\}$ as the least upper bound, therefore, we have,

$$\bigvee \{\phi \in \Phi : \phi \leq \psi\} \leq \psi.$$

ii. For any ψ in $\mathrm{Fun}(\mathcal{L}_1, \mathcal{L}_2)$ and U in $\mathcal{L}_1$, by definition of $\psi_{(U,V)}$ and by Proposition 4, there exists a $\phi \in \Phi$ such that $\phi \leq \psi$ and $\psi(U) \leq \phi(U)$, namely $\phi \stackrel{\Delta}{=} \psi_{(U,V)}$ with $V = \psi(U)$. Therefore, by the supremum definition, for any U in $\mathcal{L}_1$, $\psi(U) \leq \bigvee \{\phi(U) : \phi \in \Phi \text{ and } \phi \leq \psi\}$ that is, by the pointwise ordering definition and the pointwise union definition, $\psi \leq \bigvee \{\phi \in \Phi : \phi \leq \psi\}$.

Hence, by the anti-symmetry of $\leq$, we have $\psi = \bigvee \{\phi \in \Phi : \phi \leq \psi\}$.

4.2 Morphological sup-generating operators

We now recall another class of operators ([3], Definition 4.1).

Definition 2. *(morphological sup-generating operator) - Let $(\mathcal{L}_1, \leq)$ and $(\mathcal{L}_2, \leq)$ be two complete lattices. An operator λ is a* morphological sup-generating *operator from $(\mathcal{L}_1, \leq)$ to $(\mathcal{L}_2, \leq)$ if λ satisfies, for any nonempty subset $\mathcal{X}$ of $\mathcal{L}_1$, the following property:*

$$\lambda\left(\bigwedge \mathcal{X}\right) \wedge \lambda\left(\bigvee \mathcal{X}\right) = \bigwedge \lambda(\mathcal{X}).$$

The set of morphological sup-generating operators is denoted by Λ.

From Proposition 5.5 in [3], we know that any morphological sup-generating mapping is the intersection of an erosion and an anti-dilation.

As it has been shown in [3] (see proof of Lemma 6.1), the operators of the type $\psi_{(U,V)}$ are also examples of morphological sup-generating operators.

As for the loose sup-generating operators, the morphological sup-generating operators have the following nice property.

Proposition 6. *(decomposition in terms of morphological sup-generating operators) - The set Λ of morphological sup-generating operators from $\mathcal{L}_1$ to $\mathcal{L}_2$ is a sup-generating family for the power lattice $(\mathrm{Fun}(\mathcal{L}_1, \mathcal{L}_2), \leq)$. Equivalently, for any ψ in $\mathrm{Fun}(\mathcal{L}_1, \mathcal{L}_2)$,*

$$\psi = \bigvee \{\lambda \in \Lambda : \lambda \leq \psi\}.$$

Proof. See proof of Lemma 6.1 in [3].

At this stage, we have introduced two sup-generating families for the power lattice. We already know that they intersect each other since the operators of the type $\psi_{(U,V)}$ belong to both. Let's be more conclusive.

In order to make the proof of the next proposition simpler to write, we recall, at once, the concept of operator kernel [3].

Let ψ be an operator in $\mathrm{Fun}(\mathcal{L}_1, \mathcal{L}_2)$, a mapping $\cdot\mathcal{K}$ from $\mathrm{Fun}(\mathcal{L}_1, \mathcal{L}_2)$ to $\mathrm{Fun}(\mathcal{L}_2, \mathcal{P}(\mathcal{L}_1))$ defined by, for any Y in $\mathcal{L}_2$,

$$\cdot\mathcal{K}(\psi)(Y) \stackrel{\Delta}{=} \{X \in \mathcal{L}_1 : Y \leq \psi(X)\}.$$

is called the *left kernel mapping with respect to the complete lattice* $(\mathcal{L}_2, \leq)$. The mapping $\cdot\mathcal{K}(\psi)$ from $\mathcal{L}_2$ to $\mathcal{P}(\mathcal{L}_1)$ is called the *left kernel* (or simply, *kernel*) of ψ.

Proposition 7. *(comparison between the loose and morphological sup-generating operators) - The set Λ of morphological sup-generating operators is included in the set Φ of loose sup-generating operators, in other words $\Lambda \subset \Phi$.*

Proof. Let ψ be the operator from $\mathcal{L}_1$ to $\mathcal{L}_2$, and let $\cdot\underline{\psi}$ and $\overline{\cdot\psi}$ be the mappings from $\mathcal{L}_2$ to $\mathcal{L}_1$ defined by, for any Y in $\mathcal{L}_2$,

$$\cdot\underline{\psi}(Y) \overset{\triangle}{=} \bigwedge \cdot\mathcal{K}(\psi)(Y) \qquad \text{and} \qquad \overline{\cdot\psi}(Y) \overset{\triangle}{=} \bigvee \cdot\mathcal{K}(\psi)(Y).$$

By Theorem 5.1 in [3], for any $\lambda \in \Lambda$, there exist two mappings α and β from $\mathcal{L}_2$ to $\mathcal{L}_1$, such that $\lambda = \overline{\alpha\beta}$, namely $\alpha \overset{\triangle}{=} \cdot\underline{\lambda}$ and $\beta \overset{\triangle}{=} \overline{\cdot\lambda}$, that is, by loose sup-generating operator definition, $\lambda \in \Phi$, in other words, $\Lambda \subset \Phi$.

Actually, the previous proposition is a consequence of a property of the so-called *morphological connection* ([3], Definition 5.2). Furthermore, if $\overline{\alpha\beta}$ is a morphological sup-generating operator, then, by Theorem 5.1 in [3], α and β are, respectively, a dilation and an anti-dilation. This means that α is necessarily increasing and β is necessarily decreasing.

We have chosen to qualify the sup-generating operators in Φ as *loose* because they don't necessarily have to be derived from an increasing and a decreasing mappings. Even though two inputs are comparable, in the loose case, the α and β outputs don't need to be comparable

We have chosen to qualify the sup-generating operators in Λ as *morphological* because they are derived from a dilation and an anti-dilation which are two elementary morphological mappings.

At first glance, one could think that the morphological decomposition is better than the loose one in the sense that we don't need to consider all the loose sup-generating operators to construct any lattice operator; the morphological sup-generating operators are sufficient.

Nevertheless, one can expect that more morphological than loose sup-generating operators may be needed in a minimal construction.

4.3 Constructive decompositions

In order to state the constructive decomposition of a lattice operator, we have to recall the concept of interval function ([3], Definition 6.2).

$\mathcal{I}$ is an *interval function* from $\mathcal{L}_2$ to $\mathcal{P}(\mathcal{L}_1)$ if for any Y in $\mathcal{L}_2$, $\mathcal{I}(Y)$ is a interval of $\mathcal{P}(\mathcal{L}_1)$ or the empty set. We denote by $\mathrm{IntFun}(\mathcal{L}_2, \mathcal{P}(\mathcal{L}_1))$ the set of interval functions from $\mathcal{L}_2$ to $\mathcal{P}(\mathcal{L}_1)$.

We denote by $\cdot\mathbf{A}_\mathrm{l}(\psi)$ the interval function collection defined by, for any ψ from $\mathcal{L}_1$ to $\mathcal{L}_2$,

$$\cdot\mathbf{A}_\mathrm{l}(\psi) \overset{\triangle}{=} \{\mathcal{I} \in \mathrm{IntFun}(\mathcal{L}_2, \mathcal{P}(\mathcal{L}_1)) : \mathcal{I} \leq \cdot\mathcal{K}(\psi)\}.$$

In the above statement, $\leq$ is the pointwise ordering on $\mathrm{Fun}(\mathcal{L}_2, \mathcal{P}(\mathcal{L}_1))$, that is the extension of the inclusion $\subseteq$ on $\mathcal{P}(\mathcal{L}_1)$ to $\mathrm{Fun}(\mathcal{L}_2, \mathcal{P}(\mathcal{L}_1))$.

Let $\mathcal{I}$ be an interval function from $\mathcal{L}_2$ to $\mathcal{P}(\mathcal{L}_1)$, we denote by $\alpha_\mathcal{I}$ and $\beta_\mathcal{I}$ the mappings from $\mathcal{L}_2$ to $\mathcal{L}_1$ defined by, for any Y in $\mathcal{L}_2$,

$$\alpha_\mathcal{I}(Y) \overset{\triangle}{=} \bigwedge \mathcal{I}(Y) \qquad \text{and} \qquad \beta_\mathcal{I}(Y) \overset{\triangle}{=} \bigvee \mathcal{I}(Y).$$

We call the mappings $\alpha_{\mathcal{I}}$ and $\beta_{\mathcal{I}}$ the *extremity functions of* $\mathcal{I}$. For any Y in $\mathcal{L}_2$, we have $\alpha_{\mathcal{I}}(Y) \leq \beta_{\mathcal{I}}(Y)$ or (exclusive) $(\alpha_{\mathcal{I}}(Y) = I$ and $\beta_{\mathcal{I}}(Y) = O)$.

By observing that for any X in $\mathcal{L}_1$, $\psi(X) = \bigvee \{Y \in \mathcal{L}_2 : Y \leq \psi(X)\}$, we can prove the following theorem.

Theorem 3. *(constructive decomposition in terms of loose sup-generating operators) - Let $(\mathcal{L}_1, \leq)$ and $(\mathcal{L}_2, \leq)$ be two lattices. Any operator ψ from $\mathcal{L}_1$ to $\mathcal{L}_2$ can be decomposed by a set of loose sup-generating operators and its constructive decomposition is*

$$\psi = \bigvee \left\{ \overline{\alpha_{\mathcal{I}} \beta_{\mathcal{I}}} : \mathcal{I} \in \cdot \mathbf{A}_{\mathrm{l}}(\psi) \right\}.$$

Let Δ and Δ^{a} be, respectively, the set of dilations and anti-dilations from $\mathcal{L}_2$ to $\mathcal{L}_1$.

We denote by $\cdot \mathbf{A}_{\mathrm{m}}(\psi)$ the interval function collection defined by, for any ψ from $\mathcal{L}_1$ to $\mathcal{L}_2$,

$$\cdot \mathbf{A}_{\mathrm{m}}(\psi) \triangleq \left\{ \mathcal{I} \in \mathrm{IntFun}(\mathcal{L}_2, \mathcal{P}(\mathcal{L}_1)) : \mathcal{I} \leq \cdot \mathcal{K}(\psi), \alpha_{\mathcal{I}} \in \Delta \text{ and } \beta_{\mathcal{I}} \in \Delta^{\mathrm{a}} \right\}.$$

When using only the morphological sup-generating operators, we have the following theorem.

Theorem 4. *(constructive decomposition in terms of morphological sup-generating operators) - Let $(\mathcal{L}_1, \leq)$ and $(\mathcal{L}_2, \leq)$ be two lattices. Any operator ψ from $\mathcal{L}_1$ to $\mathcal{L}_2$ can be decomposed by a set of morphological sup-generating operators and its constructive decomposition is*

$$\psi = \bigvee \left\{ \overline{\alpha_{\mathcal{I}} \beta_{\mathcal{I}}} : \mathcal{I} \in \cdot \mathbf{A}_{\mathrm{m}}(\psi) \right\}.$$

Proof. See proof of Theorem in [3].

We now make a comparison between the loose and morphological sup-generating operators in terms of their kernels.

Let α and β be two mappings from $\mathcal{L}_2$ to $\mathcal{L}_1$, we denote by $[\alpha, \beta]$ the interval function from $\mathcal{L}_2$ to $\mathcal{P}(\mathcal{L}_1)$ defined by, for any Y in $\mathcal{L}_2$,

$$[\alpha, \beta](Y) \triangleq \begin{cases} [\alpha(Y), \beta(Y)] & \text{if } \alpha(Y) \leq \beta(Y) \\ \emptyset & \text{otherwise} \end{cases}.$$

From the definitions of $\overline{\alpha \beta}$ and morphological connection ([3], Proposition 5.2), we can prove the following proposition.

Proposition 8. *(loose and morphological sup-generating operator kernel properties) - We have the following properties.*

i) Let ψ be a loose sup-generating operator from $(\mathcal{L}_1, \leq)$ to $(\mathcal{L}_2, \leq)$, and let α and β be two mappings from $\mathcal{L}_2$ to $\mathcal{L}_1$, such that $\psi = \overline{\alpha \beta}$, then $[\alpha, \beta] \leq \cdot \mathcal{K}(\psi)$.

ii) Let ψ be an operator from $\mathcal{L}_1$ to $\mathcal{L}_2$, and let $\alpha \triangleq \cdot \psi$ and $\beta \triangleq \overline{\cdot \psi}$, then $\cdot \mathcal{K}(\psi) \leq [\alpha, \beta]$. Furthermore, ψ is a morphological sup-generating operator from $(\mathcal{L}_1, \leq)$ to $(\mathcal{L}_2, \leq)$ if and only if $\cdot \mathcal{K}(\psi) = [\alpha, \beta]$.

The above proposition shows that the equality holds for and only for the morphological sup-generating operators.

4.4 Minimal decompositions

Since $\cdot\mathbf{A}_{\mathrm{m}}\left(\psi\right)$ is included in $\cdot\mathbf{A}_{\mathrm{l}}\left(\psi\right)$, the morphological decomposition is slimmer than the loose one. Nevertheless, when the decompositions are based on the maximal interval functions of $\cdot\mathbf{A}_{\mathrm{l}}\left(\psi\right)$ and $\cdot\mathbf{A}_{\mathrm{m}}\left(\psi\right)$, then we may come to the opposite conclusion.

For the sake of simplicity, we will assume, from now on, that the sets $\mathcal{L}_1$ and $\mathcal{L}_2$ are finite. We denote by $\mathrm{Max}\mathcal{A}$ the set of maximal elements of a subset $\mathcal{A}$ of a lattice.

Let $\cdot\mathbf{B}_{\mathrm{l}}\left(\psi\right)$ and $\cdot\mathbf{B}_{\mathrm{m}}\left(\psi\right)$ be, respectively, the collection of maximal interval functions of $\cdot\mathbf{A}_{\mathrm{l}}\left(\psi\right)$ and $\cdot\mathbf{A}_{\mathrm{m}}\left(\psi\right)$, that is, $\cdot\mathbf{B}_{\mathrm{l}}\left(\psi\right) = \mathrm{Max}\mathbf{A}_{\mathrm{l}}\left(\psi\right)$ and $\cdot\mathbf{B}_{\mathrm{m}}\left(\psi\right) = \mathrm{Max}\cdot\mathbf{A}_{\mathrm{m}}\left(\psi\right)$. We call them, respectively, the *loose (left) basis* and the *morphological (left) basis* of ψ.

For any $\mathcal{I}$ and $\mathcal{J}$ in $\mathrm{IntFun}(\mathcal{L}_2, \mathcal{P}(\mathcal{L}_1))$, $\mathcal{I} \le \mathcal{J}$ implies $\overline{\alpha_{\mathcal{I}}\beta_{\mathcal{I}}} \le \overline{\alpha_{\mathcal{J}}\beta_{\mathcal{J}}}$. Hence if $\mathcal{I}$ and $\mathcal{J}$ are two interval functions in $\cdot\mathbf{A}_{\mathrm{l}}\left(\psi\right)$ (resp., $\cdot\mathbf{A}_{\mathrm{m}}\left(\psi\right)$) and $\mathcal{I} \le \mathcal{J}$, then $\overline{\alpha_{\mathcal{I}}\beta_{\mathcal{I}}}$ has no contribution in the constructive loose (resp., morphological) decomposition of ψ. Therefore, in the finite case, we have the following two theorems.

Theorem 5. *(minimal decomposition in terms of loose sup-generating operators) - Let $(\mathcal{L}_1, \le)$ and $(\mathcal{L}_2, \le)$ be two finite lattices. Any operator ψ from $\mathcal{L}_1$ to $\mathcal{L}_2$ can be decomposed by a set of loose sup-generating operators and its minimal decomposition is*

$$\psi = \bigvee \left\{ \overline{\alpha_{\mathcal{I}}\beta_{\mathcal{I}}} : \mathcal{I} \in \cdot\mathbf{B}_{\mathrm{l}}\left(\psi\right) \right\}.$$

Theorem 6. *(minimal decomposition in terms of morphological sup-generating operators) - Let $(\mathcal{L}_1, \le)$ and $(\mathcal{L}_2, \le)$ be two finite lattices. Any operator ψ from $\mathcal{L}_1$ to $\mathcal{L}_2$ can be decomposed by a set of morphological sup-generating operators and its minimal decomposition is*

$$\psi = \bigvee \left\{ \overline{\alpha_{\mathcal{I}}\beta_{\mathcal{I}}} : \mathcal{I} \in \cdot\mathbf{B}_{\mathrm{m}}\left(\psi\right) \right\}.$$

The minimal loose decomposition can still be slimmed since it may involve redundant loose sup-generating operators:

$$\psi = \bigvee \mathrm{Max} \left\{ \overline{\alpha_{\mathcal{I}}\beta_{\mathcal{I}}} : \mathcal{I} \in \cdot\mathbf{B}_{\mathrm{l}}\left(\psi\right) \right\}.$$

For any operator ψ in $\mathrm{Fun}(\mathcal{L}_1, \mathcal{L}_2)$, we denote by $\cdot\mathbf{B}\left(\psi\right)$ the mapping from $\mathcal{L}_2$ to $\mathcal{P}(\mathcal{P}(\mathcal{L}_1))$ such that, for any Y in $\mathcal{L}_2$, $\cdot\mathbf{B}\left(\psi\right)\left(Y\right)$ is the subcollection of maximal intervals of $\cdot\mathcal{K}(\psi)\left(Y\right)$, that is, $\cdot\mathbf{B}\left(\psi\right)\left(Y\right) = \mathrm{Max}\,\cdot\mathcal{K}(\psi)\left(Y\right)$

We write $\mathcal{I}\ (\in)\ \cdot\mathbf{B}\left(\psi\right)$ to mean that, for any Y in $\mathcal{L}_2$, $\mathcal{I}(Y) \in \cdot\mathbf{B}\left(\psi\right)\left(Y\right)$. We can prove that $\mathcal{I}\ (\in)\ \cdot\mathbf{B}\left(\psi\right)$ is equivalent to $\mathcal{I} \in \cdot\mathbf{B}_{\mathrm{l}}\left(\psi\right)$ and consequently, for any X in $\mathcal{L}_1$,

$$\psi(X) = \bigvee \left\{ Y \in \mathcal{L}_2 : \exists I \in \cdot\mathbf{B}\left(\psi\right)(Y) : X \in I \right\}.$$

This expression, derived from the minimal loose decomposition, is attractive because it is computationally simple. It has been extensively used in [10] and [5].

When $\mathcal{L}_2$ is simply a chain with two elements then the minimal loose decomposition becomes a minimal morphological decomposition. This is a consequence of the fact that, in this case, for any $\mathcal{I} \in \cdot\mathbf{B}_1\left(\psi\right)$, $\alpha_{\mathcal{I}} \in \Delta$ and $\beta_{\mathcal{I}} \in \Delta^a$.

In other words, the decomposition of binary image operators is a special case of the **two** decompositions presented above.

5 Gray-scale operator design

Optimizing gray-scale operators is inherently more difficult than optimizing binary operators owing to the much greater complexity of gray-scale characteristic functions; nevertheless the problem has been addressed, for quite some time, for increasing operators, and more recently for non-increasing operators [19]. Owing to space limitations, we will confine ourselves to brief descriptions of the kind of gray-scale designs that have been studied. From the standpoint of representation, the lattice representations are again involved, with lattices being discrete-integer valued rather than binary.

The optimization problem for increasing gray-scale operators has been studied in several ways. Among the first issue addressed was finding the maximal search space for the structuring elements. Later, MAE theorems were derived in different settings, and these were used in much the same way as the MAE theorem in the binary case [31], [32]. The gradient-type adaptation mentioned earlier for binary operators has also been applied.

A constrained class of increasing operators are those that satisfy the threshold decomposition property. For any random variable X, define the binary random variable X^k by $X^k \overset{\Delta}{=} 1$ if $X \geq k$ and $X^k \overset{\Delta}{=} 0$ if $X < k$. ψ satisfies the *threshold decomposition property* if there exists an increasing binary operator ζ_ψ such that

$$\psi(\mathbf{X}) = \sum_{k=1}^{m} \zeta_\psi(\mathbf{X}^k)$$

where $\mathbf{X} \overset{\Delta}{=} (X_1, X_2, ..., X_d)$ and $\mathbf{X}^k \overset{\Delta}{=} (X_1^k, X_2^k, ..., X_d^k)$. Since ψ is defined via a single binary operator, it is in effect a binary operator. Its basis representation consists of erosions with flat structuring elements. Thus, such filters are called *flat filters* or *stack filters*. The design cost for flat filters is reduced in comparison to general increasing filters, but the constraint error is increased. The increase is sufficiently severe in most cases that flat filters are essentially confined to restoring signals degraded by additive noise. In operator design it is assumed that all threshold vectors $\mathbf{X}^k$ are identically distributed. Numerous design methods have been developed for flat filters [11], [1], [21], [29].

Up till recently the statistical design of non-increasing operators had been kept to very small windows or to operators that are combinations of increasing operators. The opportunities have been expanded with the introduction of aperture filters [27]. These are windowed in both domain and range, thereby reducing design complexity resulting from the gray-scale. Owing to reduced design error and larger windows, aperture filters have been shown to outperform unconstrained gray-scale operators. Indeed, they have been shown to outperform linear inverse (Wiener) filters for restoring images corrupted by nonlinear smoothing [27]. Active research on aperture filters is continuing, with various special types of aperture filters under investigation.

Example 2. In this example[10], the noise to be filtered consists of a composition of impulse plus horizontal dropout noise. The two-side impulse noise is uniformely distributed with probability of occurrence of 10% and has amplitude 200. Values greater than 255 and smaller than 0 were saturated, respectively, to 255 and 0. The dropout noise consists of horizontal line segments of intensity 255 with probability of occurrence 0.35% . The length of these segments follows a normal distribution with mean 5 and variance 49. Figure 3 shows the observed-ideal pairs of images used for the training of a stack filter. Image (a) in Figure 4 shows another realization of the same noise process, and image (b) and image (c) show, respectively, the same image filtered by a 1 and 5-iterations stack filter designed over a 17 points window. Just for comparision pouposes, image (d) shows the result produced by the median filter by the 5×5 window in the same image. Note that the median blurs the image. The disgned operator was also tested on another image, corrupted by a realization of the same noise process, shown in Figure 5.

6 Conclusion

The study of the morphological language, initiated by Matheron in the seventies, has led to the construction of a solid algebraic framework for the design of non-linear discrete signal and image processing operators. Compared to other algebraic structures as neural networks or decision trees, this approach has an important advantage: the adequacy for modeling prior information (i.e., user's knowledge about the problem studied). This point is absolutely fundamental for practical application of the technique, otherwise the amount of training data necessary would be prohibitive. Examples of techniques for modeling prior information in morphological operators design are envelopes, multi-resolution and apertures. These approaches have been successfully applied in the design of document image processing procedures.

Usually, the basis estimation algorithms construct the basis components from which can be derived the loose canonical representation. However, this canonical representation may not be efficient for computing the operators. Trying to overcome this difficulty, recent studies present techniques for the transformation of some canonical morphological representations into more efficient morphological operators [38], [9], [37]. Even in the simple cases studied, these techniques require the solution of complex combinatorial optimization problems.

There are many challenges for future work in this field, mainly in relation to modeling of prior knowledge and to creating specialized basis estimation algorithms. Perhaps the greatest future challenge is the generalization of theese ideas to the identification of discrete-time discrete-range dynamical systems. This advance would have many applications in the modeling of genetic networks from observation of expression patterns by microarray technology, which is among the greatest scientific challenges of the twenty-first century.

References

1. J. T. Astola and P. Kuosmanen. Representation and optimization of stack filters. In E. R. Dougherty and J. T. Astola, editors, *Nonlinear Filters for Image Processing*, pages 237–279. SPIE and IEEE Press, Bellingham, 1999.
2. G. J. F. Banon and J. Barrera. Minimal Representations for Translation-Invariant Set Mappings by Mathematical Morphology. *SIAM J. Applied Mathematics*, 51(6):1782–1798, December 1991.
3. G. J. F. Banon and J. Barrera. Decomposition of Mappings between Complete Lattices by Mathematical Morphology, Part I. General Lattices. *Signal Processing*, 30:299–327, 1993.
4. J. Barrera and G. J. F. Banon. Expressiveness of the Morphological Language. In *Image Algebra and Morphological Image Processing III*, volume 1769 of *Proc. of SPIE*, pages 264–274, San Diego, California, 1992.
5. J. Barrera and E. R. Dougherty. Representation of Gray-Scale Windowed Operators. In H. J. Heijmans and J. B. Roerdink, editors, *Mathematical Morphology and its Applications to Image and Signal Processing*, volume 12 of *Computational Imaging and Vision*, pages 19–26. Kluwer Academic Publishers, Dordrecht, May 1998.
6. J. Barrera, E. R. Dougherty, and M. Brun. Hybrid human-machine binary morphological operator design. An independent constraint approach. *Signal Processing*, 80(8):1469–1487, August 2000.
7. J. Barrera, E. R. Dougherty, and N. S. T. Hirata. Design of Optimal Morphological Operators from Prior Filters. *Acta Steriologica*, 16(3):193–200, 1997. Special issue on Mathematical Morphology.
8. J. Barrera, E. R. Dougherty, and N. S. Tomita. Automatic Programming of Binary Morphological Machines by Design of Statistically Optimal Operators in the Context of Computational Learning Theory. *Electronic Imaging*, 6(1):54–67, January 1997.
9. J. Barrera and R. F. Hashimoto. Sup-compact and inf-compact representation of w-operators. *Fundamenta Informaticae*, 45(4):283–294, 2001.

10. J. Barrera, R. Terada, R. Hirata Jr, and N. S. T. Hirata. Automatic Programming of Morphological Machines by PAC Learning. *Fundamenta Informaticae*, 41(1-2):229–258, January 2000.

11. E. J. Coyle and J.-H. Lin. Stack Filters and the Mean Absolute Error Criterion. *IEEE Transactions on Acoustics, Speech and Signal Processing*, 36(8):1244–1254, August 1988.

12. E. R. Dougherty. Optimal Mean-Square N-Observation Digital Morphological Filters I. Optimal Binary Filters. *CVGIP: Image Understanding*, 55(1):36–54, January 1992.

13. E. R. Dougherty and J. Barrera. Bayesian Design of Optimal Morphological Operators Based on Prior Distributions for Conditional Probabilities. *Acta Stereologica*, 16(3):167–174, 1997.

14. E. R. Dougherty and J. Barrera. Logical Image Operators. In E. R. Dougherty and J. T. Astola, editors, *Nonlinear Filters for Image Processing*, pages 1–60. SPIE and IEEE Press, Bellingham, 1999.

15. E. R. Dougherty, J. Barrera, G. Mozelle, S. Kim, and M. Brun. Multiresolution Analysis for Optimal Binary Filters. *Mathematical Imaging and Vision*, (14):53–72, 2001.

16. E. R. Dougherty and C. R. Giardina. A digital version of the matheron representation theorem for increasing tau-mappings in terms of a basis for the kernel. In *Proc. IEEE Computer Vision and Pattern Recognition*, pages 534–536, Miami, 1986.

17. E. R. Dougherty and R. P. Loce. Precision of Morphological-Representation Estimators for Translation-invariant Binary Filters: Increasing and Nonincreasing. *Signal Processing*, 40:129–154, 1994.

18. E. R. Dougherty and R. P. Loce. Optimal Binary Differencing Filters: Design, Logic Complexity, Precision Analysis, and Application to Digital Document Processing. *Electronic Imaging*, 5(1):66–86, January 1996.

19. E. R. Dougherty and D. Sinha. Computational Mathematical Morphology. *Signal Processing*, 38:21–29, 1994.

20. E. R. Dougherty, Y. Zhang, and Y. Chen. Optimal Iterative Increasing Binary Morphological Filters. *Optical Engineering*, 35(12):3495–3507, December 1996.

21. M. Gabbouj and E. J. Coyle. Minimum Mean Absolute Error Stack Filtering with Structural Constraints and Goals. *IEEE Transactions on Acoustics, Speech and Signal Processing*, 38(6):955–968, June 1990.

22. N. R. Harvey and S. Marshall. The Use of Genetic Algorithms in Morphological Filter Design. *Signal Processing: Image Communication*, 8(1):55–71, January 1996.

23. H. J. A. M. Heijmans. *Morphological Image Operators*. Academic Press, Boston, 1994.

24. H. J. A. M. Heijmans and C. Ronse. The Algebraic Basis of Mathematical Morphology – Part I: Dilations and Erosions. *Computer Vision, Graphics and Image Processing*, 50:245–295, 1990.

25. N. S. T. Hirata, E. R. Dougherty, and J. Barrera. A Switching Algorithm for Design of Optimal Increasing Binary Filters Over Large Windows. *Pattern Recognition*, 33(6):1059–1081, June 2000.

26. N. S. T. Hirata, E. R. Dougherty, and J. Barrera. Iterative Design of Morphological Binary Image Operators. *Optical Engineering*, 39(12):3106–3123, December 2000.

27. R. Hirata Jr., E. R. Dougherty, and J. Barrera. Aperture Filters. *Signal Processing*, 80(4):697–721, April 2000.

28. V. G. Kamat, E. R. Dougherty, and J. Barrera. Multiresolution Bayesian Design of Binary Filters. *submitted*, 2000.

29. P. Kuosmanen and J. T. Astola. Optimal stack filters under rank selection and structural constraints. *Signal Processing*, 41:309–338, 1995.

30. R. P. Loce and E. R. Dougherty. Facilitation of Optimal Binary Morphological Filter Design Via Structuring Element Libraries and Design Constraints. *Optical Engineering*, 31(5):1008–1025, May 1992.

31. R. P. Loce and E. R. Dougherty. Optimal Morphological Restoration: The Morphological Filter Mean-Absolute-Error Theorem. *Visual Communication and Image Representation*, 3(4):412–432, December 1992.

32. R. P. Loce and E. R. Dougherty. Mean-Absolute-Error representation and Optimization of Computational-Morphological Filters. *Graphical Models and Image Processing*, 57(1):27–37, 1995.

33. P. Maragos. A Representation Theory for Morphological Image and Signal Processing. *IEEE Transactions on Pattern Analysis and Machine Intelligence*, 11(6):586–599, June 1989.

34. P. A. Maragos. *A Unified Theory of Translation-invariant Systems with Applications to Morphological Analysis and Coding of Images*. PhD thesis, School of Elect. Eng. - Georgia Inst. Tech., 1985.

35. G. Matheron. *Random Sets and Integral Geometry*. John Wiley, 1975.

36. A. V. Mathew, E. R. Dougherty, and V. Swarnakar. Efficient Derivation of the Optimal Mean-Square Binary Morphological Filter from the Conditional Expectation Via a Switching Algorithm for Discrete Power-Set Lattice. *Circuits, Systems, and Signal Processing*, 12(3):409–430, 1993.

37. J. B. R. F. Hashimoto and E. R. Dougherty. From the sup-decomposition to a sequential decomposition. In L. V. John Goutsias and D. S. Bloomberg, editors, *Mathematical morphology and its applications to image and signal processing*, pages 13–22, Palo Alto, 2000.

38. J. B. R. F. Hashimoto and C. E. Ferreira. A combinatorial optimization technique for the sequential decomposition of erosions and dilations. *Journal of Mathematical Imaging and Vision*, 13(1):17–33, 2000.

39. P. Salembier. Structuring element adaptation for morphological filters. *Visual Communication and Image Representation*, 3(2):115–136, 1992.

40. O. V. Sarca, E. Dougherty, and J. Astola. Two-stage Binary Filters. *Electronic Imaging*, 8(3):219–232, July 1999.

41. O. V. Sarca, E. R. Dougherty, and J. Astola. Secondarily Constrained Boolean Filters. *Signal Processing*, 71(3):247–263, December 1998.

42. J. Serra. *Image Analysis and Mathematical Morphology*. Academic Press, 1982.

43. I. Tăbuş, D. Petrescu, and M. Gabbouj. A training Framework for Stack and Boolean Filtering – Fast Optimal Design Procedures and Robustness Case Study. *IEEE Transactions on Image Processing*, 5(6):809–826, June 1996.

Morphological Decomposition Systems with Perfect Reconstruction: From Pyramids to Wavelets

Henk J.A.M. Heijmans[1] and John Goutsias[2]

[1] CWI
[2] Center for Imaging Science, The Johns Hopkins University

1 Introduction

Multiresolution methods in signal and image processing are very useful for the following reasons: (i) there is substantial evidence that the human visual system processes visual information in a 'multiresolution' fashion; (ii) often, images contain features of physically significant structure at different resolutions; (iii) sensors may provide data of the same source at multiple resolutions; (iv) multiresolution image processing algorithms offer computational advantages and, moreover, appear to be robust.

This chapter, which is based on our work in [18, 25], introduces a general signal decomposition system with perfect reconstruction. By concatenating several instances of such a system, one obtains a multistage signal decomposition scheme, which covers two well-known signal representations, namely pyramids and wavelets. We discuss both representations at length.

In a pyramid representation, every analysis operator that brings a signal from a given level to the next coarser level reduces information. This information is captured by the detail signal, which is the difference between the original signal and the approximation obtained by applying a synthesis operator on the coarser signal. In general, a pyramid representation, comprising the coarsest signal along with detail signals at all levels, is redundant.

In a wavelet representation, the detail signal lives at the same level as the coarse signal itself, and it is obtained from a second family of analysis operators. In this case, the analysis and synthesis operators need to satisfy a condition that is very similar in nature to the biorthogonality condition known from the theory of linear wavelets (note, however, that this condition is formulated in operator terms only, and does not require any sort of linearity assumption or inner product). A major property of the wavelet representation is that it is non-redundant.

To design the multiresolution signal decomposition schemes discussed in this chapter, we need to find operators that satisfy constraints characteristic to the particular scheme. It is a relatively easy and direct task to design operators that satisfy the constraints required by pyramidal signal decomposition schemes. However, designing operators that lead to legitimate wavelet decomposition schemes is a more difficult and delicate task, which requires the use of special mathematical tools. For example, to design linear wavelet decomposition schemes, the Z-transform is used to transform the required constraints into a system of polynomial equations, whose solution provides the impulse responses of the underlying linear operators [44]. On the other hand, the *lifting scheme*, introduced by Sweldens [40, 41, 42], provides a powerful tool for constructing nonlinear wavelet decompositions. The enormous flexibility and freedom that the lifting scheme offers has challenged researchers to develop various nonlinear wavelet decomposition schemes [6, 10, 7, 8, 9, 13, 15, 14, 17, 16, 22, 20, 21, 23, 24]. In this chapter, the lifting scheme is introduced for general decomposition schemes, and as such it can also be applied to pyramids.

We would like to emphasize here the enormous influence that G. Matheron's work had on the material presented in this chapter, and on virtually all work we have published in mathematical morphology. During the last 40 years G. Matheron contributed a rich collection of scientific results, which provided a solid foundation to mathematical morphology. Moreover, he produced numerous ideas and concepts that have been used by others and ourselves to further the area of mathematical morphology. In particular, our work on morphological multiresolution systems is a direct consequence of a number of concepts pioneered by G. Matheron in his seminal book on 'Random Sets and Integral Geometry' [32] and in [37], such as granulometries and morphological filters. We are deeply indebted to him and respectfully honor his memory.

2 Preliminaries

In this section, we briefly recall some concepts from mathematical morphology that we use in the sequel. We refer to [26] for a comprehensive discussion. Recall first that a *partially ordered set* or *poset* $\mathcal{L}$ is a set endowed with a partial ordering. A poset $\mathcal{L}$ is called a *lattice* if every finite subset in $\mathcal{L}$ has a supremum (least upper bound) and an infimum (greatest lower bound). It is called a *complete lattice* if every (finite or infinite) subset of $\mathcal{L}$ has an infimum and a supremum. If $\mathcal{K} \subseteq \mathcal{L}$, then we denote the supremum and infimum of $\mathcal{K}$ by $\bigvee \mathcal{K}$ and $\bigwedge \mathcal{K}$, respectively. Instead of $\bigvee \{x_1, x_2, \ldots, x_n\}$ we write $x_1 \vee x_2 \vee \cdots \vee x_n$ (same for the infimum).

A fundamental concept in mathematical morphology is that of an adjunction. Consider two partially ordered sets (posets) $\mathcal{L}, \mathcal{M}$ and two operators $\varepsilon \colon \mathcal{L} \to \mathcal{M}$ and $\delta \colon \mathcal{M} \to \mathcal{L}$. The pair (ε, δ) defines an *adjunction* between $\mathcal{L}$ and $\mathcal{M}$ if

$$\delta(y) \leq x \;\Leftrightarrow\; y \leq \varepsilon(x), \quad x \in \mathcal{L}, \; y \in \mathcal{M}.$$

It is easy to show that, in an adjunction, both operators ε and δ are increasing; i.e., $x_1 \leq x_2$ implies that $\varepsilon(x_1) \leq \varepsilon(x_2)$ (the same for δ). If (ε, δ) is an adjunction between two lattices $\mathcal{L}$ and $\mathcal{M}$, then

$$\varepsilon(x_1 \wedge x_2 \wedge \cdots \wedge x_n) = \varepsilon(x_1) \wedge \varepsilon(x_2) \wedge \cdots \wedge \varepsilon(x_n), \quad x_1, x_2, \ldots, x_n \in \mathcal{L}$$

and, dually,

$$\delta(y_1 \vee y_2 \vee \cdots \vee y_n) = \delta(y_1) \vee \delta(y_2) \vee \cdots \vee \delta(y_n), \quad y_1, y_2, \ldots, y_n \in \mathcal{M}.$$

In a complete lattice, this relationship also holds for infinite infima and suprema, respectively. Operators ε and δ, with the properties stated above, are called *erosion* and *dilation*, respectively. In the following, id denotes the identity operator. It can be easily demonstrated that, if (ε, δ) is an adjunction between two posets $\mathcal{L}$ and $\mathcal{M}$, then

$$\varepsilon\delta\varepsilon = \varepsilon \quad \text{and} \quad \delta\varepsilon\delta = \delta$$

$$\varepsilon\delta \geq \text{id} \quad \text{and} \quad \delta\varepsilon \leq \text{id}.$$

A basic adjunction on the set $\mathrm{Fun}(\mathbb{Z}^d, \mathcal{T})$ of all functions from $\mathbb{Z}^d$ to a complete lattice $\mathcal{T}$ is formed by the *flat* dilation δ_A and the *flat* erosion ε_A, given by:

$$\delta_A(x)(n) = (x \oplus A)(n) = \bigvee_{k \in A} x(n - k) \tag{1}$$

$$\varepsilon_A(x)(n) = (x \ominus A)(n) = \bigwedge_{k \in A} x(n + k). \tag{2}$$

Here, $A \subseteq \mathbb{Z}^d$ is a given set, the so-called *structuring element*.

Let ψ be an operator from a poset $\mathcal{L}$ into itself.

- ψ is called *idempotent*, if $\psi^2 = \psi$.

- If ψ is increasing and idempotent, then ψ is called a (morphological) *filter*.

- A filter ψ that satisfies $\psi \leq \text{id}$ (ψ is anti-extensive) is called an *opening*.

- A filter ψ that satisfies $\psi \geq \text{id}$ (ψ is extensive) is called a *closing*.

Now, if (ε, δ) is an adjunction between two posets $\mathcal{L}$ and $\mathcal{M}$, then, $\varepsilon\delta$ is a closing on $\mathcal{M}$ and $\delta\varepsilon$ is an opening on $\mathcal{L}$. We have seen that the pair $(\varepsilon_A, \delta_A)$, given by (1) and (2), constitutes an adjunction on $\mathrm{Fun}(\mathbb{Z}^d, \mathcal{T})$. Thus, we may conclude that the compositions

$$x \bigcirc A = (x \ominus A) \oplus A \quad \text{and} \quad x \bullet A = (x \oplus A) \ominus A$$

form an opening and closing, respectively.

An operator ν on a complete lattice $\mathcal{L}$ is a *negation*, if it is a bijection that reverses ordering (i.e., $x \leq y \Rightarrow \nu(y) \leq \nu(x)$) such that $\nu^2 = \text{id}$, the identity

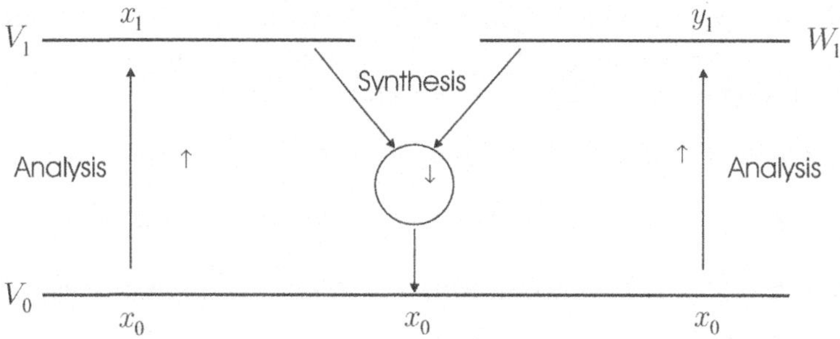

Fig. 1. *A signal decomposition scheme with perfect reconstruction (from [25]).*

operator. For example, for every $x \in \mathrm{Fun}(E, \mathcal{T})$, $\nu(x) = -x$, if $\mathcal{T} = \overline{\mathbb{R}}$, whereas $\nu(x) = N - 1 - x$, if $\mathcal{T} = \{0, 1, \ldots, N - 1\}$. Let $\mathcal{L}, \mathcal{M}$ be two complete lattices with negations $\nu_{\mathcal{L}}$, $\nu_{\mathcal{M}}$, respectively. With an operator $\psi \colon \mathcal{L} \to \mathcal{M}$, we can associate the *negative operator* $\psi^* = \nu_{\mathcal{M}} \psi \nu_{\mathcal{L}}$. When no confusion about the respective negation seems possible, we set $\psi^*(x) = [\psi(x^*)]^*$. If (ε, δ) forms an adjunction between two complete lattices $\mathcal{L}$ and $\mathcal{M}$ and if both lattices have a negation, then the pair $(\delta^*, \varepsilon^*)$ forms an adjunction between $\mathcal{M}$ and $\mathcal{L}$ as well.

3 Decomposition Systems with Perfect Reconstruction

3.1 Signal Decomposition

To analyze a signal, it is sometimes useful to decompose it into different parts in such a way that no information is removed from the signal. In Fig. 1, we depict a general scheme for the decomposition of an input signal $x_0 \in V_0$ into two parts, $(x_1, y_1) \in V_1 \times W_1$. Here, x_1 has the interpretation of an *approximation* or *simplification* of signal x_0, whereas y_1 represents a kind of *detail* or *error* signal. The operators $\psi^\uparrow \colon V_0 \to V_1$ and $\omega^\uparrow \colon V_0 \to W_1$ are called *analysis operators*, whereas the operator $\Psi^\downarrow \colon V_1 \times W_1 \to V_0$ is called the *synthesis operator*.

Our previous assumption that no information is lost by the decomposition is expressed by the condition that $\Psi^\downarrow$ is the left inverse of $\Psi^\uparrow = (\psi^\uparrow, \omega^\uparrow)$; i.e.,

$$\Psi^\downarrow(\psi^\uparrow(x_0), \omega^\uparrow(x_0)) = x_0 \,, \quad \text{for } x_0 \in V_0.$$

This condition will be referred to as the *perfect reconstruction condition*.

We do not intend to elaborate much further on a general theory of decomposition systems. In the following sections, we treat two special types of decomposition systems, namely pyramids and wavelets, and we present several examples of such decompositions. Before restricting attention to these two special cases however, we discuss a general method, called *lifting*, which

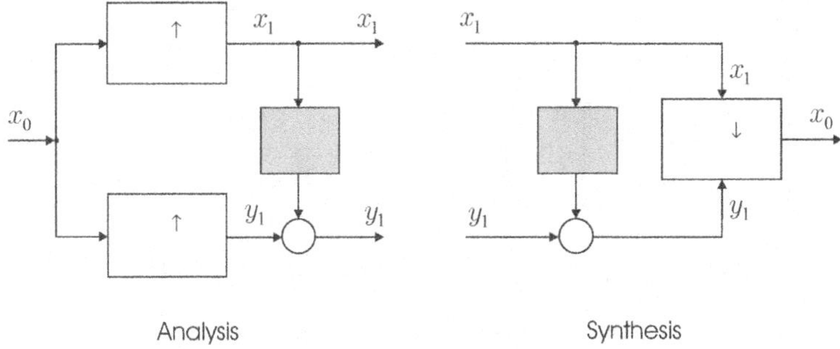

Fig. 2. *Prediction lifting scheme (from [25]).*

can be exploited to modify a given decomposition system. Subsequently, we discuss *multistage decomposition systems* obtained by concatenating several systems of the form depicted in Fig. 1.

3.2 Lifting

We now describe a general and flexible technique to modify a given decomposition system (which can be a trivial one) into another one, possibly with some improved characteristics. This technique, called *lifting*, was developed about six years ago by W. Sweldens in the context of wavelets [40, 41, 42] (see also [3], for a predecessor to this scheme, known as a 'ladder network'). As we shall see below, this technique extends readily to general decomposition systems. The formulation given here is based on our previous work in [25].

Two types of lifting schemes can be distinguished: *prediction lifting* and *update lifting*. We will treat both cases separately.

Prediction lifting.

This scheme, depicted in Fig. 2, modifies the detail analysis operator $\omega^{\uparrow}$ as well as the synthesis operator $\Psi^{\downarrow}$. We assume that W_1 is closed under addition and subtraction (see [25], for a more general formulation) and that π is a mapping from V_1 into W_1. The detail signal y_1 is modified by subtracting $\pi(x_1)$:

$$y_1' = y_1 - \pi(x_1);$$

see the left part of Fig. 2. This modification is called 'prediction,' since π is often chosen in such a way that $\pi(x_1)$ is an estimate (i.e., a prediction) of y_1; hence, their difference y_1' is a detail or error signal. Obviously, the original signal x_0 can be reconstructed from x_1 and y_1' by means of:

$$x_0 = \Psi^{\downarrow}(x_1, y_1) = \Psi^{\downarrow}(x_1, y_1' + \pi(x_1)).$$

Thus, we arrive at the modified decomposition system with analysis and synthesis operators given by (the subscript 'p' means 'prediction')

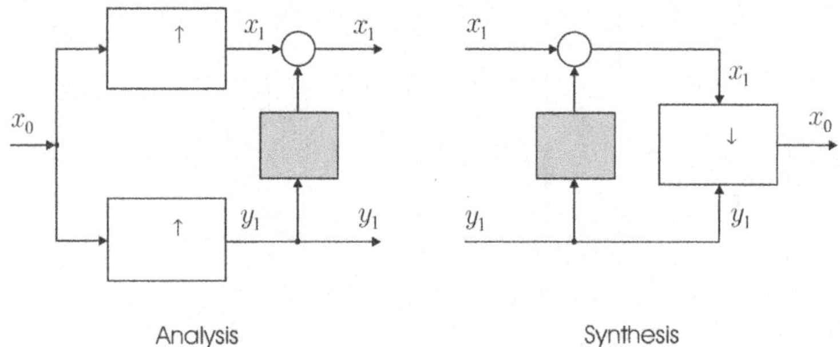

Fig. 3. *Update lifting scheme (from [25]).*

$$\psi_p^\uparrow(x) = \psi^\uparrow(x), \quad x \in V_0 \tag{3}$$

$$\omega_p^\uparrow(x) = \omega^\uparrow(x) - \pi\psi^\uparrow(x), \quad x \in V_0 \tag{4}$$

$$\Psi_p^\downarrow(x,y) = \Psi^\downarrow(x, y + \pi(x)), \quad x \in V_1,\ y \in W_1. \tag{5}$$

Update lifting.

This scheme, depicted in Fig. 3, modifies the signal analysis operator $\psi^\uparrow$ as well as the synthesis operator $\Psi^\downarrow$. We assume that V_1 is closed under addition and subtraction (see [25], for a more general formulation) and that λ is a mapping from W_1 into V_1. The approximation signal x_1 is modified by subtracting $\lambda(y_1)$:

$$x_1' = x_1 - \lambda(y_1);$$

see the left part of Fig. 3. Operator λ is called the *update operator*. In practice, the update operator is chosen in such a way that the resulting signal x_1' satisfies a certain constraint. For example, one might require that the mapping $x_0 \mapsto x_1'$ preserves a given signal attribute, such as the average or the (local) maximum. If the unmodified signal x_1 does not satisfy the constraint, we may choose λ in such a way that x_1' does.

As before, the original signal can be reconstructed from x_1' and y_1:

$$x_0 = \Psi^\downarrow(x_1, y_1) = \Psi^\downarrow(x_1' + \lambda(y_1), y_1).$$

Thus, we find the modified decomposition system with analysis and synthesis operators given by (the subscript 'u' means 'update')

$$\psi_u^\uparrow(x) = \psi^\uparrow(x) - \lambda\omega^\uparrow(x), \quad x \in V_0 \tag{6}$$

$$\omega_u^\uparrow(x) = \omega^\uparrow(x), \quad x \in V_0 \tag{7}$$

$$\Psi_u^\downarrow(x,y) = \Psi^\downarrow(x + \lambda(y), y), \quad x \in V_1,\ y \in W_1. \tag{8}$$

So far, we have shown that an existing decomposition system with perfect reconstruction can be modified by an arbitrary prediction or update lifting step. Perfect reconstruction is guaranteed by the very structure of this scheme

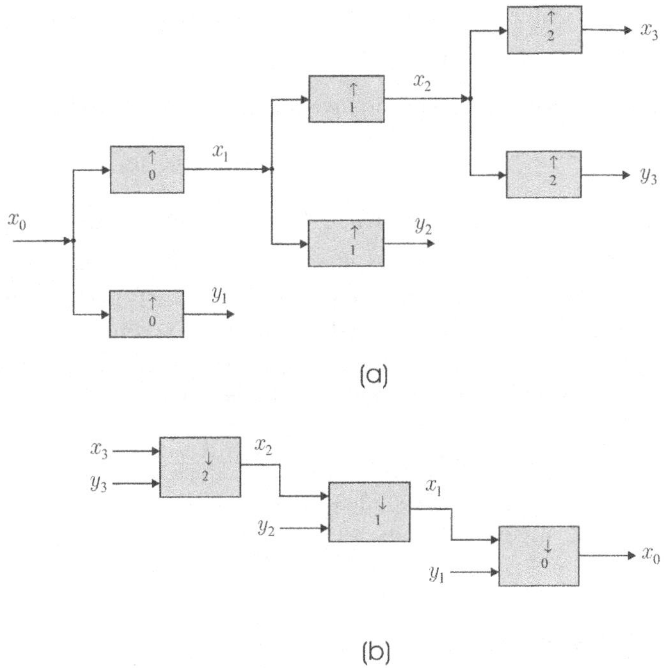

Fig. 4. *A 3-level decomposition system: (a) signal analysis, and (b) signal synthesis (from [25]).*

and does not require any particular assumptions on the lifting operators involved. Obviously, we can concatenate any number of lifting steps in order to modify a given decomposition system. In the context of linear wavelets, it has been shown that any system (using finite filters) can be decomposed into elementary lifting steps [11].

3.3 Multilevel Decomposition

As we said before, x_1 and y_1 can be interpreted as the approximation and the detail signals of a signal x_0, respectively. In other words, x_1 is a sort of a 'simplification' of x_0, inheriting many of its properties, whereas y_1 contains (at least) the information that has been discarded in order to obtain this simplification. In various signal and image processing applications, the decomposition $x_0 \mapsto (x_1, y_1)$ is only a first step towards an analysis of x_0. Subsequent steps comprise a decomposition of x_1 into x_2 and y_2, of x_2 into x_3 and y_3, and so forth.

To formalize this procedure, assume that there exists a sequence of signal spaces $V_0, V_1, V_2, \ldots$ and detail spaces $W_1, W_2, \ldots$. At each level $j \geq 0$ we have two analysis operators, $\psi_j^{\uparrow}: V_j \rightarrow V_{j+1}$ and $\omega_j^{\uparrow}: V_j \rightarrow W_{j+1}$, and a synthesis operator $\Psi_j^{\downarrow}: V_{j+1} \times W_{j+1} \rightarrow V_j$, satisfying the perfect reconstruction condition:

$$\Psi_j^{\downarrow}(\psi_j^{\uparrow}(x), \omega_j^{\uparrow}(x)) = x, \quad \text{for } x \in V_j. \tag{9}$$

A given input signal $x_0 \in V_0$ can be decomposed by the recursive scheme

$$x_0 \rightarrow \{x_1, y_1\} \rightarrow \{x_2, y_2, y_1\} \rightarrow \cdots \rightarrow \{x_k, y_k, y_{k-1}, \ldots, y_1\} \tag{10}$$

depicted in Fig. 4 (for the case when $k = 3$), where $x_{j+1} = \psi_j^{\uparrow}(x_j)$ and $y_{j+1} = \omega_j^{\uparrow}(x_j)$. We refer to this scheme as a *multilevel decomposition system*.

If the analysis operator $\psi_j^{\uparrow}$ involves sampling, then such a scheme is also called a *multiresolution* or *multiscale* decomposition scheme.

The original signal x_0 can be perfectly reconstructed from x_k and $y_1, y_2, \ldots, y_k$ by means of the following recursive synthesis scheme:

$$x_j = \Psi_j^{\downarrow}(x_{j+1}, y_{j+1}), \quad j = k-1, k-2, \ldots, 0, \tag{11}$$

where $x_0, x_1, \ldots, x_{k-1}$ are the signals reconstructed at each level. This is also depicted in Fig. 4, for the case when $k = 3$.

4 Morphological Pyramids

4.1 The Pyramid Transform

In the particular case when $W_{j+1} \subseteq V_j$ and

$$\Psi_j^{\downarrow}(x, y) = \psi_j^{\downarrow}(x) + y, \quad \text{for } x \in V_{j+1}, y \in W_{j+1}, \tag{12}$$

for some synthesis operator $\psi_j^{\downarrow}: V_{j+1} \mapsto V_j$ satisfying the perfect reconstruction condition (see (9) and (12))

$$\psi_j^{\downarrow}\psi_j^{\uparrow}(x) + \omega_j^{\uparrow}(x) = x, \quad \text{for } x \in V_j,$$

we have that

$$\omega_j^{\uparrow}(x) = x - \psi_j^{\downarrow}\psi_j^{\uparrow}(x).$$

Note that we have implicitly assumed that V_j is closed under addition and subtraction. In this case, a given input signal $x_0 \in V_0$ can be decomposed by the recursive scheme (10), where

$$\begin{cases} x_{j+1} = \psi_j^{\uparrow}(x_j) \in V_{j+1} \\ y_{j+1} = x_j - \psi_j^{\downarrow}(x_{j+1}) \in W_{j+1} \end{cases}, \quad j = 0, 1, \ldots, k-1. \tag{13}$$

The *detail signal* $y_{j+1} = x_j - \psi_j^{\downarrow}(x_{j+1})$ contains information about x_j that is not present in $\hat{x}_j = \psi_j^{\downarrow}(x_{j+1})$. Clearly, the signal $x_0 \in V_0$ can be *exactly* reconstructed from x_k and $y_1, y_2, \ldots, y_k$ by means of the backward recursion (see (11) and (12))

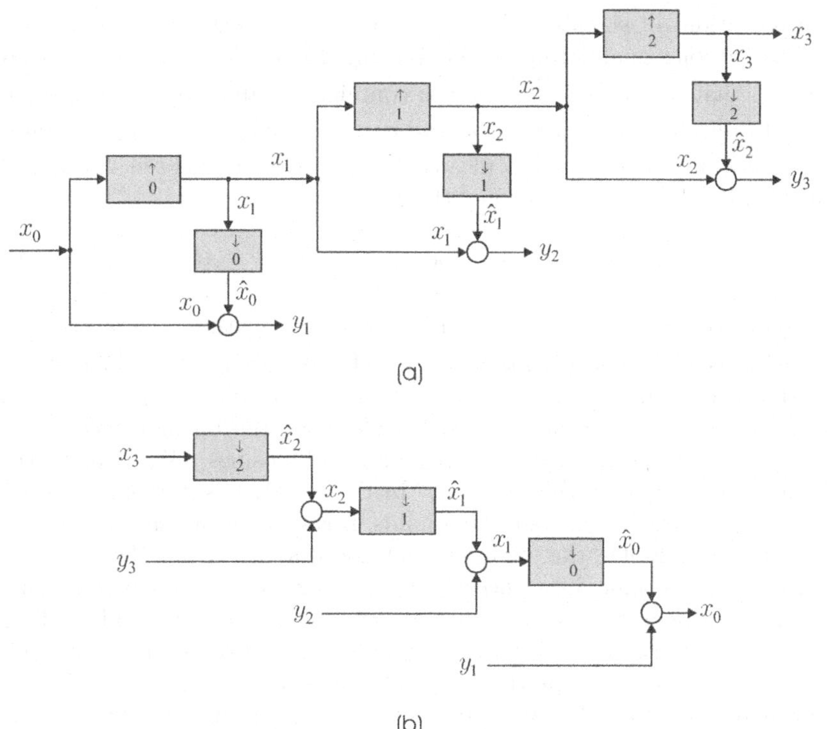

Fig. 5. *(a) A three-level pyramid transform, and (b) its inverse (from [18]).*

$$x_j = \psi_j^{\downarrow}(x_{j+1}) + y_{j+1}, \quad j = k-1, k-2, \ldots, 0. \tag{14}$$

It is not difficult to see that the multiscale signal decomposition scheme proposed by Burt and Adelson in [4] is a special case of the scheme in (13), (14) [18]. They called the sequence $\{x_j\}$ of approximation signals the Gaussian pyramid and the sequence $\{y_j\}$ of detail signals the Laplacian pyramid. We follow their nomenclature and refer to the process of decomposing a signal $x_0 \in V_0$ into $\{x_k, y_k, y_{k-1}, \ldots, y_1\}$ by means of (13) as the *pyramid transform* of x_0. On the other hand, the process of synthesizing x_0 by means of (14) is called the *inverse pyramid transform*. Block diagrams illustrating the pyramid transform and its inverse, for the case when $k = 3$, are depicted in Fig. 5.

It is worthwhile noticing here that, if $V_i^{(j)} = \mathrm{Ran}(\widehat{\psi}_{i,j})$ (i.e., the range of the *approximation operator* $\widehat{\psi}_{i,j} = \psi_i^{\downarrow}\psi_{i+1}^{\downarrow} \cdots \psi_{j-1}^{\downarrow}\psi_{j-1}^{\uparrow}\psi_{j-2}^{\uparrow} \cdots \psi_i^{\uparrow}, j > i$), then it is desirable that

$$V_i^{(j)} \subseteq V_i^{(j-1)} \subseteq V_i, \quad j > i+1. \tag{15}$$

In this case, operator $\widehat{\psi}_{i,j}$ maps the signal space V_i into nested subspaces $\cdots \subseteq V_i^{(i+2)} \subseteq V_i^{(i+1)} \subseteq V_i$, each subspace $V_i^{(j)}$ containing all 'level j' $(j > i)$

approximations of signals in V_i. Equation (15) is a basic requirement for a multiresolution signal decomposition scheme [12, 44, 30] that agrees with our intuition that the space $V_i^{(j-1)}$, which contains the approximations of signals at level i obtained by means of operator $\widehat{\psi}_{i,j-1}$, contains the approximations of signals at level i obtained by means of $\widehat{\psi}_{i,j}$ as well. It has been shown in [18] that (15) is satisfied, if we assume that

$$\psi_j^\uparrow \psi_j^\downarrow = \mathsf{id} \quad \text{on} \quad V_{j+1}. \tag{16}$$

If (16) is satisfied, then we say that the analysis and synthesis operators $\psi_j^\uparrow$ and $\psi_j^\downarrow$ satisfy the *pyramid condition*. In this chapter, we only consider pyramid transforms for which the pyramid condition is satisfied for all j. Furthermore, we focus our attention on pyramid transforms based on morphological operators (e.g., erosions, dilations, openings, and closings). We refer to these pyramids as *morphological pyramids*. In the rest of this section, we provide many examples of morphological pyramids. It is straightforward to verify that, for all these examples, the pyramid condition is satisfied (see [18]).

The lifting scheme, introduced in Section 3.2 for arbitrary decomposition systems, can be applied to pyramid decompositions. Update and prediction lifting both give rise to a modified pyramid scheme. However, it is not a priori clear whether lifting keeps the pyramid condition intact. In fact, it is not difficult to show that the pyramid condition is not invalidated by an update lifting step. In this case, the updated analysis operator is given by

$$\psi_u^\uparrow(x) = \psi^\uparrow(x) - \lambda(x - \psi^\downarrow \psi^\uparrow(x)),$$

where $\lambda\colon V_0 \to V_1$ is the update operator. The synthesis operator is not modified by the update. Obviously, the pair $(\psi_u^\uparrow, \psi^\downarrow)$ satisfies the pyramid condition if the pair $(\psi^\uparrow, \psi^\downarrow)$ does so. Unfortunately, similar results *cannot* be established in the case of a prediction lifting step: in general, the pyramid condition will no longer be valid after prediction lifting.

4.2 Pyramids without Sample Reduction

In this subsection, we present several examples of pyramid transforms based on morphological analysis and synthesis operators that preserve the number of data samples at each level. As stated before, all pyramids considered in this chapter do satisfy the pyramid condition.

Example 1 (Granulometries)

Recall that a discrete family of operators $\{\alpha_j \mid j \geq 0\}$ on the complete lattice $\mathcal{L}$ is called a *granulometry* if it satisfies the semigroup property (see [32, 26])

$$\alpha_i \alpha_j = \alpha_j \alpha_i = \alpha_j, \ j \geq i.$$

Here, we consider a discrete granulometry $\{\alpha_j \mid j \geq 0\}$ on the complete lattice $\mathcal{L} = \mathrm{Fun}(E, \mathcal{T})$, where $\mathcal{T} \subseteq \overline{\mathrm{I\!R}}$. Set $V_0 = \mathcal{L}$ and $V_{j+1} = \mathrm{Ran}(\alpha_j)$, $j > 0$, and define $\psi_j^{\uparrow} = \alpha_j$ and $\psi_j^{\downarrow} = \mathrm{id}$. It is evident that $\psi_j^{\uparrow}$ maps V_j into V_{j+1} and $\psi_j^{\downarrow}$ maps V_{j+1} into V_j, since $V_{j+1} \subseteq V_j$. Given an input signal x_0, we arrive at the signal analysis scheme (recall (13)):

$$\begin{cases} x_{j+1} = \alpha_j(x_j) \in V_{j+1} \\ y_{j+1} = x_j - x_{j+1} \end{cases}, \quad j \geq 0. \tag{17}$$

For synthesis, we find that (recall (14))

$$x_0 = \sum_{j=1}^{\infty} y_j. \tag{18}$$

Now, consider the anti-granulometry $\{\beta_j = \alpha_j^* \mid j \geq 0\}$ on $\mathcal{L} = \mathrm{Fun}(E, \mathcal{T})$. This leads to the following signal analysis and synthesis schemes (compare with (17), (18))

$$\begin{cases} x_0' = x_0 \in V_0 \\ x_{j+1}' = \beta_j(x_j') \in V_{j+1} , \quad j \geq 0, \qquad x_0 = \sum_{j=1}^{\infty} y_j'. \\ y_{j+1}' = x_{j+1}' - x_j' \end{cases}$$

In the literature, the decomposition of a signal x_0 into the detail signals $\{\ldots, y_2', y_1', y_1, y_2, \ldots\}$ is called the *discrete size transform* of x_0 [31]. If the space E is finite or countably infinite, then $\{\ldots, |y_2'|, |y_1'|, |y_1|, |y_2|, \ldots\}$, where $|x| = \sum_n |x(n)|$, is called the *pattern spectrum* of x_0 [31].

A particular case of a discrete granulometry can be obtained by the following scheme. Set $\mathcal{T} = \{0, 1\}$; in this case, $\mathcal{L} = \mathcal{P}(E)$, which is the complete Boolean lattice of all subsets of E. Assume that we are given a nested family of subspaces of $\mathcal{P}(E)$:

$$\cdots \subseteq \mathcal{C}_2 \subseteq \mathcal{C}_1 \subseteq \mathcal{C}_0 \subseteq \mathcal{P}(E), \tag{19}$$

and define the openings $\alpha_j \colon V_j \to V_{j+1}$ by

$$\alpha_j(x) = \bigcup \{c \in \mathcal{C}_j \mid c \subseteq x\}.$$

Interpreting $\mathcal{C}_j$ as 'components' at level j, we may think of $\alpha_j(x)$ as the union of all components in $\mathcal{C}_j$ that are contained inside x. The family $\{\alpha_j \mid j \geq 0\}$ forms a discrete granulometry, and V_j comprises all sets that can be obtained as a union of components in $\mathcal{C}_j$.

Let now E be the d-dimensional Euclidean space $\mathrm{I\!R}^d$. A subset $x \subseteq \mathrm{I\!R}^d$ will be called *r-connected* if any two points in x can be connected by an arc that lies entirely inside the dilation $x \oplus b_r$, where b_r is the closed ball with radius r. Recall that $x \oplus b_r$ comprises all points in $\mathrm{I\!R}^d$ that lie at distance $\leq r$

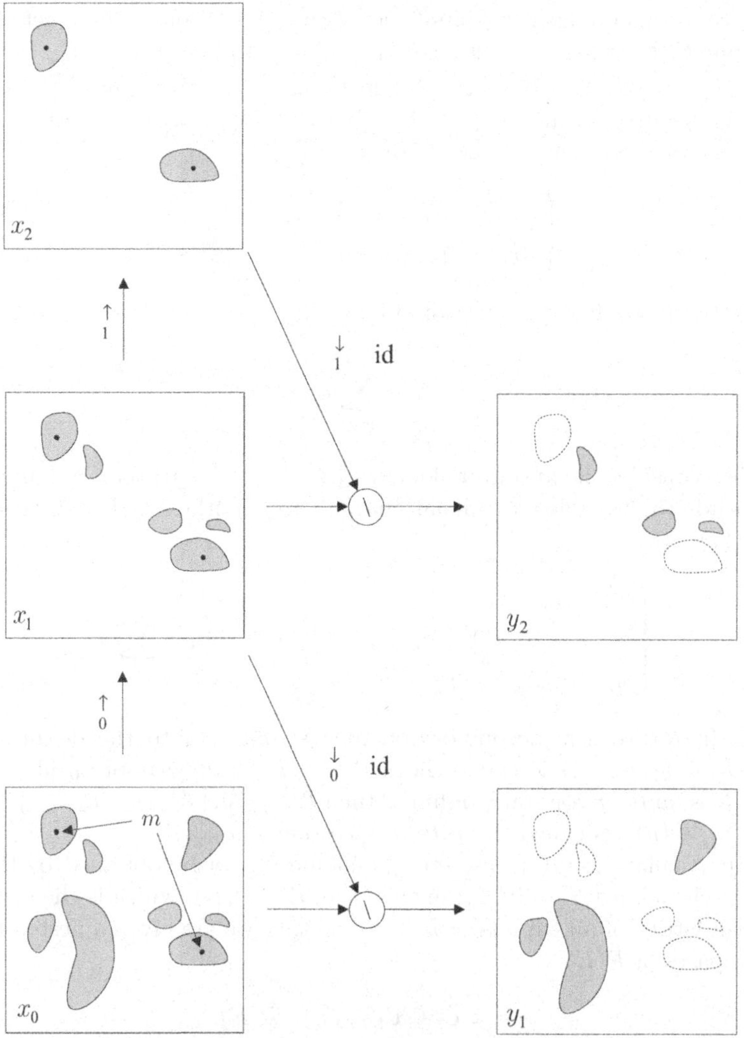

Fig. 6. *Three-levels of the object-oriented pyramid decomposition of Example 1.*

of a point in x. Suppose that we are given a non-increasing sequence $r_j \geq 0$ and a set $m \subseteq \mathbb{R}^d$ to be called the *marker set*. Let $\mathcal{C}_j$ be the r_j-connected sets that intersect the marker set m. It is obvious that $\mathcal{C}_j$ forms a nested family as in (19). The resulting discrete granulometry $\{\alpha_j \mid j \geq 0\}$, where α_j is given by (19), can be used to generate a pyramidal decomposition scheme with analysis and synthesis operators given by (17) and (18), respectively (in (17), $-$ should be replaced by the set difference $\backslash$, whereas in (18), the sum should be replaced by set union).

An example of such a pyramidal decomposition scheme is depicted in Fig. 6. Note that this is an object-oriented decomposition scheme that acts on individual objects in an image rather than on individual pixels. See [1] for more details on such decomposition schemes.

Example 2 (Morphological Skeletons)

Consider the complete lattice $\mathrm{Fun}(E,\mathcal{T})$, where $\mathcal{T} \subseteq \overline{\mathbb{R}}$, and an adjunction (ε,δ) on $\mathcal{L}$. Define $V_j = \mathrm{Ran}(\varepsilon^j)$, where $\varepsilon^0 = \mathsf{id}$ and $\varepsilon^j = \varepsilon\varepsilon\cdots\varepsilon$ (j times). Let $\psi_j^{\uparrow}\colon V_j \mapsto V_{j+1}$ and $\psi_j^{\downarrow}\colon V_{j+1} \mapsto V_j$ be given by

$$\psi_j^{\uparrow} = \varepsilon \quad \text{and} \quad \psi_j^{\downarrow} = \varepsilon^j \delta^{j+1},$$

where $\delta^j = \delta\delta\cdots\delta$ (j times). This leads to the following signal analysis scheme (recall (13)):

$$\begin{cases} x_{j+1} = \varepsilon(x_j) \in V_{j+1} \\ y_{j+1} = x_j - \varepsilon^j \delta^{j+1}(x_{j+1}) \end{cases}, \quad j \geq 0. \tag{20}$$

For synthesis, we find that (recall (14))

$$x_j = \varepsilon^j \delta^{j+1}(x_{j+1}) + y_j, \quad j \geq 0. \tag{21}$$

Notice that the detail signal y_j can be written as

$$y_{j+1} = \varepsilon^j(x) - (\varepsilon^j \delta^j)(\delta\varepsilon)\varepsilon^j(x). \tag{22}$$

On the other hand, Lantuéjoul's formula for discrete skeletons, well-known from mathematical morphology [36], produces skeleton subsets y_j from a signal $x_0 \in \mathrm{Fun}(E,\mathcal{T})$, given by (see [18])

$$y_{j+1} = \varepsilon^j(x) - (\delta\varepsilon)\varepsilon^j(x), \quad j \geq 0. \tag{23}$$

Comparing (22) with (23), we see that the decomposition scheme (20), (21) has an extra closing $\varepsilon^j \delta^j$. As a result, the detail signal y_j in (20) is never larger than the detail signal in the Lantuéjoul formula (23). Therefore, the decomposition scheme (20), (21) may give rise to a more efficient compression scheme than Lantuéjoul's skeleton. The decomposition scheme (20), (21) has been proposed earlier by Goutsias and Schonfeld in [19]. Fig. 7(c) depicts the result of applying this decomposition to the binary image in Fig. 7(a). The resulting image is different than the one depicted in Fig. 7(b) in 66 pixels. Since the image depicted in Fig. 7(b) is non-zero at 1,453 pixels, this amounts to 4.5% data reduction.

Example 3 (Curve Evolution Pyramid)

In the literature, one finds many different algorithms for curve evolution. A drawback of curve evolution methods for curve denoising is that they tend to

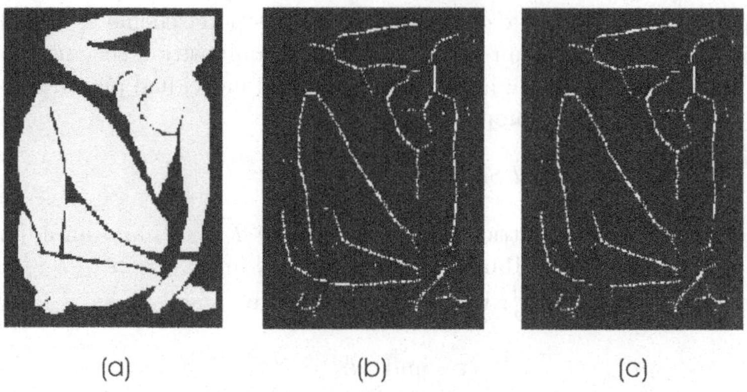

Fig. 7. *(a) A binary image, (b) the decomposition $\bigcup y_j$ obtained by means of (23), and (c) the decomposition $\bigcup y_j$ obtained by means of (20). In this case, the operation − in (20) and (23) is taken to be set difference.*

shrink the original curve. In [28, 29], it is explained in detail how morphological pyramids can help to circumvent this problem. Here, we present a condensed version of the account in [28, 29].

The basic idea is to apply inverse flow to the outcome of a forward flow. In practice, however, the inverse flow is highly unstable. Below, we see that a pyramid approach can help us to circumvent this problem. In the following, we restrict ourselves to polygons x_0 with N vertex points $\boldsymbol{p}_0(0), \boldsymbol{p}_0(1), \ldots, \boldsymbol{p}_0(N-1) \in \mathbb{R}^2$. The curve evolution algorithm introduced below can be performed most easily in the Fourier domain; hence, we use the (cyclic) Fourier transform of $\boldsymbol{p}_0$. In fact, we use the polar representation $[r_0(k), \theta_0(k)]$ of the Fourier transform of $\boldsymbol{p}_0$. Here, r_0 is the radius and θ_0 is the phase. Because of rounding effects in our computations, we assume that the radius $r_0(\cdot)$ is quantized. For simplicity, we assume that the radius is integer-valued. As we will see, quantization of the phase is not necessary.

Now, assume that $V_j = V$, for all $j \geq 0$, where V is the set of all N-vertex polygons for which the radius function $r(\cdot)$ is integer-valued. The $j+1$'th step of the (discrete-time) curve evolution considered here maps a polygon with vertex points $\boldsymbol{p}_j(\cdot)$ into another N-vertex polygon with vertices

$$\boldsymbol{p}_{j+1}(i) = \frac{1}{3}\left[\boldsymbol{p}_j(i-1) + \boldsymbol{p}_j(i) + \boldsymbol{p}_j(i+1)\right],$$

where $i \pm 1$ are taken modulo N. Denoting by $F(k)$ the frequency response of the filter $(\frac{1}{3}, \frac{1}{3}, \frac{1}{3})$, which is a real-valued function, we find that the polar representation of the Fourier transform of $\boldsymbol{p}_{j+1}$ looks as follows:

$$r_{j+1}(k) = \lfloor |F(k)| r_j(k) \rfloor \qquad (24)$$

$$\theta_{j+1}(k) = \theta_j(k) + \mathrm{phase}(F(k)) \bmod 2\pi. \qquad (25)$$

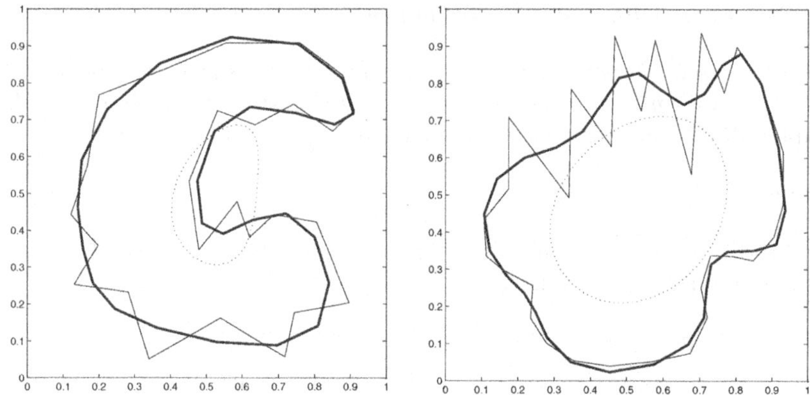

Fig. 8. *Forward and inverse curve evolution of a polygon flow (from [29]).*

Note that, since F is real-valued, its phase can only assume the values 0 and π (and hence quantization is not required here). The expression in (24), (25) defines the analysis operator $\psi^\uparrow$.

A synthesis operator $\psi^\downarrow$ can be defined by determining a solution x_j of the equation $\psi^\uparrow(x_j) = x_{j+1}$, for every $x_{j+1} \in V_{j+1}$. One such solution map is given by

$$r_j(k) = \left\lceil |\widetilde{F}^{-1}(k)| r_{j+1}(k) \right\rceil$$

$$\theta_j(k) = \theta_{j+1}(k) + \text{phase}(\widetilde{F}^{-1}(k)) \bmod 2\pi,$$

where

$$\widetilde{F}^{-1}(k) = \begin{cases} 1/F(k), & \text{if } F(k) \neq 0 \\ 0, & \text{otherwise} \end{cases}.$$

In [28], it is argued that the choice of this operator is triggered by a minimization of some energy function.

The pair $(\psi^\uparrow, \psi^\downarrow)$ forms an adjunction (relative to some partial ordering that is not specified here). Moreover, the pyramid condition is satisfied in this case as well. The resulting pyramid is referred to as the *curve evolution pyramid*. In this case, the decomposition $\{x_0, x_1, \ldots, x_k\}$, by means of $x_{j+1} = \psi^\uparrow(x_j)$, $j = 0, 1, \ldots, k-1$ is a (discrete-time) forward flow, whereas the decomposition $\{\widehat{x}_0, \widehat{x}_1, \ldots, \widehat{x}_k\}$, by means of $\widehat{x}_j = \psi^\downarrow(\widehat{x}_{j+1})$, $j = k-1, k-2, \ldots, 0$, $\widehat{x}_k = x_k$, is the corresponding 'inverse' flow.

Fig. 8 depicts the result of applying the forward and inverse flows to two 'noisy' polygons (thin, solid curves). The dotted curves result after 30 iterations of a forward flow, whereas the bold-faced curves result after 30 iterations of the 'inverse' flow applied on the dotted curves.

4.3 Pyramids with Sample Reduction

We now present several examples of pyramid transforms based on morphological analysis operators that contain, at each level, half the number of data samples. Again, all examples considered here do satisfy the pyramid condition.

Example 4 (Toet pyramid)

Let $\mathcal{T}$ be a complete chain and $t \in \mathcal{T}$. Consider the operators $\sigma^\uparrow, \sigma_t^\downarrow$, both mapping $\mathrm{Fun}(\mathbb{Z}^d, \mathcal{T})$ into itself:

$$\sigma^\uparrow(x)(n) = x(2n)$$

$$\sigma_t^\downarrow(x)(2n) = x(n) \quad \text{and} \quad \sigma_t^\downarrow(x)(m) = t, \text{ if } m \notin 2\mathbb{Z}^d,$$

Here $2\mathbb{Z}^d$ denotes all vectors in $\mathbb{Z}^d$ with *even* coordinates. Define

$$\psi_j^\uparrow = \beta\alpha\sigma^\uparrow \quad \text{and} \quad \psi_j^\downarrow = \beta\alpha\sigma_\top^\downarrow, \quad j \geq 0,$$

where α and β are the opening and closing by the structuring element $A = \{0,1\}^d$ (for the d–dimensional case), respectively, and $t = \top$ is the greatest element in $\mathcal{T}$. One can show [18] that the pyramid condition holds. The resulting pyramidal signal decomposition scheme has been suggested by Toet in [43] for use in contrast enhancement applications.

Example 5 (Median pyramids)

Assume that $\mathcal{T}$ is a complete chain, and consider a pyramid for which $V_j = \mathrm{Fun}(\mathbb{Z}, \mathcal{T})$, for every j, and the same analysis and synthesis operators are used at every level j, given by

$$\psi^\uparrow(x)(n) = \mathsf{median}\{x(2n-1), x(2n), x(2n+1)\} \tag{26}$$

$$\psi^\downarrow(x)(2n) = \psi^\downarrow(x)(2n+1) = x(n). \tag{27}$$

This leads to a 1-D pyramid that is referred to as the *median pyramid*.

An alternative 1-D median pyramid can be constructed by considering the following analysis and synthesis operators:

$$\psi^\uparrow(x)(n) = \begin{cases} x(2n), & \text{if } x(2n-1) \wedge x(2n) \wedge x(2n+1) = x(2n) \\ \mathsf{median}\{x(2n-1), x(2n), x(2n+1)\}, & \text{otherwise} \end{cases} \tag{28}$$

$$\psi^\downarrow(x)(2n) = x(n), \quad \psi^\downarrow(x)(2n+1) = x(n) \vee x(n+1). \tag{29}$$

In this case, the synthesis operator is a dilation from V_{j+1} into V_j. The median pyramid based on operators (28), (29) may provide a better approximation $\psi^\downarrow\psi^\uparrow(x)$ of x than the pyramid based on operators (26), (27), since the former pyramid utilizes more information from signal x in order to obtain the sample values $\psi^\downarrow(x)(2n+1)$ (compare (26) with (28)).

A 2-D median pyramid can be obtained by using the same analysis and synthesis operators at every level j, and by setting

$$\psi^{\uparrow}(x)(m,n) = \text{median}\{x(2m+k, 2n+l) \mid (k,l) \in A\}, \tag{30}$$

where A is the 3×3 square centered at the origin, and

$$\psi^{\downarrow}(x)(2m, 2n) = x(m,n) \tag{31}$$

$$\psi^{\downarrow}(x)(2m, 2n+1) = x(m,n) \wedge x(m, n+1) \tag{32}$$

$$\psi^{\downarrow}(x)(2m+1, 2n) = x(m,n) \wedge x(m+1, n) \tag{33}$$

$$\psi^{\downarrow}(x)(2m+1, 2n+1) = x(m,n) \vee x(m, n+1) \vee$$
$$x(m+1, n+1) \vee x(m+1, n). \tag{34}$$

An example, illustrating the resulting 2-D median pyramid, is depicted in Fig. 9. For clarity of presentation, the size of the images depicted in Fig. 9 (and subsequent figures) is larger than their actual size.

It has been suggested in [38] that median pyramids preserve details and produce decompositions that can be compressed more efficiently than other (linear) hierarchical signal decomposition schemes.

Example 6 (Morphological Haar pyramid)

Assume that $\mathcal{T}$ is a complete chain, and consider a pyramid for which $V_j = \text{Fun}(\mathbb{Z}, \mathcal{T})$, for every j, and the same analysis and synthesis operators are used at every level j, given by

$$\psi^{\uparrow}(x)(n) = x(2n) \wedge x(2n+1) \tag{35}$$

$$\psi^{\downarrow}(x)(2n) = \psi^{\downarrow}(x)(2n+1) = x(n). \tag{36}$$

It turns out that the analysis operator (35) is the morphological counterpart the analysis operator of the linear Haar wavelet decomposition scheme, given by [12, 44, 30]

$$\psi^{\uparrow}(x)(n) = \frac{1}{2}(x(2n) + x(2n+1)).$$

Moreover, the synthesis operators of both schemes are identical. For this reason, the nonlinear scheme governed by (35) and (36) is referred to as the *morphological Haar pyramid.*

A 2-D version of the morphological Haar pyramid is obtained by analysis and synthesis operators given by

$$\psi^{\uparrow}(x)(m,n) = x(2m, 2n) \wedge x(2m, 2n+1) \wedge x(2m+1, 2n+1) \wedge x(2m+1, 2n) \tag{37}$$

$$\psi^{\downarrow}(x)(2m, 2n) = \psi^{\downarrow}(x)(2m, 2n+1) = \psi^{\downarrow}(x)(2m+1, 2n+1)$$
$$= \psi^{\downarrow}(x)(2m+1, 2n) = x(m,n). \tag{38}$$

An example of this decomposition is depicted in Fig. 10. In the following section, where we deal with wavelet decompositions, we meet these operators again.

Fig. 9. *Multiresolution image decomposition based on the median pyramid (30)–(34).*

Example 7 (Symmetrized Version of Morphological Haar Pyramid)

A more interesting example than the morphological Haar pyramid is obtained by considering the case when

$$\psi^{\uparrow}(x)(n) = x(2n - 1) \wedge x(2n) \wedge x(2n + 1)$$

$$\psi^{\downarrow}(x)(2n) = x(n) \quad \text{and} \quad \psi^{\downarrow}(x)(2n + 1) = x(n) \vee x(n + 1).$$

This leads to a symmetrized version of the morphological Haar pyramid.
A 2-D version of the previous decomposition is obtained by setting

$$\psi^{\uparrow}(x)(m, n) = \bigwedge_{-1 \le k, l \le 1} x(2m + k, 2n + l), \tag{39}$$

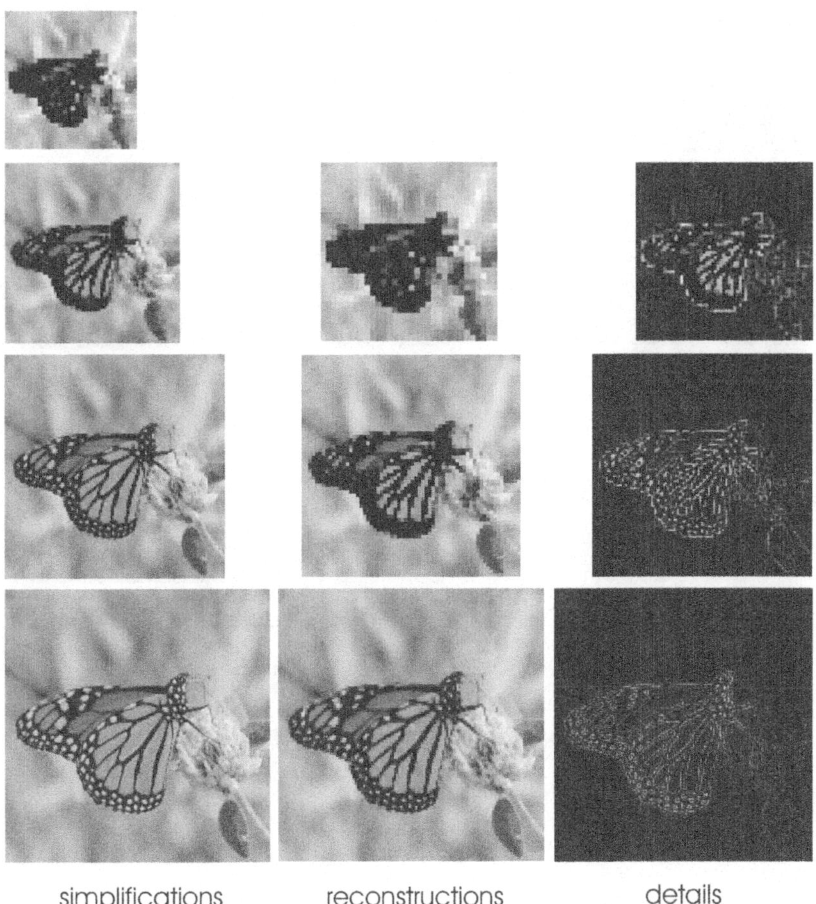

simplifications   reconstructions   details

Fig. 10. *Multiresolution image decomposition based on the morphological Haar pyramid (37), (38).*

$$\psi^{\downarrow}(x)(2m, 2n) = x(m, n) \tag{40}$$

$$\psi^{\downarrow}(x)(2m, 2n + 1) = x(m, n) \vee x(m, n + 1) \tag{41}$$

$$\psi^{\downarrow}(x)(2m + 1, 2n) = x(m, n) \vee x(m + 1, n) \tag{42}$$

$$\psi^{\downarrow}(x)(2m + 1, 2n + 1) = x(m, n) \vee x(m, n + 1) \vee$$
$$x(m + 1, n + 1) \vee x(m + 1, n). \tag{43}$$

An example of this decomposition is depicted in Fig. 11.

Fig. 11. *Multiresolution image decomposition based on the symmetrized morphological Haar pyramid (39)–(43).*

4.4 Morphological Adjunction Pyramids

Examples 6–7 are particular cases of the family of adjunction pyramids, which has been discussed in [18]. Here, we briefly recall one of the results derived in that paper.

Assume that $V_0 = V_1 = \mathrm{Fun}(\mathbb{Z}^d, \mathcal{T})$. We are interested in analysis and synthesis operators $\psi^\uparrow$, $\psi^\downarrow$ with the following three properties:

1. The pair $(\psi^\uparrow, \psi^\downarrow)$ is an adjunction (see Section 2); hence, $\psi^\uparrow$ is an erosion and $\psi^\downarrow$ is a dilation.

2. The operators $\psi^\uparrow$ and $\psi^\downarrow$ are both *flat* (see [26]), meaning that their structuring elements are sets rather than grayscale functions.

3. The operators $\psi^\uparrow$ and $\psi^\downarrow$ are translation invariant in the following sense: for every translation operator $\tau = \tau_k$, with $k \in \mathbb{Z}^d$, we have

$$\psi^\uparrow \tau^2 = \tau \psi^\uparrow \quad \text{and} \quad \psi^\downarrow \tau = \tau^2 \psi^\downarrow.$$

Recall that the translation operator τ_k is given by $(\tau_k x)(n) = x(n - k)$, for $k, n \in \mathbb{Z}^d$.

This last property means that the analysis operator $\psi^\uparrow$ involves subsampling by 2 in every spatial direction. In [18], it is shown that $\psi^\uparrow$ and $\psi^\downarrow$ can only be of the following form:

$$\psi^\uparrow(x)(n) = \bigwedge_{k \in A} x(2n + k) \tag{44}$$

$$\psi^\downarrow(x)(k) = \bigvee_{n \in A[k]} x(\frac{k - n}{2}), \tag{45}$$

where $A \subseteq \mathbb{Z}^d$ is the *structuring element*. Here, we use the following notation: for $n \in \mathbb{Z}^d$, set

$$\mathbb{Z}^d[n] = \{k \in \mathbb{Z}^d \mid k - n \in 2\mathbb{Z}^d\},$$

where $2\mathbb{Z}^d$ denotes all vectors in $\mathbb{Z}^d$ with *even* coordinates. The sets $\mathbb{Z}^d[n]$ form a disjoint partition of $\mathbb{Z}^d$ into 2^d parts. For $A \subseteq \mathbb{Z}^d$ and $n \in \mathbb{Z}^d$, we define

$$A[n] = A \cap \mathbb{Z}^d[n],$$

which yields a partition of A.

In order that the pyramid condition is satisfied for the pair in (44), (45), it is necessary that the structuring element A satisfies the following condition:

$$A[a] = \{a\}, \text{ for some } a \in A, \tag{46}$$

or, equivalently,

$$\exists a \in A : \quad \forall k \in \mathbb{Z}^d \setminus \{0\}, \quad a + 2k \notin A.$$

In the 1-D version of Example 7, we have $d = 1$ and $A = \{-1, 0, 1\}$. Thus, in this case, $A[0] = \{0\}$, meaning that the condition in (46) is satisfied.

In [28], it is argued that, in many cases, the pyramid condition means that the analysis and synthesis operators form an adjunction. For example, to show that $(\psi_j^\uparrow, \psi_j^\downarrow)$ in Example 1 defines an adjunction between V_j and V_{j+1}, we must show the following relation:

$$y \leq \psi_j^\uparrow(x) \Leftrightarrow \psi_j^\downarrow(y) \leq x, \quad \text{for } x \in \text{Ran}(\alpha_{j-1}) \text{ and } y \in \text{Ran}(\alpha_j).$$

Indeed, writing $x = \alpha_{j-1}(x')$ and $y = \alpha_j(y')$, we find that

$$y \leq \psi_j^{\uparrow}(x) \Leftrightarrow \alpha_j(y') \leq \alpha_j \alpha_{j-1}(x')$$
$$\Leftrightarrow \alpha_j(y') \leq \alpha_{j-1}(x')$$
$$\Leftrightarrow \psi_j^{\downarrow}(y) \leq x.$$

If the analysis and synthesis operators $(\psi_j^{\uparrow}, \psi_j^{\downarrow})$ of a pyramidal decomposition scheme form an adjunction, then the pyramid is called a *morphological adjunction pyramid* [18]. The analysis operator $\psi_j^{\uparrow}$ of a morphological adjunction pyramid is necessarily an erosion, whereas the synthesis operator $\psi_j^{\downarrow}$ is necessarily a dilation. It has been shown in [18] that, in this case, the pyramid condition is satisfied if and only if $\psi_j^{\uparrow}$ is surjective or $\psi_j^{\downarrow}$ is injective. Moreover, $\psi_j^{\downarrow}\psi_j^{\uparrow}$ is an opening and hence $\psi_j^{\downarrow}\psi_j^{\uparrow} \leq \mathrm{id}$, which, together with (13), implies that $y_{j+1} \geq 0$, for all $j \geq 0$. This property is important in signal compression and coding applications, since it implies that the detail signals assume no negative values, thus saving one bit of information, required for coding the sign of y_j.

Finally, the analysis and synthesis operators used in the previous examples map integers into integers. This is a very desirable property, since these operators preserve the integer form of signals in most data compression and coding applications [5].

5 Morphological Wavelets

5.1 The Wavelet Transform

This section is concerned with another interesting family of decomposition systems with perfect reconstruction, the *wavelet decomposition*. Later, we show that every wavelet decomposition includes a pyramid decomposition (or rather, an entire family of pyramid decompositions).

A general wavelet decomposition has the same structure as in Fig. 1. But, besides the perfect reconstruction condition

$$\Psi^{\downarrow}(\psi^{\uparrow}(x_0), \omega^{\uparrow}(x_0)) = x_0, \quad \text{for } x_0 \in V_0, \tag{47}$$

it also satisfies the additional constraints

$$\psi^{\uparrow}(\Psi^{\downarrow}(x_1, y_1)) = x_1, \quad \text{for } x_1 \in V_1, \, y_1 \in W_1 \tag{48}$$

$$\omega^{\uparrow}(\Psi^{\downarrow}(x_1, y_1)) = y_1, \quad \text{for } x_1 \in V_1, \, y_1 \in W_1, \tag{49}$$

which guarantees that the wavelet decomposition is non-redundant. Note that (47)–(49) imply that the analysis operator $\Psi^{\uparrow} = (\psi^{\uparrow}, \omega^{\uparrow})$ and the synthesis operator $\Psi^{\downarrow}$ are inverses. Concatenation of a series of analysis steps, as in Section 3.3, yields a multilevel decomposition which, in the literature, is called

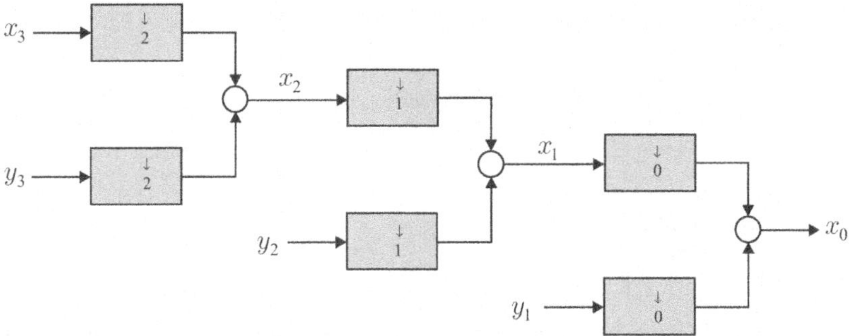

Fig. 12. *The signal synthesis part of a 3-level uncoupled wavelet decomposition (from [25]).*

the *wavelet transform*. In the examples to be discussed later, the analysis operators involve downsampling (or decimation), which is 'undone' by upsampling (or interpolation) in the synthesis part.

Fixing an element $\overline{y} \in W_1$ and defining the operator $\psi^{\downarrow} \colon V_1 \to V_0$ as

$$\psi^{\downarrow}(x) = \Psi^{\downarrow}(x, \overline{y}), \quad x \in V_1,$$

we derive from (48) the pyramid condition $\psi^{\uparrow}\psi^{\downarrow}(x) = x$, for $x \in V_1$. This proves our claim that every wavelet decomposition 'includes' a pyramid decomposition.

Of particular interest is the case when the synthesis operator $\Psi^{\downarrow}$ is of the special form

$$\Psi^{\downarrow}(x, y) = \psi^{\downarrow}(x) + \omega^{\downarrow}(y), \quad x \in V_1, \; y \in W_1,$$

in which case we speak of an *uncoupled wavelet decomposition*. Now, conditions (47)–(49) reduce to

$$\psi^{\downarrow}\psi^{\uparrow}(x) + \omega^{\downarrow}\omega^{\uparrow}(x) = x, \quad x \in V_0 \tag{50}$$

$$\psi^{\uparrow}(\psi^{\downarrow}(x) + \omega^{\downarrow}(y)) = x, \quad x \in V_1, \; y \in W_1 \tag{51}$$

$$\omega^{\uparrow}(\psi^{\downarrow}(x) + \omega^{\downarrow}(y)) = y, \quad x \in V_1, \; y \in W_1. \tag{52}$$

We refer to $\psi^{\downarrow}, \omega^{\downarrow}$ as the *signal synthesis* and the *detail synthesis operator*, respectively. In the multilevel case, the corresponding synthesis scheme is depicted in Fig. 12.

A trivial example of an uncoupled wavelet decomposition is the *lazy wavelet transform* that splits a 1-D discrete signal x into its odd and even samples. Here, the analysis operators are given by

$$\psi^{\uparrow}(x)(n) = x(2n)$$
$$\omega^{\uparrow}(x)(n) = x(2n + 1).$$

Reconstruction is achieved through the synthesis operators

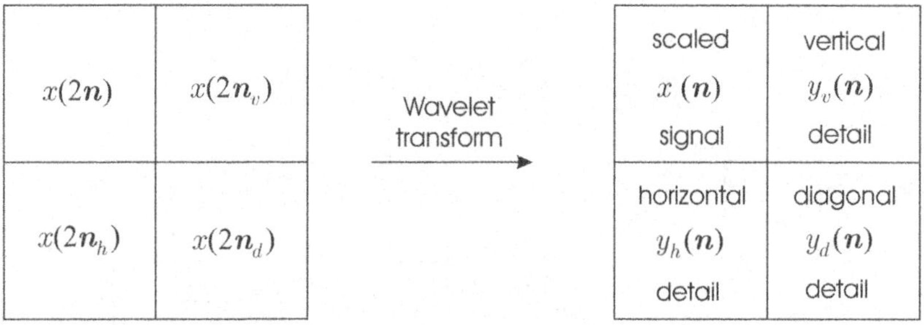

			scaled	vertical
$x(2\boldsymbol{n})$	$x(2\boldsymbol{n}_v)$	Wavelet transform $\longrightarrow$	$x'(\boldsymbol{n})$ signal	$y_v(\boldsymbol{n})$ detail
$x(2\boldsymbol{n}_h)$	$x(2\boldsymbol{n}_d)$		horizontal $y_h(\boldsymbol{n})$ detail	diagonal $y_d(\boldsymbol{n})$ detail

Fig. 13. *The 2-D morphological Haar wavelet transforms an input signal x to a scaled signal x' and the vertical, horizontal, and diagonal detail signals y_v, y_h, y_d, respectively (from [25]).*

$$\psi^{\downarrow}(x)(2n) = x(n) \quad \text{and} \quad \psi^{\downarrow}(x)(2n+1) = 0$$

$$\omega^{\downarrow}(y)(2n) = 0 \quad \text{and} \quad \omega^{\downarrow}(y)(2n+1) = y(n).$$

This decomposition is better known in the signal processing community as the *polyphase transform of order* 2 [44]. The lazy wavelet transform is not of great interest by itself; however, it is often used as a starting point for the lifting scheme to be discussed below.

A more interesting example is the decomposition that we have called the *morphological Haar wavelet*, since it resembles the classical linear Haar wavelet. The major difference is that the linear signal analysis filter of the latter is replaced by a morphological operator (e.g., by an erosion). Let $V_0 = V_1 = W_1 = \mathbb{R}^{\mathbb{Z}}$ be the lattice of doubly infinite real-valued sequences. Define the analysis operators

$$\psi^{\uparrow}(x)(n) = x(2n) \wedge x(2n+1)$$

$$\omega^{\uparrow}(x)(n) = x(2n) - x(2n+1),$$

and the synthesis operators

$$\psi^{\downarrow}(x)(2n) = \psi^{\downarrow}(x)(2n+1) = x(n)$$

$$\omega^{\downarrow}(y)(2n) = y(n) \vee 0 \quad \text{and} \quad \omega^{\downarrow}(y)(2n+1) = -(y(n) \wedge 0).$$

It is rather straightforward to verify that conditions (50)–(52) are satisfied. Observe that the operators $\psi^{\uparrow}, \psi^{\downarrow}$ form the *morphological Haar pyramid* discussed in Example 6.

The previous example can be easily extended to two dimensions. Indeed, let us define a simple 2-D wavelet. Let V_0 and V_1 consist of all functions from $\mathbb{Z}^2$ into $\mathbb{R}$ and let W_1 consist of all functions from $\mathbb{Z}^2$ into $\mathbb{R}^3$. We introduce the following notation. By $\boldsymbol{n}, 2\boldsymbol{n}$, we denote the points $(m, n), (2m, 2n) \in \mathbb{Z}^2$, respectively, and by $2\boldsymbol{n}_h, 2\boldsymbol{n}_v, 2\boldsymbol{n}_d$, we denote the points $(2m+1, 2n), (2m, 2n+1), (2m+1, 2n+1)$, respectively; see Fig. 13. Define

$$\psi^\uparrow(x)(\boldsymbol{n}) = x(2\boldsymbol{n}) \wedge x(2\boldsymbol{n}_h) \wedge x(2\boldsymbol{n}_v) \wedge x(2\boldsymbol{n}_d) \tag{53}$$

$$\omega^\uparrow(x)(\boldsymbol{n}) = (\omega_v(x)(\boldsymbol{n}), \omega_h(x)(\boldsymbol{n}), \omega_d(x)(\boldsymbol{n})), \tag{54}$$

where $\omega_v, \omega_h, \omega_d$ represent the vertical, horizontal, and diagonal detail signals, given by:

$$\omega_v(x)(\boldsymbol{n}) = \tfrac{1}{2}(x(2\boldsymbol{n}) - x(2\boldsymbol{n}_v) + x(2\boldsymbol{n}_h) - x(2\boldsymbol{n}_d)) \tag{55}$$

$$\omega_h(x)(\boldsymbol{n}) = \tfrac{1}{2}(x(2\boldsymbol{n}) - x(2\boldsymbol{n}_h) + x(2\boldsymbol{n}_v) - x(2\boldsymbol{n}_d)) \tag{56}$$

$$\omega_d(x)(\boldsymbol{n}) = \tfrac{1}{2}(x(2\boldsymbol{n}) - x(2\boldsymbol{n}_h) - x(2\boldsymbol{n}_v) + x(2\boldsymbol{n}_d)). \tag{57}$$

The synthesis operators are given by

$$\psi^\downarrow(x)(2\boldsymbol{n}) = \psi^\downarrow(x)(2\boldsymbol{n}_h) = \psi^\downarrow(x)(2\boldsymbol{n}_v) = \psi^\downarrow(x)(2\boldsymbol{n}_d) = x(\boldsymbol{n}),$$

and

$$\omega^\downarrow(y)(2\boldsymbol{n}) = (y_v(\boldsymbol{n}) + y_h(\boldsymbol{n})) \vee (y_v(\boldsymbol{n}) + y_d(\boldsymbol{n})) \vee (y_h(\boldsymbol{n}) + y_d(\boldsymbol{n})) \vee 0$$

$$\omega^\downarrow(y)(2\boldsymbol{n}_h) = (y_v(\boldsymbol{n}) - y_h(\boldsymbol{n})) \vee (y_v(\boldsymbol{n}) - y_d(\boldsymbol{n})) \vee (-y_h(\boldsymbol{n}) - y_d(\boldsymbol{n})) \vee 0$$

$$\omega^\downarrow(y)(2\boldsymbol{n}_v) = (y_h(\boldsymbol{n}) - y_v(\boldsymbol{n})) \vee (-y_v(\boldsymbol{n}) - y_d(\boldsymbol{n})) \vee (y_h(\boldsymbol{n}) - y_d(\boldsymbol{n})) \vee 0$$

$$\omega^\downarrow(y)(2\boldsymbol{n}_d) = (-y_v(\boldsymbol{n}) - y_h(\boldsymbol{n})) \vee (y_d(\boldsymbol{n}) - y_v(\boldsymbol{n})) \vee (y_d(\boldsymbol{n}) - y_h(\boldsymbol{n})) \vee 0,$$

where we have written $y \in W_1$ as $y = (y_v, y_h, y_d)$. It is not difficult to show that conditions (50)–(52) are all satisfied. The analysis operators $\psi^\uparrow$ and $\omega^\uparrow$ in (53), (54) map a quadruple of signal values, as the ones depicted on the left hand-side of Fig. 13, to the quadruple at the right hand-side; here $x' = \psi^\uparrow(x)$ and $y_v = \omega_v(x)$ (the same for y_h, y_d).

An example, illustrating one step of this decomposition, is depicted in Fig. 14. As in the 1-D case, the minimum in the expression for $\psi^\uparrow$ can be replaced by a maximum.

In fact, it has been shown in [25] that the minimum can also be replaced by an arbitrary *positive Boolean function* without destroying the perfect reconstruction property. As an example, we consider a 2-D wavelet transform on binary images, where the 'low-pass' analysis operator is the median operator:

$$\psi^\uparrow(x)(\boldsymbol{n}) = \mathsf{median}(x(2\boldsymbol{n}), x(2\boldsymbol{n}), x(2\boldsymbol{n}_h), x(2\boldsymbol{n}_v), x(2\boldsymbol{n}_d)). \tag{58}$$

Take $\omega^\uparrow$ as in (54), with

$$\omega_v(x)(\boldsymbol{n}) = x(2\boldsymbol{n}) \,\triangle\, x(2\boldsymbol{n}_v) \tag{59}$$

$$\omega_h(x)(\boldsymbol{n}) = x(2\boldsymbol{n}) \,\triangle\, x(2\boldsymbol{n}_h) \tag{60}$$

$$\omega_d(x)(\boldsymbol{n}) = x(2\boldsymbol{n}) \,\triangle\, x(2\boldsymbol{n}_d), \tag{61}$$

where $\triangle$ denotes 'exclusive or.' In [25], it has been shown that the inverse synthesis operators are given by

(a)   (b)

Fig. 14. *Multiresolution image decomposition based on the 2-D morphological Haar wavelet transform: (a) An image x, and (b) its decomposition into the scaled image $\psi^\uparrow(x)$, given by (53), and the detail images $\omega_v(x)$, $\omega_h(x)$ and $\omega_d(x)$, given by (55)–(57).*

$$\omega^{\downarrow}(y)(2\boldsymbol{n}) = y_v(\boldsymbol{n}) \wedge y_h(\boldsymbol{n}) \wedge y_d(\boldsymbol{n})$$

$$\omega^{\downarrow}(y)(2\boldsymbol{n}_h) = y_h(\boldsymbol{n}) \,\triangle\, (y_v(\boldsymbol{n}) \wedge y_h(\boldsymbol{n}) \wedge y_d(\boldsymbol{n}))$$

$$\omega^{\downarrow}(y)(2\boldsymbol{n}_v) = y_v(\boldsymbol{n}) \,\triangle\, (y_v(\boldsymbol{n}) \wedge y_h(\boldsymbol{n}) \wedge y_d(\boldsymbol{n}))$$

$$\omega^{\downarrow}(y)(2\boldsymbol{n}_d) = y_d(\boldsymbol{n}) \,\triangle\, (y_v(\boldsymbol{n}) \wedge y_h(\boldsymbol{n}) \wedge y_d(\boldsymbol{n})).$$

Clearly, conditions (50)–(52) are satisfied. An example, illustrating one step of this decomposition, is depicted in Fig. 15.

5.2 Lifting

In Section 3.2, we have described the technique of lifting for general decomposition systems. This technique can also be applied for wavelet decompositions. In [25], it has been shown that the modified scheme resulting from lifting, i.e., (3)–(5) for prediction and (6)–(8) for update, satisfies the additional constraints for a wavelet scheme, i.e., (48), (49). It is important to note, however, that lifting may turn a wavelet decomposition that is uncoupled into one that does no longer have this property. Again, refer to [25] for a precise statement. Here, we present one rather simple example. In the next subsection, we discuss a rather special wavelet decomposition scheme, called the *max-lifting scheme*, which can be obtained by two lifting steps.

Example 8 (Lifting based on the median operator)

In this example, we lift the lazy wavelet by means of the prediction operator

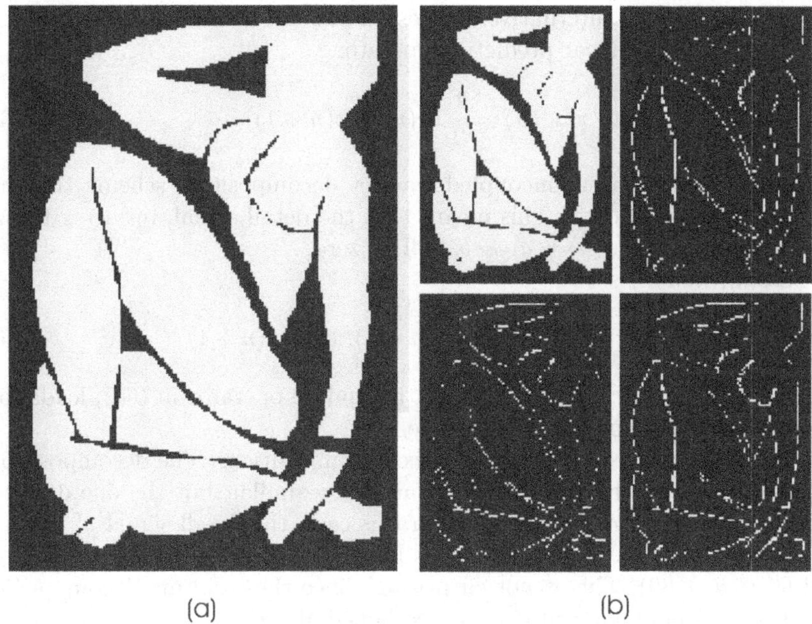

(a)   (b)

Fig. 15. *Multiresolution binary image decomposition based on the 2-D median wavelet transform: (a) A binary image x, and (b) its decomposition into the scaled image $\psi^\uparrow(x)$, given by (58), and the detail images $\omega_v(x)$, $\omega_h(x)$ and $\omega_d(x)$, given by (59)–(61).*

$$\pi(x)(n) = x(n)\,, \tag{62}$$

and the update operator

$$\lambda(y)(n) = -\mathsf{median}(0, y(n-1), y(n))\,. \tag{63}$$

Thus, we arrive at the *uncoupled* wavelet decomposition with analysis operators

$$\psi^\uparrow(x)(n) = x(2n) + \mathsf{median}(0, x(2n-1) - x(2n-2), x(2n+1) - x(2n))$$
$$\omega^\uparrow(x)(n) = x(2n+1) - x(2n)\,,$$

and synthesis operators

$$\psi^\downarrow(x)(n) = x(n)$$
$$\omega^\downarrow(y)(2n) = -\mathsf{median}(0, y(n-1), y(n))$$
$$\omega^\downarrow(y)(2n+1) = y(n) - \mathsf{median}(0, y(n-1), y(n))\,.$$

It is not difficult to see that this wavelet decomposition is invariant under grey-scale translations and multiplications (also with respect to negative values).

We mention two alternative choices for the prediction operator π. First, we may choose the linear prediction operator

$$\pi(x)(n) = \frac{1}{2}\left(x(n) + x(n+1)\right). \tag{64}$$

This choice leads to an uncoupled wavelet decomposition scheme that has two 'vanishing moments.' This means that the detail signal, resulting from a 'linear' input signal $x(n) = an + b$, will be zero.

Second, we may set

$$\pi(x)(n) = \mathsf{median}(x(n-1), x(n), x(n+1)). \tag{65}$$

This last choice, in combination with the update operator in (63), leads to a coupled wavelet decomposition scheme.

Fig. 16 depicts examples of all three decompositions. The decompositions depicted in Fig. 16(a) and Fig. 16(c) are quite similar, but the one depicted in Fig. 16(b) needs more attention. In this case, the detail signal y_1' is small at points where the input signal x_0 is linear-like (e.g., at points $0 \leq n \leq 18$ and $60 \leq n \leq 80$). This is not surprising, since the resulting decomposition has two 'vanishing moments,' as we explained above.

5.3 Max-Lifting

In this subsection, we construct a particular wavelet decomposition scheme using one prediction and one update lifting steps; it is called the *max-lifting scheme*, since the scheme has the intriguing property that its analysis operator preserves local maxima. We first give the most general formulation of this scheme. Afterwards, we show how it can be applied in the 1-D and the 2-D cases.

Consider a set S of samples, which is the disjoint union of two other sets Q and R. Assume that we have a symmetric binary adjacency relation $\sim$ on S such that $p \sim p'$ is never satisfied if p, p' lie both in Q or R. Such a relation defines a so-called *bi-graph* with vertex sets Q and R; see Fig. 17. If $q \sim r$, then we say that q and r are neighbors. Every sample is assumed to have at most finitely many neighbors. We write $r \sim\sim p$, if there exists an element q such that $r \sim q$ and $q \sim p$; see Fig. 17. In particular, $r \sim\sim r$, if r possesses at least one neighbor.

An input signal x_0 on S is first split into the approximation signal x_1, defined on R, and a detail signal y_1, defined on Q; i.e.,

$$\begin{cases} x_1(r) = x_0(r), \ r \in R \\ y_1(q) = x_0(q), \ q \in Q \end{cases}.$$

Consider the coupled wavelet transform obtained by applying a prediction and an update lifting step. The prediction step looks as follows:

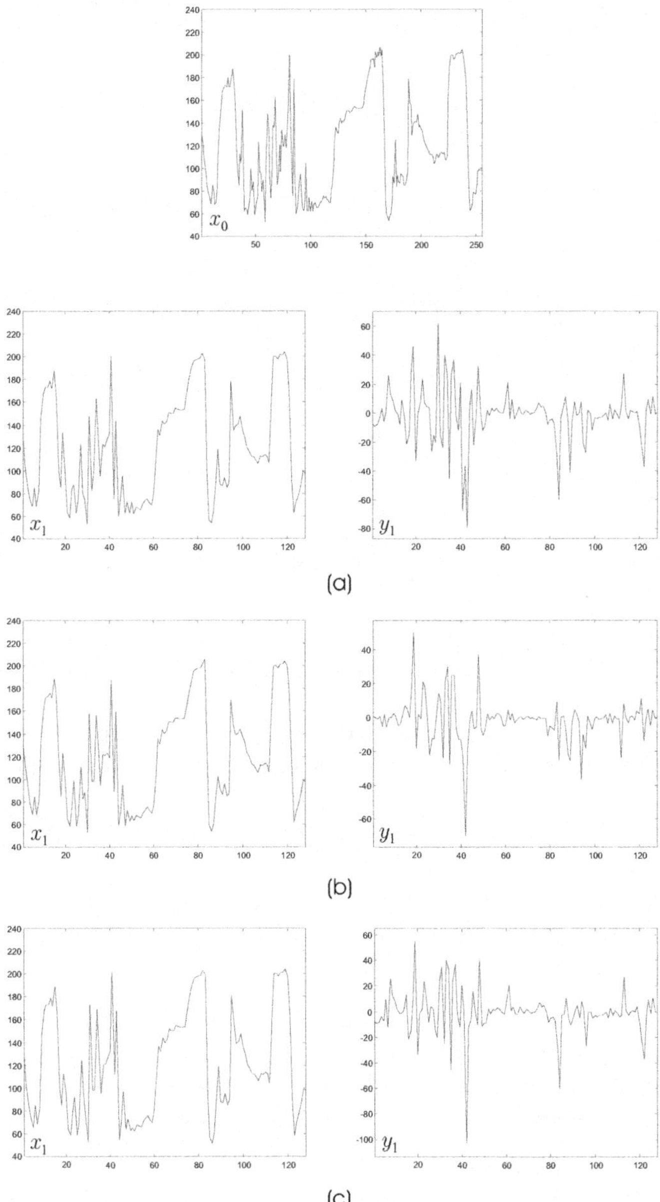

Fig. 16. *One-stage decomposition of a signal x_0 by means of two lifting steps. The prediction operators are given by: (62) in (a), (64) in (b), and (65) in (c). The subsequent update step is given by (63) in all three cases. Observe that the detail signal in (b) is small at points where the input signal is linear-like. This is due to the fact that, in (b), the decomposition has two 'vanishing moments.'*

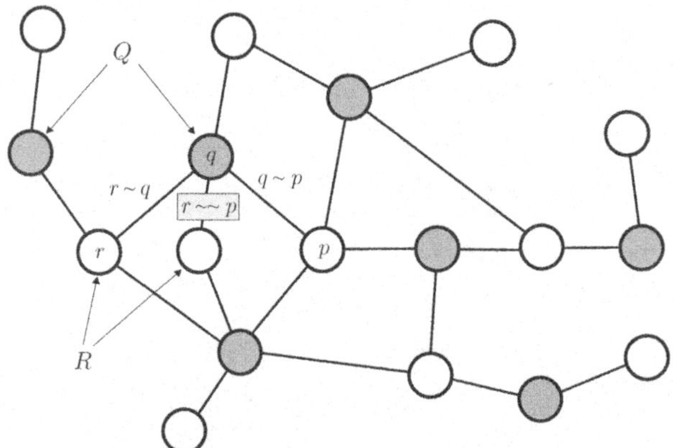

Fig. 17. *A bi-graph with vertex sets R (white nodes) and Q (gray nodes). Neighbors are connected with edges.*

$$\pi(x_1)(q) = \bigvee_{r:\, r \sim q} x_1(r), \quad q \in Q.$$

This means that the prediction of the signal at a point q is given by the maximum of its neighbors, which, by definition, all lie in R. The update operator is chosen in such a way that local maxima of the input signal x_0 are mapped to the next level x_1'. It turns out that such a property, a more precise formulation of which will be given below, can be achieved if we choose the update lifting step

$$\lambda(y_1)(r) = -\max\{0, \bigvee_{q:\, q \sim r} y_1(q)\}, \quad r \in R.$$

Thus, we arrive at the decomposition of a signal x_0 with domain S into an approximation signal x_1' on R and a detail signal y_1' on Q, given by

$$x_1'(r) = x_1(r) - \lambda(y_1')(r),\ r \in R \quad \text{and} \quad y_1'(q) = y_1(q) - \pi(x_1)(q),\ q \in Q.$$

For a point $p \in S$, we denote by $A(x_0 \mid p)$ the set of neighbors p' of p such that $x_0(p') \geq x_0(p'')$, for all neighbors p'' of p. Note that this set is nonempty, since p has finitely many neighbors. In [25], we have shown the following results:

(a) $x_0(r) \leq x_1'(r) \leq \max\{x_0(q) \mid q = r \text{ or } q \sim r\}$, for $r \in R$.

(b) $x_0(q) \leq \max\{x_1'(r) \mid r \sim q\}$, for $q \in Q$.

(c) Assume that $q \in Q$ is such that $x_0(q) \geq x_0(r)$, for $r \sim q$, and $r \sim\sim q$; then, $x_1'(p) = x_0(q)$, for every $p \in A(x_0 \mid q)$.

Indeed, this result expresses the fact that local maxima of x_0 'survive' to the next level. If x_0 has a local maximum at $r \in R$, in the sense that $x_0(r) \geq x_0(q)$, for $q \sim r$, then $x_1'(r) = x_0(r)$. If x_0 has a local maximum at $q \in Q$, in the sense that $x_0(q) \geq x_0(r)$, for $r \sim q$ and $r \sim\sim q$, then

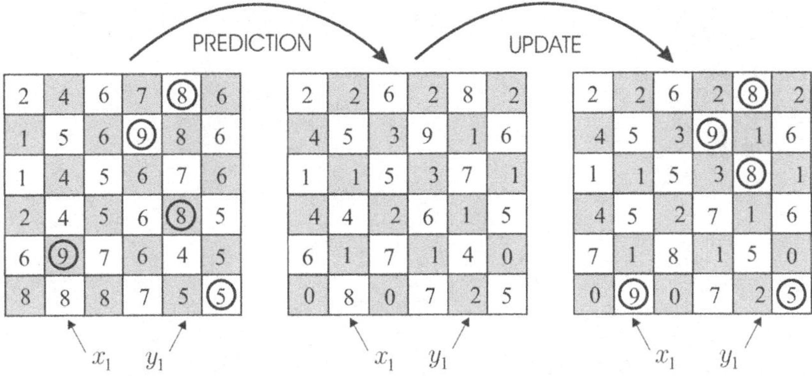

Fig. 18. *A diagram illustrating the 2-D max-lifting scheme. The white nodes contain the scaled signal x_1 (resp. x_1'), whereas, the gray nodes contain the detail signal y_1 (resp. y_1'). In this scheme, local maxima (indicated by circles) are preserved.*

$x_1'(p) = x_0(q)$, for all 'largest neighbors' p of q. In this case, 'largest neighbor' means that $x_0(p) \geq x_0(p')$, for all other neighbors p' of p.

Not only does the max-lifting scheme preserve local maxima, we can also show that this scheme will never create new maxima. To be more specific, assume that the approximation signal x_1' has a local maximum at $r \in R$; i.e., $x_1'(r) \geq x_1'(r')$, for $r' \in R$, with $r' \sim\sim r$. Then, x_0 has a local maximum at some $p \in S$, with $p = r$ or $p \sim r$, and $x_0(p) = x_1'(r)$. For a detailed proof, we again refer to [25].

In the 1-D case, where $S = \mathbb{Z}$, we can take for R, Q the even and odd samples, respectively, and define $r \sim q$ if $|r - q| = 1$. Here, we restrict ourselves to the 2-D case, and more specifically to the *quincunx sampling* scheme. In this case, R and Q consist of points $(m, n) \in \mathbb{Z}^2$, with $m + n$ even and odd, respectively. Moreover, $(m_1, n_1) \sim (m_2, n_2)$ if and only if $|m_1 - m_2| + |n_1 - n_2| = 1$.

An important feature of lifting schemes is that they allow in-place calculations. In this case, the original signal values can be replaced by the transformed ones without having to allocate additional memory. This is clearly illustrated in Fig. 18, where we apply the 2-D max-lifting scheme on the quincunx lattice to a 6×6 square matrix. This provides also an illustration of our previous assertion that the max-lifting scheme preserves local maxima; in Fig. 18, these local maxima are indicated by circles.

Note that R can be mapped onto S through a rotation of the form $(r_1, r_2) \mapsto \frac{1}{2}(r_1 - r_2, r_1 + r_2)$, after which the same scheme can be applied again. In Fig. 19, we apply the 2-D max-lifting scheme to a particular image. Here, the scaled signal x_1 and the detail signal y_1 are both defined on a quincunx grid. To properly depict these signals, we perform a $45°$ counterclockwise rotation.

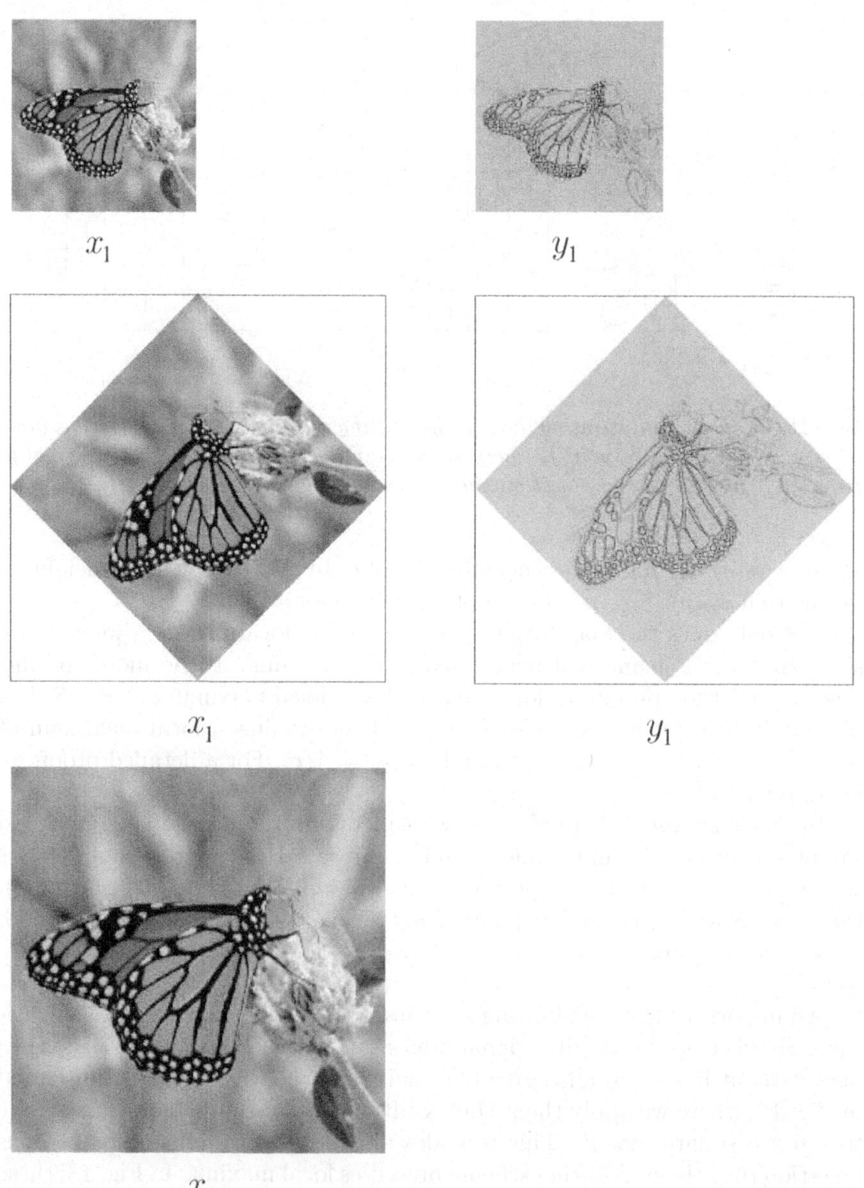

Fig. 19. *Image decomposition based on the two-dimensional max-lifting scheme with quincunx sampling. Bottom row: original image x_0. Middle row: transformed images x_1' and y_1' (after $45°$ counterclockwise rotation). Top row: images x_1'' and y_1'' obtained by applying max-lifting to x_1'. Notice that the detail image may contain positive (bright) as well as negative (dark) values.*

6 Conclusions

In this study of morphological decomposition systems with perfect reconstruction, we have presented an axiomatic treatise on pyramid and wavelet decomposition schemes. Many multiresolution signal decomposition schemes proposed in the literature are special cases of the general schemes discussed here.

The nonlinear schemes discussed as examples in this chapter enjoy some useful and attractive properties:

- Implementation can be done extremely fast by means of simple operations (e.g., addition, subtraction, max, min, median, etc.). This is partially due to the fact that only integer arithmetic is used in calculations and that use of prediction/update steps in the decomposition produces computationally efficient implementations.

- If the input to the proposed schemes is integer-valued, the output will be integer-valued as well. Clearly, these schemes can avoid quantization, an attractive property for lossless data compression.

- The proposed schemes can be easily adopted to the case of binary images. This is of particular interest in document image processing, analysis, and compression applications, but it is also important on its own right (e.g., see [39, 27], for works on constructing wavelet decomposition schemes for binary images).

- Due to the nonlinear nature of the proposed signal analysis operators, important geometric information (e.g., edges) is well preserved at lower resolutions.

Morphological pyramids and wavelets comprise a relatively new and largely unexplored research area. In this chapter, we have only been able to give the reader a glimpse of this field and the underlying mathematics. For more comprehensive discussions, one may refer to [18, 25]. In [27], Kamstra introduces an even more general framework for finite-valued wavelets. One of his major conclusions is the combinatoric explosion of the number of wavelets with finite filter length. In two recent papers, Piella, Pesquet, and Heijmans [34, 35] present results on constructing adaptive wavelets based on the lifting scheme.

Finally, the material presented in this chapter forms the basis for further developments in nonlinear multiscale signal decomposition schemes and more particularly for the construction of morphological scale-spaces [29, 2].

References

1. Braga-Neto, U., Goutsias, J.: A multiscale approach to connectivity. Computer Vision and Image Understanding, **89**, 70-107 (2003)
2. Braga-Neto, U., Goutsias, J.: Supremal multiscale analysis. SIAM Journal of Mathematical Analysis, **36**, 94-120 (2004)

3. Bruekers, F. A. M. L., van den Enden, A. W. M.: New networks for perfect inversion and perfect reconstruction. IEEE Journal on Selected Areas in Communications, **10**, 130–137 (1992)

4. Burt, P. J., Adelson, E. H.: The Laplacian pyramid as a compact image code. IEEE Transactions on Communications, **31**, 532–540 (1983)

5. Calderbank, A. R., Daubechies, I., Sweldens, W., Yeo, B.-L.: Wavelet transforms that map integers to integers. Applied and Computational Harmonic Analysis, **5**, 332–369 (1998)

6. Cha, H., Chaparro, L. F.: Adaptive morphological representation of signals: Polynomial and wavelet methods. Multidimensional Systems and Signal Processing, **8**, 249–271 (1997)

7. Claypoole, R. L., Baraniuk, R. G., Nowak, R. D.: Adaptive wavelet transforms via lifting. In: Proceedings of the IEEE International Conference on Acoustics, Speech, and Signal Processing. Seattle, Washington (1998)

8. Claypoole, R. L., Baraniuk, R. G., Nowak, R. D.: Lifting construction of nonlinear wavelet transforms. In: Proceedings of the IEEE-SP International Symposium on Time-Frequency and Time-Scale Analysis. Pittsburgh, Pennsylvania (1998).

9. Claypoole, R. L., Baraniuk, R. G., Nowak, R. D.: Adaptive wavelet transforms via lifting. Technical Report 9304, Department of Electrical and Computer Engineering, Rice University. Houston (1999)

10. Claypoole, R. L., Davis, G., Sweldens, W., Baraniuk, R. D.: Nonlinear wavelet transforms for image coding. In: Proceedings of the 31st Asilomar Conference on Signals, Systems, and Computers. **1**, 662–66 (1997)

11. Daubechies, I., Sweldens, W.: Factoring wavelet transforms into lifting steps. Journal of Fourier Analysis and Applications, **4**, 245–267 (1998)

12. Daubechies, I.: Ten Lectures on Wavelets. Society for Industrial and Applied Mathematics. Philadelphia (1992)

13. de Queiroz, R. L., Florêncio, D. A. F., Schafer, R. W.: Nonexpansive pyramid for image coding using a nonlinear filterbank. IEEE Transactions on Image Processing, **7**, 246–252 (1998)

14. Egger, O., Li, W., Kunt, M.: High compression image coding using an adaptive morphological subband decomposition. Proceedings of the IEEE, **83**, 272–287 (1995)

15. Egger, O., Li, W.: Very low bit rate image coding using morphological operators and adaptive decompositions. In: Proceedings of the IEEE International Conference on Image Processing. Austin (1994)

16. Florêncio, D. A. F., Schafer, R. W.: A non-expansive pyramidal morphological image coder. In: Proceedings of the IEEE International Conference on Image Processing. Austin (1994)

17. Florêncio, D. A. F.: A New Sampling Theory and a Framework for Nonlinear Filter Banks. PhD Thesis, Georgia Institute of Technology. Atlanta (1996)

18. Goutsias, J., Heijmans, H. J. A. M.: Nonlinear multiresolution signal decomposition schemes – Part I: Morphological pyramids. IEEE Transactions on Image Processing, **9**, 1862–1876 (2000)

19. Goutsias, J., Schonfeld, D.: Morphological representation of discrete and binary images. IEEE Transactions on Signal Processing, **39**, 1369–1379 (1991)

20. Hampson, F. J., Pesquet, J.-C.: A nonlinear subband decomposition with perfect reconstruction. In: Proceedings of the IEEE International Conference on Acoustics, Speech, and Signal Processing. Atlanta (1996)

21. Hampson, F. J., Pesquet, J.-C.: M-band nonlinear subband decompositions with perfect reconstruction. IEEE Transactions on Image Processing, **7**, 1547–1560 (1998)
22. Hampson, F. J.: Méthodes Non Linéaires en Codage d'Images et Estimation de Mouvement. PhD thesis, l'Université Paris XI Orsay. Paris (1997)
23. Heijmans, H. J. A. M., Goutsias, J.: Morphology-based perfect reconstruction filter banks. In: Proceedings of the IEEE-SP International Symposium on Time-Frequency and Time-Scale Analysis. Pittsburgh (1998)
24. Heijmans, H. J. A. M., Goutsias, J.: Constructing morphological wavelets with the lifting scheme. In: Proceedings of the Fifth International Conference on Pattern Recognition and Information Processing. Minsk, Belarus (1999)
25. Heijmans, H. J. A. M., Goutsias, J.: Nonlinear multiresolution signal decomposition schemes – Part II: Morphological wavelets. IEEE Transactions on Image Processing, **9**, 1897–1913 (2000)
26. Heijmans, H. J. A. M.: Morphological Image Operators. Academic Press, Boston (1994)
27. Kamstra, L.: Discrete wavelet transforms over finite sets which are translation invariant. Research Report PNA-R0112, CWI. Amsterdam (2001)
28. Keshet (Kresch), R., Heijmans, H. J. A. M.: Adjunctions in pyramids and curve evolution. In: Kerckhove. M. (ed) Scale-Space and Morphology in Computer Vision. Springer-Verlag, Berlin (2001)
29. Keshet (Kresch), R., Heijmans, H. J. A. M.: Adjunctions in pyramids, curve evolution and scale-spaces. International Journal of Computer Vision. **52**, 139–151 (2003)
30. Mallat, S.: A Wavelet Tour of Signal Processing. Academic Press, San Diego (1998)
31. Maragos, P.: Pattern spectrum and multiscale shape representation. IEEE Transactions on Pattern Analysis and Machine Intelligence, **11**, 701–716 (1989)
32. Matheron, G.: Random Sets and Integral Geometry. John Wiley, New York (1075)
33. Piella, G., Heijmans, H. J. A. M.: Adaptive lifting schemes with perfect reconstruction. Research Report PNA-R0104, CWI. Amsterdam (2001)
34. Piella, G., Heijmans, H. J. A. M.: Adaptive lifting schemes with perfect reconstruction. IEEE Transactions on Signal Processing, **50**, 1620–1630, 2002.
35. Piella, G., Pesquet-Popescu, B., Heijmans, H.: Adaptive update lifting with a decision rule based on derivative filters. IEEE Signal Processing Letters, **9**, 329–332, 2002.
36. Serra, J.: Image Analysis and Mathematical Morphology. Academic Press, London (1982)
37. Serra, J.: Image Analysis and Mathematical Morphology. Volume 2: Theoretical Advances. Academic Press, London (1988)
38. Song, X., Neuvo, Y.: Image compression using nonlinear pyramid vector quantization. Multidimensional Systems and Signal Processing, **5**, 133–149 (1994)
39. Swanson, M. D., Tewfik, A. H.: A binary wavelet decomposition of binary images. IEEE Transactions on Image Processing, **5**, 1637–1650 (1996)
40. Sweldens, W.: The lifting scheme: A new philosophy in biorthogonal wavelet constructions. In: Lain, A. F., Unser, M. (eds) Wavelet Applications in Signal and Image Processing III. Proceedings of SPIE, **2569**, 68–79 (1995)
41. Sweldens, W.: The lifting scheme. A custom-design construction of biorthogonal wavelets. Applied and Computational Harmonic Analysis, **3**, 186–200 (1996)

42. Sweldens, W.: The lifting scheme. A construction of second generation wavelets. SIAM Journal of Mathematical Analysis, **29**, 511–546 (1998)
43. Toet, A.: A morphological pyramidal image decomposition. Pattern Recognition Letters, **9**, 255–261 (1989).
44. Vetterli, M., Kovačević, J.: Wavelets and Subband Coding. Prentice Hall, Englewood Cliffs (1995)

Morphological segmentation revisited

Fernand Meyer

Centre de Morphologie Mathématique, Ecole des Mines de Paris

1 Introduction

Morphological segmentation is now 25 years old, and is presented in textbooks and software libraries. It relies first on the watershed transform to create contours and second on markers to select the contours of interest [12]. Separating the segmentation process into two parts constitutes its main interest: finding the markers is the intelligent part, since it requires an interpretation of the image, but the difficulty is counterbalanced by the fact that the resulting segmentation is largely independent of the precise shape or position of the markers. Furthermore, the tedious part, that is finding the contours is entirely automatic. The advantages of the method are good localization, invariance to lighting conditions, absence of parameters, high sensitivity as strongly or weakly contrasted objects are equally segmented. It has been used with success in many circumstances, in any number of dimensions and has become extremely popular. Searching the webpages containing the word watershed and segmentation, Google finds more than 15000 pages ! Another reason for its popularity is the speed of the watershed transform : hierarchical queues allow to mimic the flooding of a topographic surface from a set of markers, and require only one pass through the image [3, 16, 22]. The watershed has also successfully been implemented on dedicated hardware, on DSPs and on parallel architectures [8, 4]. Morphological segmentation being a success story, is it necessary to devote a new paper to a method so widely known ? Indeed yes, for the research in the domain is more active than ever. The present paper aims at presenting some main streams of this research.

Morphological segmentation may be reinterpreted as a 2-stage process. First, a hierarchy, that is a series of nested partitions, is created. From finer to coarser partitions only fusions of tiles take place, which means that a contour present in a coarse partition is also present in any finer partition. For this reason, the strength of a piece of contour is equal to the level of coarseness at which it disappears. During the second stage of the segmentation the strongest contours separating the markers are selected. This analysis indicates

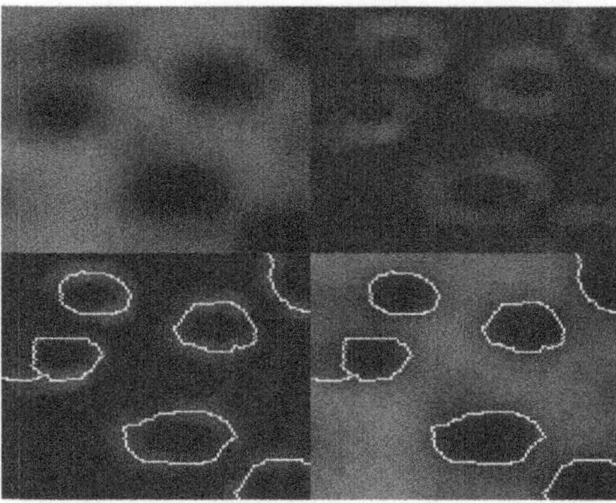

Fig. 1. 11 : Image of bubbles to be segmented ; 12: Modulus of the gradient
21: Watershed line of the gradient ; 22: Final contour

two directions for generalizing the approach, which will both be outlined in
the present paper. First create new types of hierarchies, better adapted to
particular types of images to segment. For instance, a hierarchy which is optimal for segmenting microcalcifications in breast mammographies will yield
poor results for segmenting the speaker in a videoconference sequence. Microcalcifications are tiny contrasted spots, and their optimal segmentation would
work poorly for most other types of images. A second direction will explore
new ways for extracting useful contours from the hierarchy.

2 Classical morphological segmentation

2.1 Birth of the watershed transform for segmentation

The watershed transform was first used to segment dark bubbles like those represented in fig.1 [7]. Thresholding such an image is impossible due to blurred
contours and varying background. On the gradient image, seen as a topographic surface, each blob boundary appears as a chain of mountains, with
a regional minimum inside the blob and another outside. Thresholding the
gradient image to get the contour does not lead to the solution either: a too
low threshold produces a thick contour, whereas a too high threshold misses
parts of the contour, where the boundary of the blob is somewhat fuzzy.
The watershed line appears to be the good solution: if a drop of water falls
on a topographic surface, it follows a line of steepest descent and reaches a
regional minimum. The attraction zone of a regional minimum is called its
catchment basin (CB). Two adjacent catchment basins have in common a
portion of the watershed line. Hence a drop of water falling on the watershed
line has as much chance to reach one or another of the adjacent minima. In

simple cases like these, the watershed line is a thin line, following the expected contour, independently of the intensity or contrast of the bubble. The advantages of the watershed is good localization of the contour, invariance to lighting conditions, absence of parameters, high sensitivity since strongly or weakly contrasted objects are equally segmented.

2.2 Watershed and Topographic Distance

In a digital framework, images are represented on a regular graph (square or hexagonal grid for 2D images) where the nodes represent the pixels and the edges the neighborhood relations. A connected component of uniform grey tone is called a plateau. A plateau without lower (resp. higher) neighbors is a regional minimum (resp. maximum). Consider for the moment a topographic surface g without other plateaus than the regional minima. A drop of water falling on g glides along a path of steepest descent until it reaches a regional minimum. If the altitude of a pixel x is $g(x)$, the altitude of its lowest neighbor defines the erosion $\varepsilon(g)(x)$ of size 1 at pixel x. Hence the altitude of the steepest descending slope at pixel x is $\text{slope}(x) = g(x) - \varepsilon(g)(x)$.

If π is a path $(x = p_1, p_2, ..., y = p_n)$ between two pixels x and y, we define the topographic variation along the path π as the sum $\sum_{i=2,n} \text{slope}(p_i)$ of the elementary topographic variations along the path π. The topographic distance between two pixels x and y is defined as the minimal topographic variation along all paths between x and y. Obviously the lines of steepest descent are the geodesics of the topographic distance. Putting all regional minima of g to the same altitude does not change its CBs.

Definition 1. *We call catchment basin* $\text{CB}(m_i)$ *of a regional minimum* m_i *the set of pixels which are closer to* m_i *than to any other regional minimum for the topographic distance.*

In this framework the construction of the catchment basins becomes a shortest path problem, i.e., finding the path between a marker and an image point that corresponds to the minimum weighted distance. Computing this minimum weighted distance at all image points from any marker is also equivalent to finding the *grey-weighted distance transform (GWDT)* of the image. There are several types of discrete algorithms to compute the GWDT which include iterated (sequential or parallel) min-sum differences [20] and hierarchical queues[3]. For images with plateaus, or images defined on a continuous domain, see [14, 11].

As long one stays within the catchment basin of a minimum m, of altitude 0, the set of pixels at a topographic distance smaller or equal to h are all pixels at an altitude below h. For this reason, the most efficient implementation of the watershed, is based on the idea of flooding a topographic surface. The relief is flooded from sources placed at each regional minimum. The flood level is uniform all over the relief and increases, as shown on fig.2a. The pixels

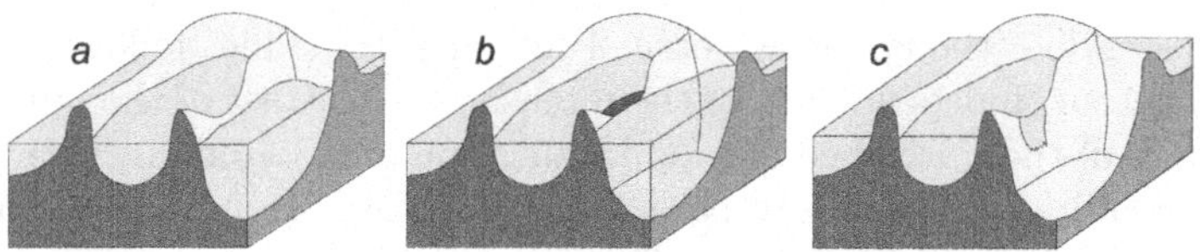

Fig. 2. a) flooding from all minima b) the watershed line is the line where adjacent lakes meet c) a catchment basin without source is flooded through a neighboring basin.

where two lakes meet are a the same topographic distance of two minima: hence they belong to the watershed line (fig.2b).

Flooding with markers

Let us now apply the method which was successful for the bubbles on a more complex image, represented by a grey tone function f and illustrated in fig.3a. For segmenting f, first its edges are highlighted by computing its gradient magnitude (see fig.3b), approximated by the discrete morphological gradient $\delta(f) - \varepsilon(f)$, where $\delta(f) = f \oplus B$ is the flat dilation of f by a small disk B and $\varepsilon(f) = f \ominus B$ is the flat erosion of f by B. The watershed line is presented in fig.3c, showing a severe oversegmentation. Even the sky, apparently rather homogeneous in the initial image is cut into multiple small pieces. As a matter of fact, the gradient image is extremely sensitive to noise and its minima are extremely numerous ; each minimum will give birth to a catchment basin, leading to an oversegmentation. In order to overcome this oversegmentation one could imagine retaining only the watershed lines on top of high gradient zones. This solution would be unsatisfactory in general: the buildings visible in the background of the cameraman image produce only low gradient intensities.

To select the contours to be retained, the best solution is to mark the objects of interest, including the background. Each marker becomes a source for flooding the topographic surface. The flood of the sources is tuned in such a way that the flooding level is uniform all over the topographic surface and grows with constant speed. As the flood level becomes higher, lakes formed from different sources meet, and a dam is erected in order to avoid that they mix [3]. In this case, catchment basins without sources are flooded from already flooded neighboring regions (see fig.2c). Segmenting the cameraman image with two markers (one of the markers is formed by two connected components) yields the segmentation of fig.6c.

2.3 A hierarchy of contours

In this section we analyze more precisely which contours are selected when markers are used. As a matter of fact, the strongest contours of the gradient image between the markers have been selected. Fig.4 shows the simplest

Fig. 3. a) initial image ; b) morphological gradient ; c) watershed of the gradient image

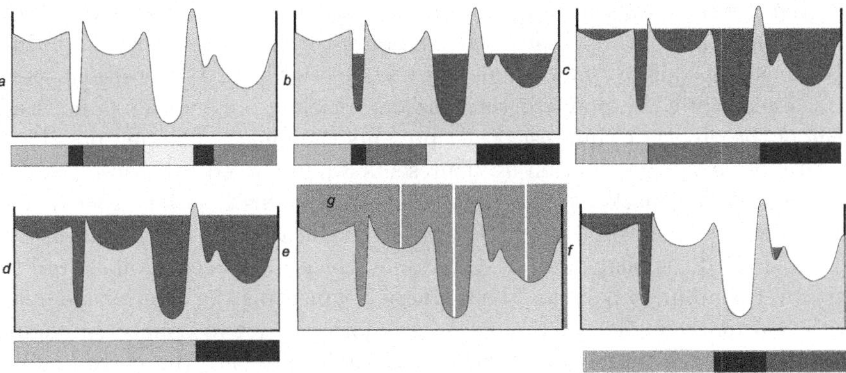

Fig. 4. a : initial topographic surface
b-d : 3 levels of flooding and delimitation of the catchment basins below
e : g is a function maximal everywhere except in the position of 3 markers.
f : largest flooding of f below g: the remaining minima correspond to the markers

way to assign a strength to the contours of a gradient image. We present a topographic surface before any flooding and 3 increasing levels of uniform flooding. For each resulting surface the catchment basins are presented below the corresponding figure with different shades of grey ; as the level of flooding increases, neighboring catchment basins merge, yielding a coarser partition. In fig.4, associated to 4 levels of increasing flooding, we have 4 decreasing partitions with respectively 6, 5, 3 and 2 regions. Such nested partitions where, from a finer level to a coarser level, only fusions of regions take place, is called a hierarchy. For this reason, a piece of contour present in a given partition is also present in all coarser partitions ; it can be weighted by the flooding level for which it disappears.

Applied to the cameraman image, the same process of flooding the gradient image yields the partitions illustrated by fig.5. On the left, the flooded gradient image, in the centre we find a mosaic image obtained by replacing each catchment basin by its mean grey value in the initial image, on the right the contours ; the whole process being repeated for 3 levels of flooding. It appears that the strongest contours of this hierarchy only select small contrasted regions, which are not the most important from a perceptual point of view. This is a general situation: the lowest level of the hierarchy presents all possible contours, generally an oversegmentation. Higher levels of the hierarchy present far less regions, but not necessarily the regions of interest.

Using markers constitutes an efficient tool to select the sole objects one is interested in and construct their contours. Let us consider again the figures 4 a to d, where the increasing levels of flooding induce a hierarchy of nested partitions, with respectively 6, 5, 3 and 2 regions. We would like 3 regions, but not the 3 regions present in fig.4c. In fig.4e, we have indicated which regions we are interested in by creating a function $g \geq f$, identical with f at three positions and equal to the maximal grey value elsewhere. The positions where $f = g$ exactly correspond to the markers, marking an internal zone inside each of the objects of interest. Fig.4 presents the highest flooding possible of f entirely below the function g : it presents 3 minima, corresponding exactly to the 3 chosen markers. The lakes produced by this flooding have varying altitudes. The associated catchment basins, delineated below fig.4 correspond to the desired segmentation. The contours which have been retained are the strongest contours separating the markers. Segmenting the cameraman image with two markers (one of the markers is formed by two connected components) yields the segmentation of fig.6c. Fig.6b presents the corresponding flooding of the gradient image, a topographic surface with only two minima, corresponding to the markers.

The remaining part of the paper will be entirely devoted to the use of hierarchies for morphological segmentation.

3 On partitions and hierarchies

The preceding section has shown that morphological segmentation relies on hierarchies of partitions. We will now define both terms and establish their lattice structure.

3.1 The lattice of partitions

We are interested in segmenting images, that is functions of Fun$(E, \mathcal{T})$ where E represents the support of the images (a continuous domain or a discrete grid, in any number of dimensions) with value in a lattice $\mathcal{T}$ (in practice the set of reals or integers). The power set $\mathcal{P}(E)$ of E contains all subsets of E. The result of any segmentation of an image f of Fun$(E, \mathcal{T})$ will be a partition $\mathfrak{S}$

Fig. 5. Catchment basins associated to increasing levels of flooding. Left: flooded gradient image. Center: Mosaic image. Right : segmentation

Fig. 6. Left : 3 markers placed on the cameraman image
Center : result of swamping
Right : resulting segmentation

of E, that is a family (C_i) of elements of $\mathcal{P}(E)$ verifying : $C_i \wedge C_j = \varnothing$ for $i \neq j$ and $\bigcup C_j = E$.

Partitions as equivalence classes

Equivalence relations are the easiest way to define partitions. They allow to shift from a local to a global point of view and vice-versa.

Definition 2. *An equivalence relation R on E is defined as a binary relation on $E \times E$ verifying:*
(i) Reflexivity: for all $x \in E$: xRx
(ii) Symmetry: for all $x, y \in E$: $xRy \Rightarrow yRx$
(iii) Transitivity: for all $x, y, z \in E$: xRy and $yRz \Rightarrow xRz$

To any partition $\mathfrak{S}$ of E, we may associate an equivalence relation R defined by $xRy \Leftrightarrow \exists C_i \in \mathfrak{S}$ such that $x, y \in C_i$.

Conversely, to any equivalence relation may be associated the partition of the equivalence classes associated to R. The equivalence class of $x \in E$ being the set of all elements y of E such that xRy.

The complete lattice of partitions

Partitions of a same set E may be more or less coarse. Two partitions $\mathfrak{S}_1$ and $\mathfrak{S}_2$ are nested if each tile of $\mathfrak{S}_2$ is a union of tiles of $\mathfrak{S}_1$; or equivalently, each tile of $\mathfrak{S}_1$ is included in a tile of $\mathfrak{S}_2$. In this case the partition $\mathfrak{S}_1$ is said to be finer than the partition $\mathfrak{S}_2$, which is coarser. We write $\mathfrak{S}_1 \leq \mathfrak{S}_2$

To be finer is an order relation : it is reflexive, transitive and anti-symmetric : $\mathfrak{S}_1 \leq \mathfrak{S}_2$ and $\mathfrak{S}_1 \geq \mathfrak{S}_2$ implies $\mathfrak{S}_1 = \mathfrak{S}_2$. The corresponding equivalence relations verify : for all $x, y \in E$: $xR_1y \Rightarrow xR_2y$.

The infimum of a family of partitions

There is one partition which is finer than all others : it is the partition made of the singletons $\{x\}$ of E. Hence the family of partitions which are finer than all partitions of $(\mathfrak{S}i)_{i \in I}$ is not empty. This family has a largest element, called infimum of $(\mathfrak{S}i)_{i \in I}$. It is characterized by $R_{\min} = \bigwedge\limits_{i \in I} R_i$ where $R_1 \wedge R_2$ means R_1 *and* R_2. It is easy to verify that $R_{\min}$ is still an equivalence relation.

The supremum of a family of partitions

There is one partition which is coarser than all others : it is the partition that is equal to E itself. Hence the family of partitions which are coarser than all partitions of $(\mathfrak{S}i)_{i \in I}$ is not empty. This family has a smallest element, called supremum of $(\mathfrak{S}i)_{i \in I}$. The problem now is that if R_1 and R_2

are two equivalence relations, then $\{R_1 \text{ or } R_2\}$ is no longer transitive. In order to obtain $R_{\max}$ one has to construct the transitive closure of $\bigvee_{i \in I} R_i$. We will write $R_{\max} = \widehat{\bigvee_{i \in I} R_i}$. The transitive closure is obtained as follows :

$$x \; \widehat{\bigvee_{i \in I} R_i} \; y. \Leftrightarrow \text{there exists a sequence } (x = z_1, z_2, ..., z_n = y) \text{ such that}$$
$$z_k \; \bigvee_{i \in I} R_i \; z_{k+1} \text{ for all } k.$$

3.2 On hierarchies, trees and ultrametric distances

We now give an axiomatic definition of hierarchies [1] and study their properties. Hierarchical classification has become famous with the classification of species. On one hand one defines partitions such that two elements of a same class are closer or more similar than two elements belonging to distinct classes. On the other hand one defines a classification tree, or dendrogram, that is a series of partitions compatible with each other: the classes of a partition are included in the classes of any coarser partition. In other words, the partitions are nested.

Definition of a dendrogram and its elements

Let $\mathcal{A}$ be a subset of $\mathcal{P}(E)$, on which we consider the inclusion order relation. $\mathcal{A}$ is a dendrogram if the following axiom is verified :

Axiom 1 *(Dendrogram axiom)* $A, U, V \in \mathcal{A}$:
$A \subset U \text{ and } A \subset V \Rightarrow U \subset V \text{ or } V \subset U$

We remark that the dendrogram axiom is weaker than the axiom defining partitions: if U, V are classes of a partition and there exists a set A included in both U and V then $U = V$.

If $\mathcal{A}$ is a dendrogram, we may define :
- the summits : $\text{Sum}(\mathcal{A}) = \{A \in \mathcal{A} \mid \forall B \in \mathcal{A} : A \subset B \Rightarrow A = B\}$
- the leaves : $\text{Leav}(\mathcal{A}) = \{A \in \mathcal{A} \mid \forall B \in \mathcal{A} : B \subset A \Rightarrow A = B\}$
- the nodes : $\text{Nod}(\mathcal{A}) = \mathcal{A} - \text{Leav}(\mathcal{A})$

$\mathcal{A}$ is a hierarchy, if the two following axioms are verified:

Axiom 2 *(Intersection axiom)* : *two elements of $\mathcal{A}$ which are not comparable for the inclusion order have an empty intersection: $A, B \in \mathcal{A}$:*
$A \cap B \in \{A, B, \emptyset\}$

Axiom 3 *(Union axiom)* *Any element A of $\mathcal{A}$ is the union of all other elements of $\mathcal{A}$ contained in A:*
$\forall A \in \mathcal{A} : \bigcup \{B \in \mathcal{A} \mid B \subset A ; B \neq A\} = \{A, \emptyset\}$

Proposition 1. *The intersection axiom implies that A is a dendrogram for the inclusion order.*

Proof. If $A \neq \emptyset$, $A \subset U$ and $A \subset V$, then $U \cap V \neq \emptyset$, implying that $U \cap V = U$ or $U \cap V = V$, that is $V \subset U$ or $U \subset V$ showing that the dendrogram axiom is satisfied.

Grouping all tiles belonging to all partitions in a series of nested partitions $(\mathfrak{S}_i)$, obviously yields a hierarchy $\mathcal{A}$.

Stratified hierarchies, ultrametric distances and nested partitions

$\mathcal{A}$ is a stratified hierarchy, if it is equipped with an index function st from $\mathcal{A}$ into $\mathbb{R}$ which is strictly increasing with the inclusion order : $\forall A, B \in \mathcal{A}$ $A \subset B$ and $B \neq A \Rightarrow \mathrm{st}(A) < \mathrm{st}(B)$. Stratification offers the possibility of thresholding a hierarchy: the elements A of a hierarchy verifying $\mathrm{st}(A) \geq \lambda$ are all coarser than λ.

Given a stratified hierarchy $\mathcal{A}$, verifying $\mathrm{st}(A) = 0$ for each $A \in \mathrm{Leav}(\mathcal{A})$, a distance between the elements of $\mathcal{P}(E)$ is defined by:
$\forall C, D \in \mathcal{P}(E), \, d(C, D) = \inf\{\mathrm{st}(A) \mid A \in \mathcal{A} : C \subset A \text{ and } D \subset A\}.$
Properties : d is an ultrametric distance :

$$\forall A, B \in \mathcal{A} \quad d(A, B) = 0 \Rightarrow A = B$$
$$\forall C, D \in \mathcal{P}(E) \quad d(C, D) = d(D, C)$$
$$\forall B, C, D \in \mathcal{P}(E) \quad d(C, D) \leq \max\{d(C, B), d(B, D)\}$$

This last inequality is called ultrametric inequality, it is stronger than the triangular inequality. It expresses that the index of the smallest tile containing C and D is smaller or equal than the index or the smallest tile containing all three elements B, C and D.

For $X \in \mathcal{P}(E)$ the closed ball of centre X and radius ρ is defined by $\mathrm{Ball}(X, \rho) = \{D \in \mathcal{P}(E) \mid d(X, D) \leq \rho\}.$

Hierarchies as balls of an ultrametric distance

The balls of an ultrametric distance verify rather peculiar properties:

- Two closed balls $\mathrm{Ball}(X, \rho)$ and $\mathrm{Ball}(Y, \rho)$ with the same radius are either disjoint or identical.
- Each element of a closed ball $\mathrm{Ball}(X, \rho)$ is centre of this ball
- The diameter of a ball is smaller or equal to its radius !

For this reason, given an ultrametric distance index d, the closed balls of radius λ form a partition. For increasing values of λ, these partitions are nested and become coarser and coarser and form a stratified hierarchy.

Hierarchies associated with a dissimilarity index

Any partition $\mathcal{A}$ for which a dissimilarity between adjacent regions has been defined can be represented as a region adjacency graph (RAG) $G = (X, \Theta)$, where X is the set of nodes and Θ is the set of edges. The nodes represent regions of the partition. Adjacent regions i and j are linked by an edge $u = (i, j)$ with a weight s_{ij} expressing the dissimilarity between them. As an example, in case of a topographic surface we may choose as partition the set of its catchment basins, the dissimilarity between two adjacent basins being the altitude of the pass separating them. A path $\mu = (i_1, i_2, ...i_k)$ is a sequence of neighboring nodes. The adjacency matrix $A = (\alpha_{ij})$ of the graph is defined by:

$$\alpha_{ij} = \begin{cases} s_{ij} & \text{if } (i, j) \in \Theta \\ \infty & \text{if not} \end{cases}$$

For any $\lambda \geq 0$, one defines a derived graph $G_\lambda = [X, \Theta_\lambda]$ with the same nodes but only a subset of edges : $\Theta_\alpha = \{(i, j) \mid \alpha_{ij} \leq \lambda\}$. The connected components of this graph create a partition of the objects X into classes. They constitute the classes of the classification at level λ. If $L = (i, i_2, ...i_p, j)$ is this path, the maximal dissimilarity along L verifies $\max(\alpha_{ii_1}, \alpha_{i_1 i_2}, ...\alpha_{i_p j}) \leq \lambda$. Two nodes belong to the same class at level λ if and only if there exists a path in G linking these two nodes along which all dissimilarity indices are below λ: $\alpha_{ij}^* = \min_{L \in C_{ij}} (\max(\alpha_{ii_1}, \alpha_{i_1 i_2}, ...\alpha_{i_p j})) \leq \lambda$ where C_{ij} is the set of all paths between i and j. Going back to morphological flooding : two catchment basins belong to the same class at level λ if and only if the corresponding minima belong to the same lake at flooding level λ. It is easy to verify that α_{ij}^* is an ultrametric distance, called max-distance, verifying for any i, j, k : $\alpha_{ik}^* \leq \max\left(\alpha_{ij}^*, \alpha_{jk}^*\right)$. If k and l are neighboring nodes, the shortest path connecting them is (k, l) itself, with a weight a_{kl}. Hence $\alpha_{kl}^* = \min_{L \in C_{kl}} (\max(\alpha_{ki_1}, \alpha_{i_1 i_2}, ...\alpha_{i_p l})) \leq a_{kl}$. The ultrametric distance α_{kl}^* is the largest ultrametric distance below α_{kl}, it is called the subdominant ultrametric distance associated with a_{ij}. The closed balls $\text{Ball}(i, \rho) = \{j \in X \mid \alpha_{ij}^* \leq \rho\}$ form the classes of the partition at level ρ. For increasing levels of ρ one obtains coarser classes.

Minimum spanning tree and forests

We will now extract the minimum spanning tree from the graph G . We start from a graph (X, T) made only of isolated nodes: T is initially an empty set of edges. We consider all edges of Θ in increasing order of weight:

- if e_{ij} is the current edge, and the nodes i and j are not yet linked in T, add e_{ij} to T. Otherwise discard this edge
- stop when T becomes a tree, spanning all nodes of X.

The resulting tree is the minimum spanning tree (MST), and the algorithm for constructing it is due to Bohuslav (1927). Among all trees spanning the nodes of X, the sum of the weights of T is minimal. The construction of the minimum spanning tree defines a distance between the nodes: the distance between two nodes k and l is the weight of the edge whose adjunction to T has created a connected path between them within T. In other words, it is the weight of the highest edge on the unique path connecting k and l within T. One can easiliy see that it is exactly the subdominant ultrametric distance associated to a_{ij} defined in the previous section. The MST conveys the same information as the RAG itself with respect to the hierarchy: cutting all edges with a threshold higher than λ in the MST or in the RAG yields the same hierarchy. Using the MST rather than the RAG is economical in processing speed and memory requirements since it has $n - 1$ edges for n nodes.

From a weighted tree to a hierarchy

Conversely, let us consider a spanning tree Θ. To any distribution of weights $W = (w_{jk})$ on the edges of Θ is associated an ultrametric distance $d_W(x_i, x_j)$, equal to the weight of the highest edge on the unique path between x_i and x_j.

3.3 The lattice of hierarchies

It is often interesting to combine several hierarchies, in order to combine various criteria or merge the information obtained from various sources (color or multispectral images for instance).

Supremum and infimum of two hierarchies

Let $\mathcal{A}$ and $\mathcal{B}$ be two stratified hierarchies, with their associated distances: $d_{\mathcal{A}}$ and $d_{\mathcal{B}}$. The following relation defines an order relation between the hierarchies: $\mathcal{B} < \mathcal{A} \Leftrightarrow \forall C, D \in \mathcal{P}(E) \quad d_{\mathcal{A}}(C, D) \leq d_{\mathcal{B}}(C, D)$.

With this order relation the stratified hierarchies of $\mathcal{P}(E)$ form a complete lattice. The maximal element is the hierarchy having E as only element and the smallest hierarchy contains only singletons $\{x\}$.

The infimum of two hierarchies $\mathcal{A}$ and $\mathcal{B}$ is written $\mathcal{A} \wedge \mathcal{B}$ and is defined by its ultrametric distance $d_{\mathcal{A} \wedge \mathcal{B}} = d_{\mathcal{A}} \vee d_{\mathcal{B}}$. Its balls are defined by :
$\mathrm{Ball}_{\mathcal{A} \wedge \mathcal{B}}(X, \rho) = \mathrm{Ball}_{\mathcal{A}}(X, \rho) \wedge \mathrm{Ball}_{\mathcal{B}}(X, \rho)$

The supremum of two hierarchies $\mathcal{A}$ and $\mathcal{B}$ is written $\mathcal{A} \vee \mathcal{B}$ and is the smallest hierarchy larger than $\mathcal{A}$ and $\mathcal{B}$; as $d_{\mathcal{A}} \wedge d_{\mathcal{B}}$ is not an ultrametric distance, $d_{\mathcal{A} \vee \mathcal{B}}$ is the subdominant ultrametric distance associated to $d_{\mathcal{A}} \wedge d_{\mathcal{B}}$. If $\mathcal{A}_{\lambda}$, $\mathcal{B}_{\lambda}$ and $\mathcal{A}_{\lambda} \vee \mathcal{B}_{\lambda}$ are the partitions obtained by taking the balls of radius λ in each of the three hierarchies, then the boundaries of $\mathcal{A}_{\lambda} \vee \mathcal{B}_{\lambda}$ are all boundaries existing in both $\mathcal{A}_{\lambda}$ and $\mathcal{B}_{\lambda}$. The infimum and supremum of two hierarchies are illustrated in fig.7. In fig.8 to segment a color image, a separate hierarchy has been created for two color components. The infimum of both provides a better segmentation than each taken separately.

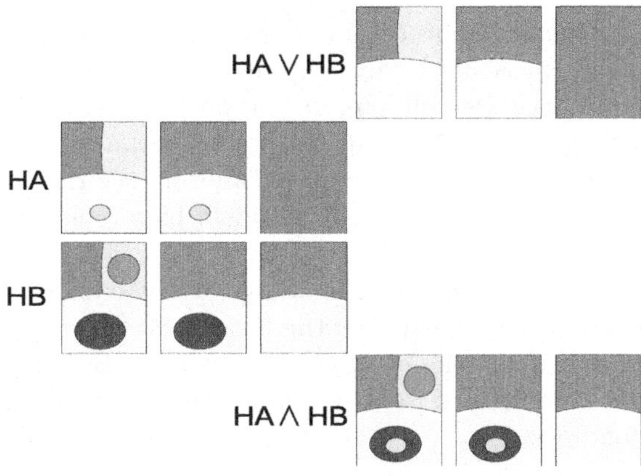

Fig. 7. Two hierarchies HA and HB and their derived supremum and infimum

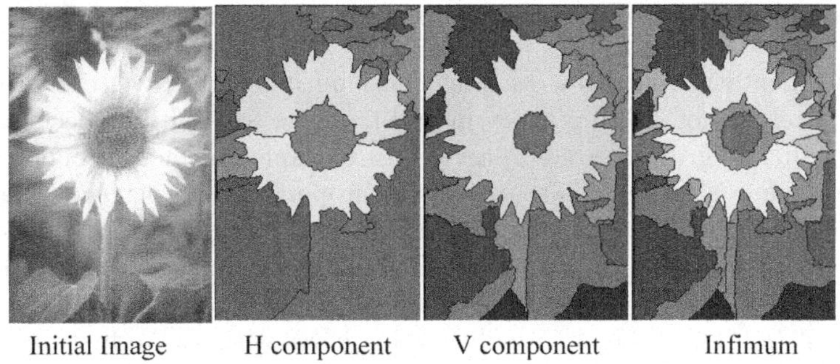

Fig. 8. Infimum of two hierarchies.

4 Creating hierarchies

4.1 Agglomerative and divisive techniques

A hierarchy being a series of nested partitions, it may obviously be constructed by two opposite approaches. In a top-down approach we start from a coarse segmentation and refine it by successive regions splitting in order to produce finer levels of hierarchy. This approach has been used by Philippe Salembier in order to produce an object oriented coder [17]: the coarse segmentation is encoded first and resegmented in order to encode additional details at finer levels. Far more common is the bottom-up approach in which a fine segmentation is produced first ; coarser segmentations are produced by merging the most similar adjacent regions. Criteria of similarity may vary as the segmentation becomes coarser. In his thesis Olivier Monga proposed a number of criteria and an optimal order in which to use them [13]. Béatriz Marcotegui used a similar bottom-up approach in order to segment a video sequence [9]. Many aggregative approaches have been published in the literature.

In the present paper we will explore the possibilities offered by levelings toconstruct hierarchies. We have already met floodings : the watershed of increasing floodings produces a hierarchy. Tailoring the rhythm of the flooding of the various basins to our need will yield hierarchies with the desired properties.

Levelings are more general than floodings. They symmetrically erase the peaks and flood the valleys, enlarging the flat-zones. We will see how we may associate a hierarchy to a series of successive levelings. Before presenting both methods to create hierarchies we will recall a number of useful properties of levelings and floodings.

4.2 Quasi-flat zones, contours and levelings.

From quasi-flat zones to the definition of levelings

Image segmentation partitions an image into mutually exclusive subsets, called regions, each one of which is uniform and homogeneous with respect to some property such as grey-tone, color, texture or motion and whose property value differs in some significant way from the property value of each neighboring regions". Extracting homogeneous regions is indeed the most elementary method of segmentation. Two homogeneous zones are separated by a transition zone. A down transition between the neighboring pixels p and q happens if $(g_p > g_q + \lambda)$; its amplitude increases with λ. We first define the negation of the up relation : $Not\,[g_p > g_q + \lambda] \Leftrightarrow g_p \leq g_q + \lambda$. The relation $\begin{vmatrix} g_p \leq g_q + \lambda \\ g_q \leq gp + \lambda \end{vmatrix} \Leftrightarrow (|g_p - g_q| \leq \lambda) \Leftrightarrow (g_p \approx g_q)$ is symmetrical. The transitive closure of this relation is an equivalence relation whose equivalence classes are the quasi-flat zones of slope λ of function f.

We wish to construct a filter Φ able to produce a simplification of f, such that the partition in quasi-flat zones of Φf is coarser than the partition of the quasi-flat-zones of f. Such a filter Φ should transform an image f into an image g with less details and simpler to segment. Furthermore, the contours of any segmentation produced on g should exactly match the contours of the same objects as seen in f. In other words, there should be no displacement of the contours when one goes from f to g. Suppose now that $g_p > g_q + \lambda$. As we require that no contour is displaced when going from f to g, a similar contour (by similar we mean that to an up transition should correspond an up transition) should exist between pixels p and q for the image f. This basic requirement is at the heart of the definition of levelings:

Definition 3. *A function g is a leveling of a function f if and only if: for any couple of neighboring pixels (p, q): $g_p > g_q + \lambda \Rightarrow f_p \geq g_p$ and $g_q \geq f_q$*

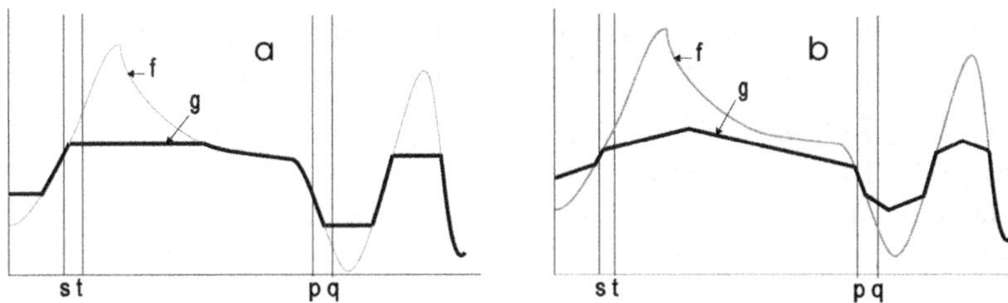

Fig. 9. The function g is a leveling of function f. On the left, a flat leveling is obtained for $\lambda = 0$; whereas $\lambda = 1$ produces a slope leveling on the right.

Hence if g is a leveling of f, then to any transition $g_p > g_q + \lambda$ corresponds an even bigger transition, since the interval $[g_q, g_p]$ is included in the interval $[f_q, f_p]$. This may be verified in fig.9 for the couples of pixels (s, t) and (p, q). Fig.9a represents a flat leveling, obtained for $\lambda = 0$, whereas fig.9b represents a slope leveling, obtained for $\lambda = 1$

The parameter λ is equal to the maximal slope of the quasi-flat zones: a transition verifying $|g_p - g_q| \leq \lambda$ is considered to be smooth. For $\lambda = 0$, the smooth zones are flat. Calling α the extensive dilation $[g \vee (\delta g - \lambda)]$ and β the anti-extensive erosion $[g \wedge (\varepsilon g + \lambda)]$, we obtain a criterion characterizing levelings $f \wedge \alpha g \leq g \leq f \vee \beta g$ [10]
Levelings verify the following algebraic properties:

-   if h is a leveling of f, then $f \vee h$ and $f \wedge h$ also are levelings of f
-   if both g and h are levelings of f, then $g \vee h$, $g \wedge h$ are also levelings of f

The relation "to be a leveling" is a preorder relation, that is a reflexive and transitive relation. In particular if h is a leveling of g and g is a leveling of f, then h is a leveling of f.

Levelings of f generated from a marker function

After describing the desirable properties of levelings, we now present how to construct them. The class of levelings of a function f is extremely huge: in fact we may associate to any function g a leveling of f. The result will be particularly interesting if we take for g an already simplified function of f, such as a low pass filter or an alternate sequential morphological filter. In this case, the leveling will retain the major simplification characteristics of the marker function but at the same time restore the boundaries sharpness and position so as to coincide with the boundaries of f. Taking a function g as a marker, we will progressively transform it into a leveling of f. An arbitrary function g will most surely not be a leveling of f, hence it will not verify the criterion:

$$f \wedge \alpha g \leq g \leq f \vee \beta g \qquad (1).$$

We will modify g as little as possible, until condition (1) is satisfied everywhere. Repeat until convergence the following modification:

-   on $\{g > f\}$ replace g by $f \vee \beta g$
-   on $\{g < f\}$ replace g by $f \wedge \alpha g$
    The algorithm converges as the values of g become closer and closer to f. Repeating until convergence $g = (f \wedge \alpha g) \vee \beta g$ produces the same result.

Fig.9 shows two levelings obtained respectively for $\lambda = 0$ on the left and $\lambda = 1$ on the right ; for $\lambda = 0$, we obtain flat zones, and for $\lambda = 1$, quasi-flat zones with a slope equal to 1. We start with the reference function f and the marker function g. The function g is transformed into a leveling g' of f. On $\{g < f\}$, the leveling raises g as little as possible until a flat zone is created or the function g' hits function f : hence on $\{g < f\}$, the function g' is quasi-flat. On $\{g > f\}$, the leveling reduces g as little as possible until a flat zone is created or the function g' hits function f : hence on $\{g > f\}$, function g' also is quasi-flat.

To simplify image 10_{11}) before segmentation, we first simplify it using an alternate sequential filter (fig.10_{12}). The contours of this image are displaced, specially in the corners. But if we use this image as a marker and construct a leveling ($\lambda = 1$), we keep the simplification of the image but restore the contours (fig.10_{13})

Useful properties of levelings

All these characteristics make the levelings vary interesting as preprocessing filters for the segmentation:

-   they are auto-dual, as they treat white and black objects in a similar way
-   they simplify the image : all connected particles where $g < f$ or where $g > f$ are quasi flat.
-   they extend the quasi-flat zones of f
-   the remaining contours in image g also correspond to contours present in image f.

Flat levelings, obtained for $\lambda = 0$ have another very interesting property: they do not generate regional minima or maxima. That means if X is a regional minimum (resp. maximum) of g, g being a leveling of f, then X contains a set Z which is a regional minimum (resp. maximum) of f.

Creating hierarchies with levelings

Levelings are particularly suited for constructing hierarchies because they enlarge the quasi-flat zones, and enlarge or suppress minima and maxima without ever creating new extrema. We will use both properties to construct hierarchies.

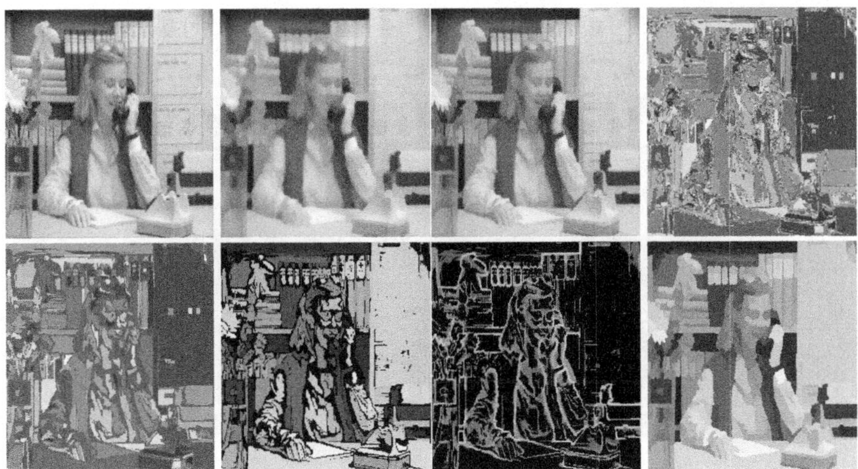

Fig. 10. Line 1 : initial image ; marker ; leveling; flat zones
Line2 : Large-flat zones in false color ; large flat zones with mean grey value ;
gradient ; watershed

Construction of a hierarchy based on the quasi-flat zones.

Since levelings enlarge the quasi flat zones, the quasi-flat zones of a family
of increasing levelings itself form a hierarchy. However, as fig.10_{13} shows, the
quasi-flat zones have two different natures: on one hand large homogeneous
zones, and in the transition zones of high gradient tiny quasi-flat zones. For
this reason, a more useful hierarchy is obtained if one gets rid of these transition zones. This may be carried out in two stages. First a fine partition is
constructed without transition zones. Then coarser partitions are constructed
by merging regions of the finest level.

A slope leveling is applied to the initial image (fig.10_{11}) using as marker
an alternate sequential filter (fig.10_{12}). On the resulting leveling (fig.10_{13})
there are still 12532 quasi flat zones (fig.10_{14}), but they obviously are of two
different natures : tiny quasi flat zones within the transition zones, and larger
quasi flat zones within the objects of interest. Let Z be the union of all smaller
flat zones, hence Z represents the contour zones of the image on one hand,
and some tiny details of the image on the other hand. In fig.10_{21} (false color)
and fig.10_{22} (mean grey value in each region) only the largest flat zones have
been retained ; they will serve as markers for flooding a topographic surface
(fig.10_{23}), equal to the restriction to Z of the modulus of the gradient. The
resulting watershed segmentation (fig.10_{24}) is a tesselation $\mathfrak{S}_0$ in which each
large flat zone has given rise to a region.

The remaining levels of the hierarchy ($\mathfrak{S}_i$) are constructed as follows. Recall that Z represents the contour zone of the image ; Ξ_i represents the quasi
flat zones of the leveling $\Lambda_i f$. We associate to Ξ_i a new partition Ξ_i' obtained

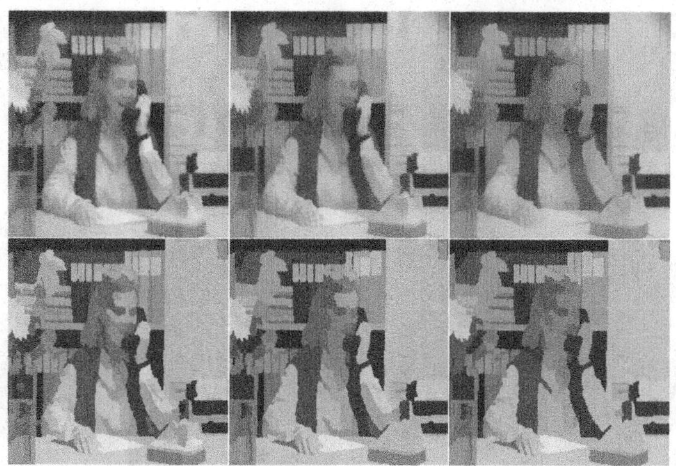

Fig. 11. Hierarchy associated to increasing levelings.
Line1 : 3 increasing levelings
Line2 : Associated increasing partitions

by replacing each pixel x of the contour zone Z by a singleton $\{x\}$. As a result, we obtain a partition which is coarser than Ξ_0 on $\overline{Z}$, but finer on Z. The partition union of Ξ_i' with $\mathfrak{S}_0$ yields a partition $\mathfrak{S}_i$ which is coarser than both Ξ_i' and $\mathfrak{S}_0$. This construction is a good illustration of the use of partition unions. Fig.11 presents in the first row 3 increasing slope levelings associated to alternate sequential filters of sizes 3, 6 and 9 and in the second row the associated segmentations.

Construction of a hierarchy based on the extrema

Extrema play a particular perceptual role : they are the top of the peaks and bottom of the wells in an image. Sometimes they carry the essential information, like the minima in electrophoretic gels, or the maxima for the detection of microcalcification in mammographies. In such cases, it is interesting to construct hierarchies based on these extrema.

Levelings enlarge or suppress minima and maxima without ever creating new extrema. Hence we may construct a hierarchy based on minima, maxima or extrema (union of minima and maxima) used as markers for a classical watershed segmentation.

Let us take as an example a construction based on the extrema. Suppose that we have applied to f a series of increasing levelings Λ_i. From one leveling to the next one some extrema have vanished, others have been enlarged, covering sometimes several extrema of the previous levels. This behavior of the extrema allows to construct a hierarchy (Ξ_i). If we use the extrema as markers for a watershed segmentation, the following changes will occur from one level to the next one:

-   if an extremum of $\Lambda_i f$ has vanished, the corresponding catchment basin of Ξ_i will merge with its closest (in terms of flooding distance) neighboring catchment basins and together they form a region of Ξ_{i+1}.
-   if an extremum of $\Lambda_{i+1} f$ covers several extrema of $\Lambda_i f$, the corresponding catchment basins of Ξ_i will be merged in one region of Ξ_{i+1}.

The next section comes back to floodings, already introduced in the first section. Flooding a gradient image is certainly the most versatile tool of morphological segmentation. To each particular progression of the flood will correspond a particular hierarchy of catchment basins.

4.3 Catchment basins and floodings

Definition of a flooding

Floodings are anti-extensive levelings : g is a flooding of f if and only if g is a leveling of f and $g \geq f$. In what follows we will use $\lambda = 0$, implying that the quasi-flat zones are really flat. The following definition of floodings is easily derived:

Definition 4. *A function g is a flooding of a function f if and only if $g \geq f$ and for any couple of neighboring pixels (p, q) : $g_p > g_q \Rightarrow g_p = f_p$*

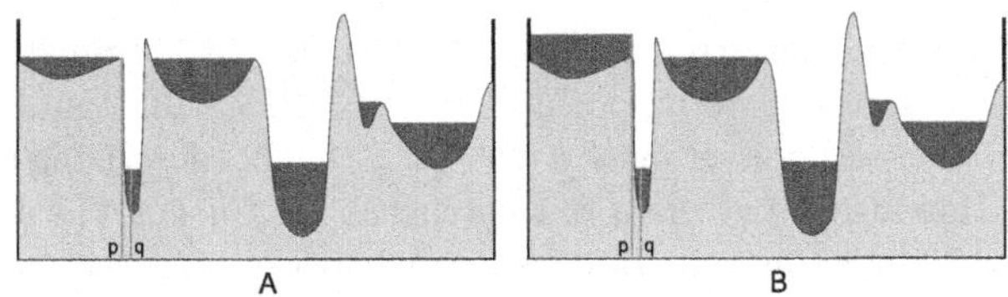

Fig. 12. A: a physically possible flooding ; B : an impossible flooding, where a lake is limited by a wall of water at position p

Any flooding g of a function f creates a number of lakes on the topographic surface of f. All connected components where $g > f$ are flat, as the following property immediately derived from the definition shows:

for any couple of neighboring pixels (p, q) : $\begin{vmatrix} g_q > f_q \\ g_p > f_p \end{vmatrix} \Rightarrow g_p = g_q$

Let L be such a lake. If all neighbors of L have a higher altitude, then L is a regional minimum. On the contrary, if L has a lower neighbor it is called full lake: there exists a couple of neighboring pixels (p, q), p belonging to L and $g_p > g_q$. According to the definition of floodings, this implies that $g_p = f_p$, meaning that the level of the flooding g and the level of the ground f are the same at pixel p ; hence the interpretation of the definition is simply that a lake

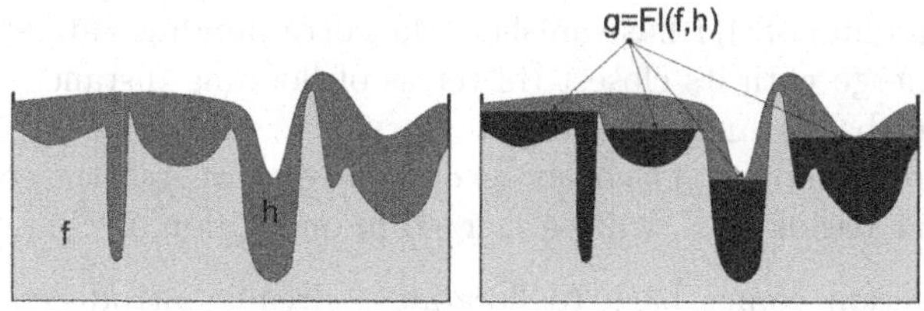

Fig. 13. $\mathrm{Fl}(f,h)$ is the flooding of g (blue function) constrained by the function h (red function)

cannot form a wall of water without solid ground in front to hold the water. This is clearly illustrated in fig.12, where the right figure cannot be a valid flooding, whereas the left figure is a valid one. The pixel p is then necessarily a pass point of g: the altitude of g decreases from p to the outside and the inside of the lake, and increases in both directions along the watershed line.

It is easy to check using their definition that:

* If g and h are two floodings of f, then $g \vee h$ and $g \wedge h$ also are floodings of f

* If g and h are floodings of f and $g \geq h$ then g is also a flooding of h.

* The relation $\{g$ is a flooding of $f\}$ is reflexive, antisymmetric and transitive: it is an order relation.

In particular, if f and h are two functions such that $f \leq h$, then the family of floodings $\left(g^i\right)$ of f verifying $g^i \leq h$ form a complete lattice for this order relation. The smallest element is f itself. The largest is called flooding of f constrained by h and is written $\mathrm{Fl}(f,h)$ (see fig.13). It is obtained by repeating the geodesic erosion of h above f : $h^{(n+1)} = f \vee \varepsilon h^{(n)}$ until stability, that is until $h^{(n+1)} = h^{(n)}$. At convergence $h^\infty = f \vee \varepsilon h^\infty$, characterizing the floodings f. This operation also is known as reconstruction of f using h as a marker [21]. We have already met this type of flooding, when we imposed a set of minima to a topographic surface. In fig.4e, we have indicated which regions we are interested in by creating a function $g \geq f$, identical with f at three positions and equal to the maximal grey value elsewhere. The positions where $f = g$ exactly correspond to the markers, marking an internal zone inside each of the objects of interest. Fig.4 presents the highest flooding possible of f entirely below the function g : it presents 3 minima, corresponding exactly to the 3 chosen markers. This operation is known as swamping in the literature [3].

These properties allow various constructions of increasing families of floodings $\left(g^i\right)$: it is necessary and sufficient that g^j is a flooding .of g^{j-1}.

Hierarchy associated to an ordered series of floodings

Floodings are an easy and flexible way to construct hierarchies. If g is a flooding of f, then the partition of catchment basins of g is coarser than the partition of catchment basins of f, as we will establish now.

Let f be a function and g a flooding of f. Consider a regional minimum m_1 of f and its associated catchment basin X. Let x be the smallest pass point separating X from its neighboring catchment basins. x will be the pass point between X and the catchment basin Y associated to a regional minimum m_2. Let us examine what happens with X with respect to g, a flooding of f. Three cases are to be considered :

- X contains no lake at all or contains a lake which is a regional minimum of g. In this case the corresponding catchment basin for g contains X.
- X contains a lake whose altitude is f_x of the pass point x. This full lake is no longer a regional minimum of g and X will have merged at least with the catchment basin Y of the minimum m_2.
- X contains a lake with an altitude larger than f_x. This lake covers several regional minima of f whose catchment basins have all merged. Hence the catchment basins of the family (g^i) are nested and form a hierarchy.

Useful families of floodings

The principal and most useful families of floodings used in morphological and multiscale segmentations will now be described. As a matter of fact, the quality of segmentation will depend to a great extent on the family of floodings it is built on. Uniform flooding is the simplest : it is the family f^λ of floodings where the level of water grows uniformly and is equal to λ. It is implicitly used in any classical morphological segmentation with markers, as we have established earlier. However, as illustrated in fig.4 and 5 it selects only small and contrasted regions in the coarsest levels of hierarchy and is not well suited to multimedia applications. Size oriented floodings permit to select regions which make more sense in such applications.

Size oriented flooding

Size oriented flooding may be visualized as a process where sources are placed at each minimum of a topographic surface and pour water in such a way that all lakes share some common measure (height, volume or area of the surface). As the flooding proceeds, some lakes finally get full, when the level of the lowest pass point is reached. Let L be such a full lake. The source of L stops pouring water and its lake is absorbed by a neighboring catchment basin X, where an active source is still present. Later the lake included in X will reach the same level as L, both lakes merge and continue growing together. Finally only one source remains active until the whole topographic surface is flooded. The series of floodings, indexed by the measure of the lakes, generates a size-oriented hierarchy.

In fig.14, a flooding starts from all minima in such a way that all lakes always have a uniform depth, as long as they are not full. The resulting hierarchy is called dynamics in case of depth driven flooding and was first introduced by M.Grimaud [6]. Deep catchment basins correspond to contrasted objects ;

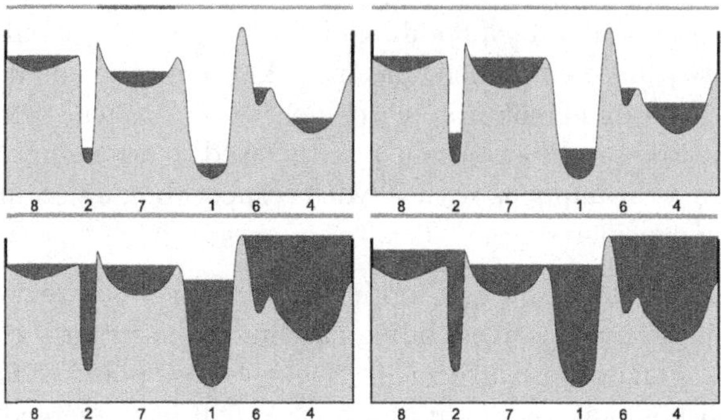

Fig. 14. Example of a height synchronous flooding. Four levels of flooding are illustrated ; each of them is topped by a representation of the corresponding catchment basins.

it will take time before they are absorbed by a neighboring catchment basin. The most contrasted one will absorb all others. This criterion obviously only takes the contrast of objects into account and not their size. If the flooding is controlled by the area or the volume of the lakes, the size of the objects is also considered. This method was introduced by C. Vachier for detecting opacities in mammographies [18]; in multimedia applications [23], good results are often obtained by using the volume of the lakes as measure, as if each source would pour water with à constant flow. This is illustrated on fig.15. The topographic surface to be flooded is a color gradient of the initial image (maximum of the morphological gradients computed in each of the R, G and B color channels). Synchronous volumic flooding has been used, and 3 levels of fusion are represented, corresponding respectively to 15, 35 and 60 regions.

Fig. 15. Initial image. Volume oriented flooding and 3 levels of the hierarchy with 15, 35 and 60 regions

In short, the depth criterion ranks the regions according to their contrast, the area, according to their size. The volume offers a nice balance between size and contrast as depicted in fig.16 where we have illustrated the differences between the criteria used to control the progression of the lakes. The initial image and its gradient are illustrated on the top row. Three types of synchronous flooding are compared. In the first one (bottom left) the lakes grow with uniform depth, resulting in a pyramid where the most contrasted regions survive longest. In the second one(bottom center) the area is used and the largest regions are favoured. In the last one(bottom right) the volume of the lakes is used, offering a good balance between size and contrast of the regions. For each hierarchy the partition with 70 tiles is selected and each tile is replaced by its mean grey tone, for the purpose of illustration.

Fig. 16. Top : initial image and inverted gradient image
Bottom : 3 partitions with 70 regions each. 3 different geometric criteria have been used during synchronous flooding : on the left, the depth of the lakes, in the centre the area and on the right the volume of the lakes.

Tailored flooding ro favor some types of regions

In some cases, while using one of the size criteria, it may also be desirable to favor some particular regions. This happens when one knows beforehand which regions are particularly important. As an example: in many cases, the topographic surface to be flooded is the gradient image $\|\nabla h\|$ of an image h. The catchment basins of $\|\nabla h\|$ correspond to flat zones in h, which may be regional minima, maxima or step zones. However minima and maxima of h are perceptually more important than transition flat zones. For this reason, it may be worthwhile to push minima and maxima of h higher in the hierarchy.

It is easy to obtain this result during synchronous flooding: by reducing the rate of flow in the corresponding minima. The more important a region

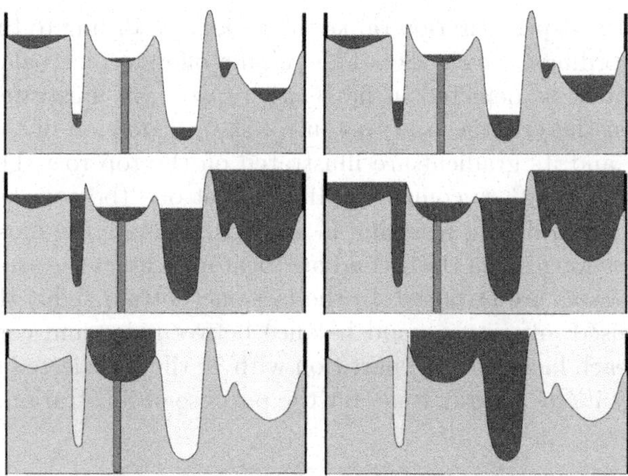

Fig. 17. 4 levels of tailored synchronous flooding, where the minimum marked red is slowed down by a factor 5. As a result we show the corresponding segmentation into 3 regions compared to the segmentation into 3 regions if no source is slowed down.

is, the more the flow of its minima has to be reduced. In fig.17 we have a case where depth synchronous flooding is performed. However the depth of the minimum marked by a black bar grows five times slower than the depth in the other catchment basins. For this reason, this particular minimum survives any absorption much longer. As a result the coarsest segmentation into 3 regions is completely different in the presence or absence of this slowed down flooding. See also fig.18, where two segmentations without and with slowing down the flooding are compared. A fine partition is created first ; the flat zones are detected and the largest of them serve as markers for flooding a gradient image (upper right picture). Then a second gradient image is constructed on the boundaries of the fine partition and this new image is flooded according volumic criteria. The result is illustrated by the lower pictures. On the left, the rate of flood is the same in all minima, on the right, regions have been selected by hand in the faces of the angels, and their rate of flow reduced by a factor 50. Then 2 partitions have been selected in the hierarchy with the same number of regions showing that the faces of the angels merge with the background if their flooding is not slowed down.

Flooding in the presence of markers

Markers are a limit case of the preceding situation. One wishes that the marked regions are present at the top of the hierarchy. This will be the case if the rate of flow in the marked minima is infinitely slowed down ; in other terms such minima have no source at all. Hence they remain minima for ever, and catch their neighboring basins as illustrated in fig.19 If there are N minima,

Fig. 18. Top row : Initial image and fine segmentation
Bottom row : On the right, the flooding in the regions corresponding to the faces of
the angels has been slowed down. Both partitions have the same number of regions.

cutting the $N-1$ highest edges of the MST yields a partition of N regions, each
containing a marker. Cutting more than $N-1$ edges shows how the regions are
further subdivided into finer segmentations ; in this case, the criterion used for
controlling the flooding of the basins without markers (depth, area or volume
of the lakes) has an effect on finer segmentations. It is interesting to observe
the resulting flooding at convergence: the only remaining minima are the
marked minima, all others are full lakes. Finally, size oriented flooding, tailored
flooding and flooding with markers may be grouped: each minimum may be
considered as a fuzzy marker, by assigning it a fuzzy level: 1 means a hard
marker, where no source is placed ; 0 means no marker at all, and the source
is not slowed down ; λ means a fuzzy marker, and the corresponding source
is slowed down by a factor λ. Fuzzy markers allow to establish a continuum
between traditional multiscale segmentation and segmentation with markers.

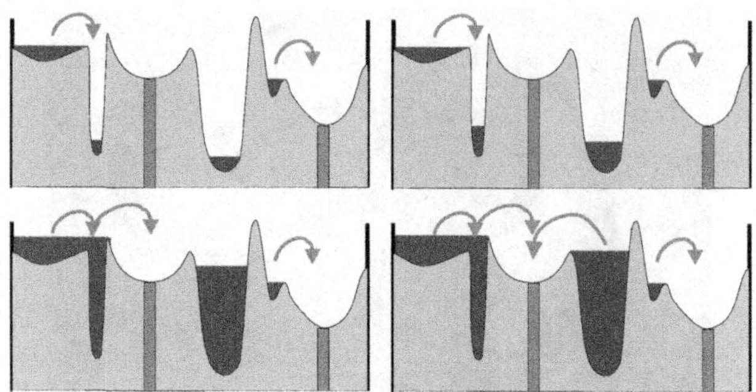

Fig. 19. Flooding in the presence of markers. The catchment basins with markers have no source at all. The arrows show in which order the catchment basins are absorbed one by another.

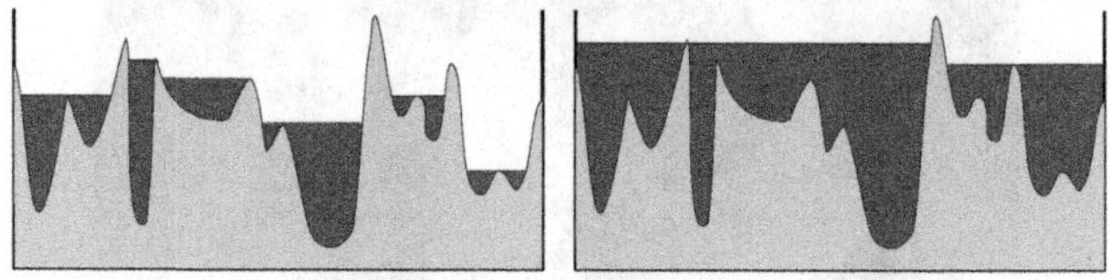

Fig. 20. Two levels of cataclysmic flooding of a topographic surface

Cataclysmic floodings or waterfalls

A flooding g of a function f is cataclysmic if each catchment basin of f is occupied by a full lake. Some of these lakes are regional minima of g ; others are not. The catchment basins of g constitute the first level of the hierarchy (see fig.20). The resulting function g itself may then be submitted to a new cataclysmic flooding and again the number of catchment basins will be strongly reduced. Repeating this flooding in sequence a few times produces an image where only one region remains. A cataclysmic flooding of an image f is easy to produce through a constrained flooding. The constraining function is equal to f on the watershed line of f and equal to ∞ everywhere else. The process is illustrated in 1 dimension in fig.21. Repeating again the same extremal flooding on the result of the first extremal flooding will drastically reduce the number of catchment basins. This process may then be repeated until a partition is created with only one catchment basin. In this way we obtain a series of nested partitions which decreases extremely rapidly. Serge Beucher introduced this type of hierarchy and illustrated the flooding process by using the image of waterfalls ; this hierarchy is also named waterfall hierarchy [2]. He used it to segment the road in videosequences : the first level of hierarchy already brings a dramatic improvement in the segmentation (see fig.22).

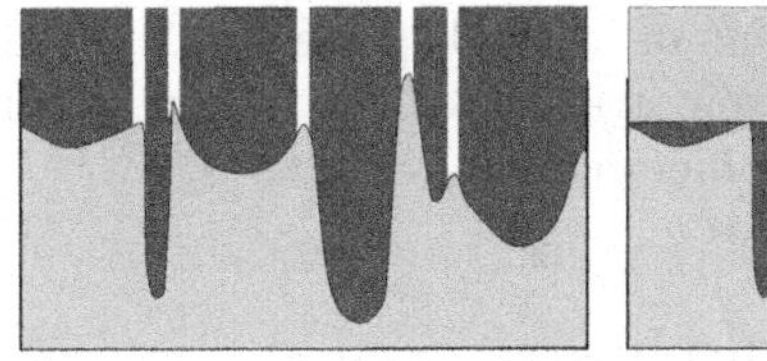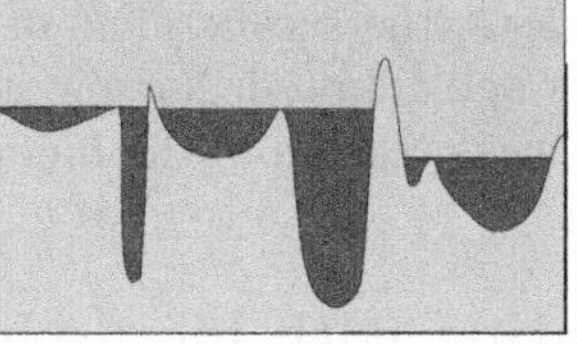

Fig. 21. Constrained flooding for producing a cataclysmic flooding.

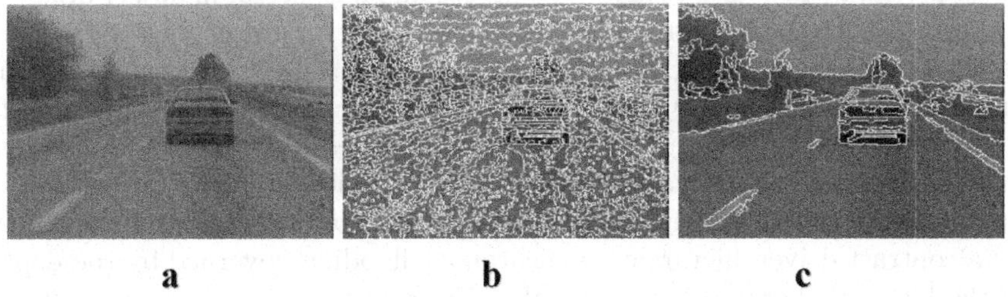

Fig. 22. a) initial image ; b) watershed of the gradient ; c) first level of waterfalls hierarchisation

Which hierarchy to use ?

We have presented a number of means for constructing hierarchies. There is obviously no one best solution, one which would be optimal for each segmentation problem. Given a problem, one has to choose the hierarchy which is best suited. The preceding section shows that we have a large choice to tailor a hierarchy well adapted to our problem.

We have seen three major classes of hierarchies. The first one starts with an initial fine partition and proceeds with progressive fusions of similar regions. This approach is extremely versatile, as we may choose any criterion for governing the successive fusions, including for instance similarity of texture or of motion. Furthermore, one may change criteria as one moves to coarser regions.

The last class, based on floodings is extremely fast: one flooding of a gradient image allows to construct a complete hierarchy. It is also versatile, as one may adjust the rate of flooding so as to optimize the segmentation for size, contrast, or a balance between size and contrast. Furthermore, it is possible to interfere with the flooding by slowing down the rate of flooding in areas of particular interest. These areas are pushed towards higher levels of the hierarchy. There is however a drawback to this method, since it is based on the gradient image: it is unable to detect narrow details. A detail which is too narrow will not produce a minimum in the gradient image but a thick gradient zone. Hence it will disappear from the final segmentation.

In such cases, the second class of methods, based on the quasi-flat zones of levelings is well adapted. Narrow stripes are detected as flat zones and thus will be present as regions in the hierarchy.

5 Segmenting with hierarchies

The simplest use of the hierarchies is choosing a number of regions or a level of subdivision which is satisfactory. This technique is useful in object oriented coding as it allows to adapt the level of detail to the targeted compression ratio. See for instance fig.23 presenting an image and 3 levels in a volume oriented flooding hierarchy. This technique is particularly useful if the objects to be segmented are characterized by some features which may be embodied into the hierarchy. This was the case in the detection of microcalcification in mammographies [18]. They are tiny contrasted spots which may be detected by a contrast driven hierarchy : synchronous flooding governed by the depth of the lakes. If there are no microcalcification in the 20 most prominent objects detected in the image, one may discard it as negative. Such situations are nevertheless rare, and most often one has to combine various levels of a hierarchy in order to construct a satisfactory result. The following sections present various means to interactive segmentation and have been incorporated in an interactive segmentation program [23] .

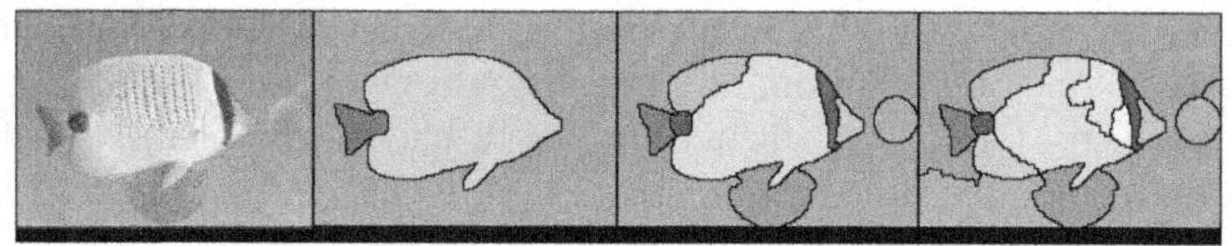

Fig. 23. Initial image and 3 levels of a hierarchy based on volume oriented flooding

5.1 Local resegmentations or mergings

The simplest refinement consists in adapting locally the coarseness of the representation. In fig.24, a first level of the hierarchy is not satisfactory as the background is oversegmented and the person undersegmented. For this reason, one chooses a coarser level of the hierarchy within the background region and on the contrary a finer level within the person. This may be recursively applied to all regions until a good result is obtained.

5.2 Magic wand

To extract a region with uniform color, most drawing/painting software packages have a function called "magic wand". For each position of the mouse, the

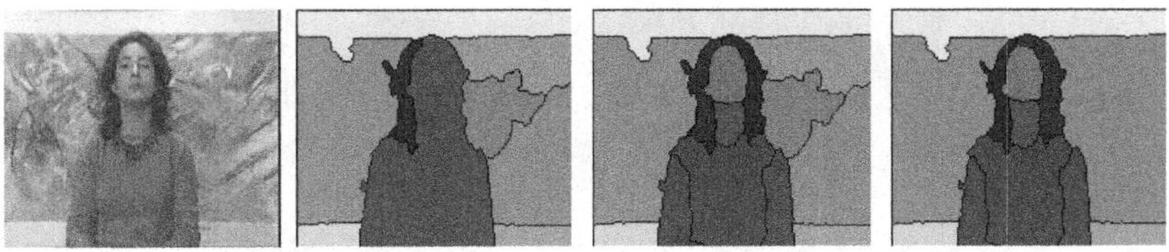

Fig. 24. Initial image followed by a progressive refinement, local resegmentations or mergings

color is determined and the connected region composed of all pixels with more or less the same color, depending on some tolerance threshold, is selected. This procedure is often helpful, but fails in some situations, when there is a progressive change of color shade, as is the case with the yellow apple in fig.25. The darker part of the apple is not selected and an irregular contour produced. On the contrary, using a hierarchy has the advantage of providing well defined contours. The hierarchy based magic wand selects the largest region in the hierarchy such that its mean color remains within some predefined limits.

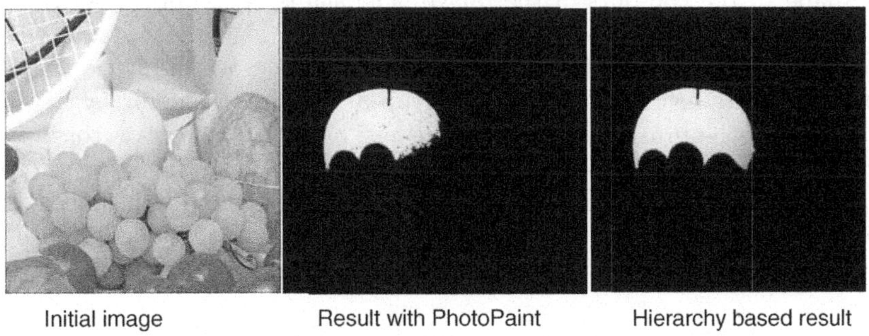

Initial image Result with PhotoPaint Hierarchy based result

Fig. 25. On the left, initial image ; center: all pixels which are within a colour tolerance of an initial pixel. On the right, result of the magic wand.

5.3 Lasso

Another classical interactive tool is the lasso : the user draws an approximate contour around the real contour as shown in fig.26a. The classical solution consists in applying the magic wand defined above to each pixel belonging to the approximate contour. For each such position one gets a piece of the background. The union of all such pieces constitutes the background. As shown on fig.26b, the result is not very satisfactory. Using a hierarchy, one may select the union of all regions of the hierarchy contained in the contour yielding a much better result as shown on fig.26c.

Fig. 26. Construction of all tiles of a hierarchy entirely included in an approximate outside contour

5.4 Intelligent brush

An intelligent brush segments an image by "painting" it: it first selects a zone of interest by painting. Contrary to conventional brushes, the brush adapts its shape to the contours of the image. The shape of the brush is given by the region of the hierarchy containing the cursor. Moving from one place to another changes the shape of the brush, when one goes from one tile of a partition to its neighboring tile. Going up and down the hierarchy modifies the shape of the brush. In fig.27, on the left, one shows the trajectory of the brush ; in the centre, the result of a fixed size brush, and on the right the result of a self-adapting brush following the hierarchy. This method has been used with success in a package for interactive segmention of organs in 3D medical images.

Fig. 27. Comparison of the drawing with a fixed size brush and a self adaptive brush.

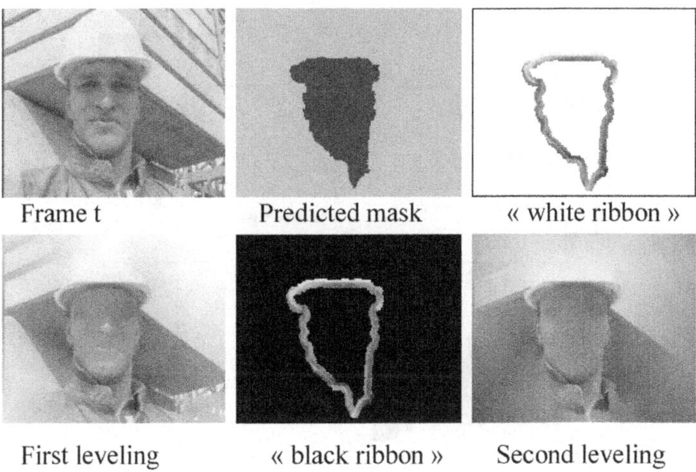

Fig. 28. Image simplification by levelings

6 Automatic tracking of the speaker in a videosequence

Our last section presents the algorithms developed within the Eureka Medea project "Multimedia for Mobile (M4M)", in which we developed a demonstrator able to track the speaker in a video-conference sequence under realistic conditions : poor lighting conditions, moving camera, moving background [5]. Fig.29 presents the flowchart of the algorithm in steady state. Inputs of the algorithms are the frames $t - 1$ and t and the mask produced for frame $t - 1$. Motion analysis and compensation allow to predict the position of the mask at time t. The contours of the real mask are within a distance ρ of the contours of the predicted mask. We will use this knowledge for simplifying the image t to segment. Two successive levelings illustrated in fig.28 will suppress almost all information in frame $t + 1$ except the contours of the new region of interest. We produce a ribbon like mask around the boundary of the mask: we take all pixels which are within a distance ρ of this boundary. The marker of both levelins is obtained by cutting out the content of frame t+1 within the ribbon ; the first marker takes a white value (255) outside the ribbon and the second marker a black value (0). The white leveling uses the white mask. The second leveling is applied on the result of the first leveling, using the black mask. As result, we obtain an image, where only the contour of the object of interest appears with its original strength, whereas the contours of all other objects of the scene have vanished or their contrast has been drastically reduced.

A hierarchical segmentation is applied to the leveled image, using volumic criteria. Marker segmentation is applied to this hierarchy for producing the segmentation of frame t. The inside marker is obtained by eroding the pre-

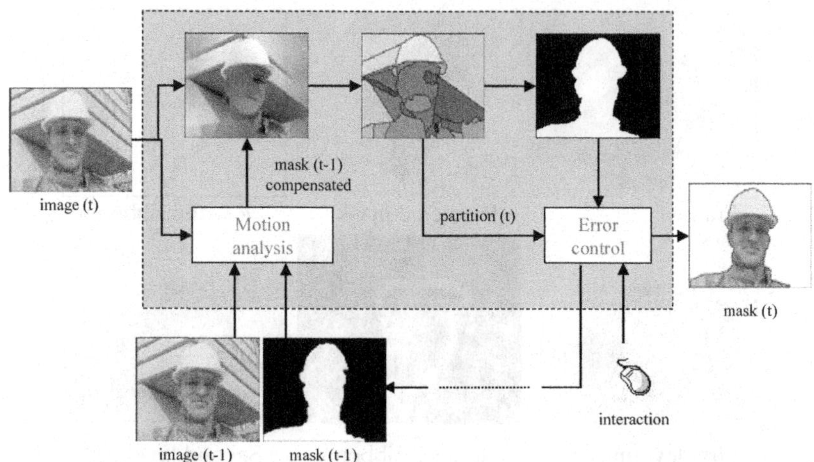

Fig. 29. Flow chart of the tracking algorithms

dicted mask and the outside mask by dilating it, using as structuring element a disk of radius ρ. The algorithm performs well and is able to process a sequence at a rate of 10 QSIF images per second on a 500 MHz Pentium III laptop PC.

7 Conclusion

The morphological approach to image segmentation based on the watershed and markers has proved to be useful and has become a sound theory and an efficient practice.

It is extremely fast as the construction of a complete hierarchy based on flooding may be obtained with only one pass through the image. It allows to inject knowledge in various ways in the process of segmentation, by using markers or by adapting the hierarchy to the type of objects to be segmented. It also permit regularizing the contours which are produced [19, 15]. Since the segmentation is based on the finding the strongest contours in a hierarchy, it is extremely robust to changes of contrast or illumination.

The method may be used in the same way whatever the number of dimensions. Moreover, the segmentation process evolves rapidly from the pixel and image level and does most of the work on a graph and tree level, which further accelerates the computation.

References

1. J. P. Benzécri. *L'analyse des données 1. La taxinomie*, chapter 3, pages 119–153. Dunod, 1973.

2. S. Beucher. *Segmentation d'Images et Morphologie Mathématique.* PhD thesis, E.N.S. des Mines de Paris, 1990.

3. S. Beucher and F. Meyer. The morphological approach to segmentation: the watershed transformation. In E. Dougherty, editor, *Mathematical morphology in image processing*, chapter 12, pages 433–481. Marcel Dekker, 1993.

4. E. Dejnozkova. *Architecture dédiée au traitement d'image basé sur les équations aux dérivées partielles.* PhD thesis, E.N.S. des Mines de Paris, 2004.

5. C. Gomila. *Mise en correspondance de partitions en vue du suivi d'objets.* PhD thesis, E.N.S. des Mines de Paris, 2001.

6. M. Grimaud. New measure of contrast : dynamics. *Image Algebra and Morphological Processing III, San Diego CA, Proc. SPIE*, 1992.

7. C. Lantuéjoul and S. Beucher. On the use of the geodesic metric in image analysis. *J. Microsc.*, 1981.

8. F. Lemonnier. *Architecture électronique dédiée aux algorithmes rapides de segmentation basés sur la Morphologie Mathématique.* PhD thesis, E.N.S. des Mines de Paris, 1996.

9. B. Marcotegui and F. Meyer. Bottom-up segmentation of image sequences for coding. *Annales des Télécommunications*, 7, 1996.

10. G. Matheron. Les nivellements. Technical report, Centre de Morphologie Mathématique, 1997.

11. F. Meyer. Topographic distance and watershed lines. *Signal Processing*, pages 113–125, 1994.

12. F. Meyer and S. Beucher. Morphological segmentation. 1(1):21–46, Sept. 1990.

13. O. Monga. *Segmentation d'Images par Croissance Hiérarchique de Régions.* PhD thesis, Université Paris Sud. Centre d'Orsay, 1988.

14. L. Najman and M. Schmitt. Watershed of a continuous function. *Signal Processing*, 38:99–112, July 1994.

15. H.-T. Nguyen, M. Worring, and R. van de Boomgaard. Watersnakes:energy-driven watershed segmentation. *PAMI*, 25:330–342, 1993.

16. Roerdink and Meijster. The watershed transform: Definitions, algorithms and parallelization strategies. *FUNDINF: Fundamenta Informatica*, 41:187–228, 2000.

17. P. Salembier. Morphological multiscale segmentation for image coding. *EURASIP Signal Processing, 359-386*, 38(3):359–386, august 1994.

18. C. Vachier. *Extraction de Caractéristiques, Segmentation d'Image et Morphologie Mathématique.* PhD thesis, E.N.S. des Mines de Paris, 1995.

19. C. Vachier and F. Meyer. The viscous watershed transform. *JMIV: in press*, 2004.

20. P. Verbeek and B. Verwer. Shading from shape, the eikonal equation solved by grey-weighted distance transform. *Pattern Recogn. Lett.*, 11:618–690, 1990.

21. L. Vincent. Morphological grayscale reconstruction in image analysis: Applications and efficient algorithms. *IEEE Trans. in Image Procesing*, pages 176–201, 1993.

22. L. Vincent and P. Soille. Watersheds in digital spaces: An efficient algorithm based on immersion simulations. *IEEE PAMI, 1991*, 13(6):583–598, 1991.

23. M. F. Zanoguera, B. Marcotegui, and F. Meyer. An interactive colour image segmentation system. In *Wiamis'99 : Workshop on Image Analysis for Multimedia Interactive Services*, pages 137–141. Heinrich-Hertz Institut Berlin, 1999.

Ubiquity of the Distance Function in Mathematical Morphology

Michel Schmitt

Ecole des Mines de Paris

1 Distance Function: a Historical Tool

Distance has been the key notion for usual Euclidean geometry and the definition of geometrical objects. For instance, a circle is the locus of points at the same distance of a given one, an ellipsis is the locus of points whose sum of distances to two given points is a constant, a parabola is the locus of points whose sum of distances to a given point and a given straight line is a constant. Also, the shortest path between two points is the line segment.

Mathematical morphology deals with set (or function) transformations, whose main applications include image processing, graph analysis, motion planning... In this framework, one very natural tool is the distance to sets, defined as the minimum of distances to any point in the set. Given a set X, we call *distance function* the distance to the complementary set to X, denoted by X^c. This function takes value 0 outside X and positive values on X (0 on the boundary of X).

In order to illustrate the way mathematical morphology uses this distance function, consider the archetypal problem of coffee beans separation (see fig. 1). How many beans can be counted in this picture? By sight, the answer is almost unambiguous. However, due to the overlaps of beans, it is not sufficient to count the number of connected components (number of objects). The solution to the problem took many years of research in order to design a robust algorithm.

Common sense ideas may not work. For instance, let us assume that the beans are approximate disks and let us take the largest disks included in X. Which meaning do we associate with the word "largest"? They are maximal in the mathematical sense, *i.e.* they are not contained in another disk. Fig. 2 shows that some maximal disks are not an approximation of a single bean. This problem is not due to the approximation of the beans by disks: the idea does not work even if the beans are perfect (overlapping) circles.

However, if we examine the distance function (fig. 1.b), we see that it takes larger values near the "center" of the beans. So, if we define the centers

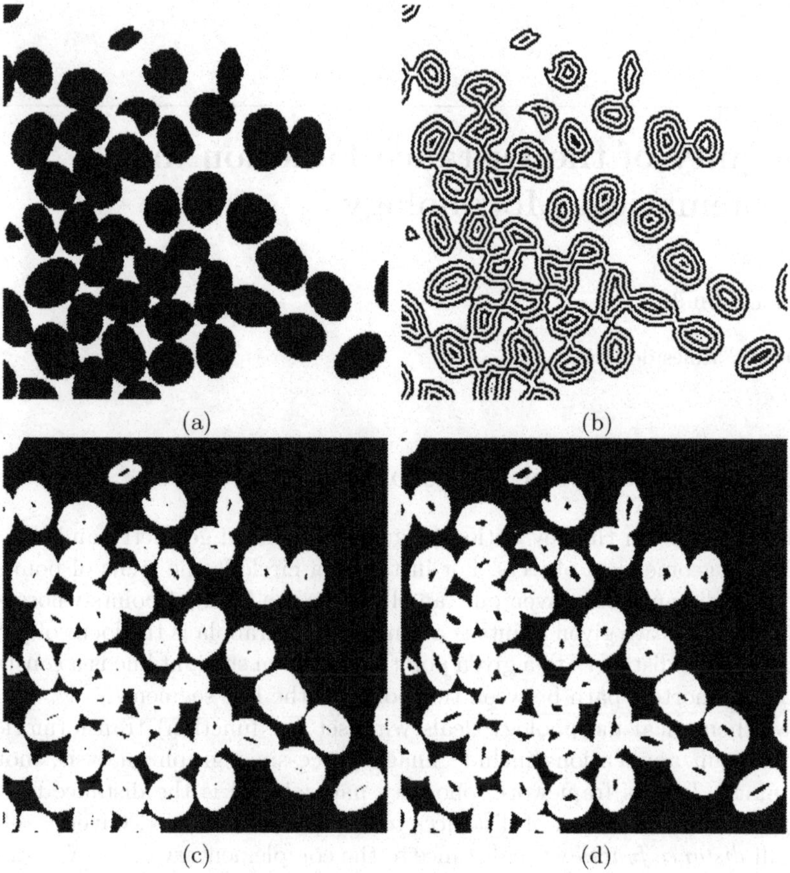

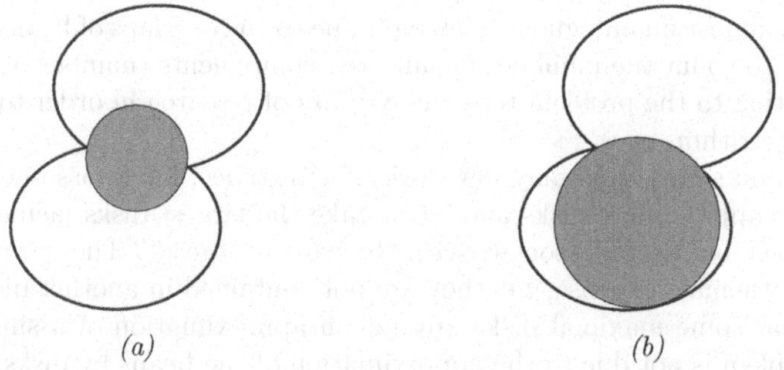

Fig. 1. Coffee beans separation: (a) original image, (b) distance function, (c) regional maxima of distance function, (d) regional maxima of the transformed distance function.

Fig. 2. Coffee beans separation: the maximal disk in *(a)* does not approximate a single bean, whereas the maximal disk in *(b)* does.

of the beans as the local maxima of our distance function, we observe that each maximum precisely corresponds to the center of one bean. Counting the number of maxima yields the number of beans. The link with the common sens idea of maximal disks is very close. At a point x, the distance function corresponds to the radius of the largest disk centered at x and totally included in X (for a rigorous assertion, one has to suppose the disks and set X to be topologically closed). A local maximum of the distance function corresponds to a largest locally centered disk. In other words, if one wants to displace a disk from one bean to another, the radius of the disk has first to decrease in order to travel through the gap and then to increase again. Note that during the whole travel, the disk always remains maximal.

From a digital point of view, things are not so simple. Disks are squares, octagons (square sampling lattice) or hexagons (hexagonal sampling lattice). As seen on fig. 1.c, some beans may contain many local maxima of the distance function. First, we must consider *regional maxima* instead of *local maxima*. Regional maxima correspond to flat regions on the distance function which are surrounded by pixels with strictly lower values. But this distinction is not sufficient: some beans still exhibit "many centers". In fact, this phenomenon is due to the digital nature of the images, where the breadth of a bean may exhibit a variation of one pixel, limit of the image precision. If we agree that this variation corresponds to sampling noise, a robust solution is obtained by decreasing the height of the regional maxima by one. Then the resulting regional maxima of the transformed distance function really describe centers of beans.

This "coffee beans separation" algorithm has been used in many similar situations. It is also known as the *perceptual graph* [14]. The purpose is here to connect a set, typically a contour, which has been broken in some places. If we name X the set, connecting X corresponds to disconnecting X^c, which can be achieved by the preceding algorithm.

Another illustration of the use of distance function in mathematical morpholology is the concept of *granulometries* [6, 24, 11]. To the question, what is the typical size of a coffee bean and what is the size distribution of the beans, one could answer the following way. Let again be X the set representing the beans. A given bean size r is obtained by use of a disk with radius r and X is replaced by the union of all disks of radius r included in X, say X_r (morphological opening of X by a disk of radius r). The beans of size less than r are those which disappear in X_r. The important point to notice is that this statement remains true even if the beans are pairwise overlapping (see fig. 3). Then the function

$$\sigma_X(r) = Area\,(X_{r+1}) - Area\,(X_r) \tag{1}$$

gives a precise idea of the size distribution of the objects in X. Large values correspond to typical sizes of particles, like 20 (*i.e.* 41 pixels in diameter) in the example presented in fig. 3.c. This curve is called *granulometric curve.*

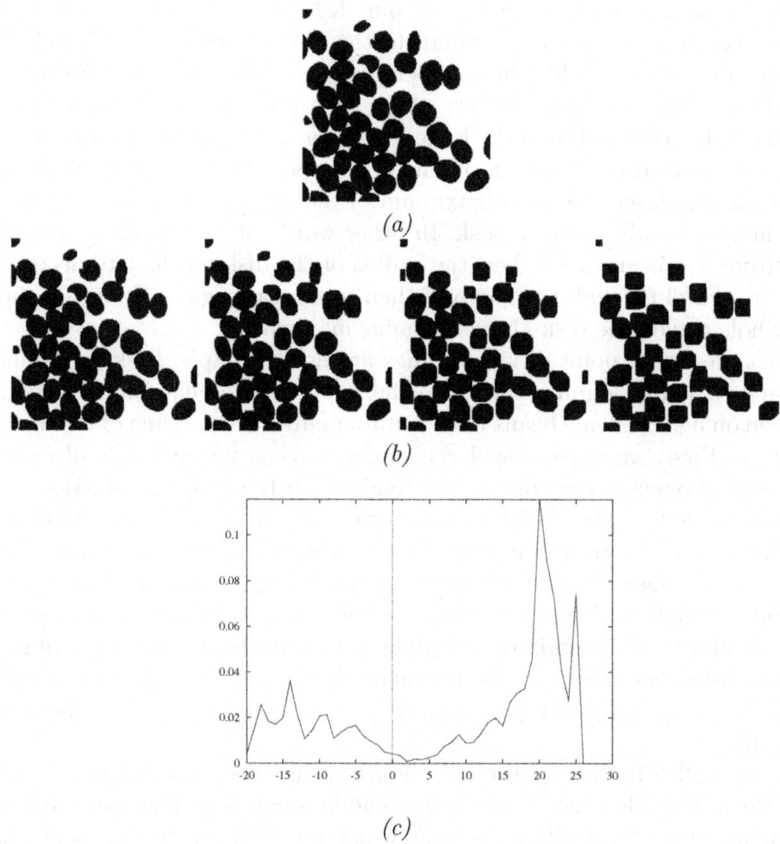

Fig. 3. Coffee beans size: (a) original image, (b) some openings of different sizes, (c) granulometric curve. Here, for computing purpose, the disk has been replaced by a square.

For a mathematical definition, see [6, 24], for characterizations, see [10]. This granulometric curve can also be used for image discrimination, see [12].

This paper aims at illustrating some of the concepts which have emerged with the notion of distance function. It is organized as follows: first, we define the distances we will use in a digital setting and link it with graph theory. Then we present some algorithms which have completely changed the computing of morphological operators and allowed the use of morphology on large scale applications. A short section illustrates the link between skeletons and distance functions. Afterward, we introduce the geodesic framework and show the antagonism that may exist between connectivity and distances on a lattice, due to its digital nature. Finally, some new insight into distance functions as solutions of evolutionary equations is addressed.

2 Distance, Erosions and Weighted Graphs

We first investigate the links between erosions and distance functions in the continuous plane and then address the discrete case, which is the natural framework for digital images.

2.1 Distance Function and Erosions

Suppose, the plane (or more generally an Euclidean vector space of any finite dimension) is equipped with a distance $d(x, y)$, non necessarily the Euclidean one. The *ball of radius r and center x*, denoted by $B(x, r)$, is defined as the set of points y such that $d(x, y) \leq r$. This distance leads to the definition of a family of erosions:

$$\varepsilon_r(X) = \{x, \; d(x, X^c) \geq r\}. \tag{2}$$

The interesting case arises when these erosions build up a semi-group, *i.e.* $\varepsilon_r(\varepsilon_{r'}(X)) = \varepsilon_{r+r'}(X)$. Then all the balls defined by the distance are convex sets and if we add the translation invariance, the usual notion of *structuring element* appears and the balls have the same shape anywhere in the space: $B(x, r) + y = B(x + y, r)$.

The interest in this link is first the algorithmical part (fast distance algorithms yield fast erosion algorithms) and second the theoretical part (an abundant mathematical literature on distances exists).

2.2 Digital Distances and Weighted Graphs

A digital lattice (square, hexagonal or other) can always be considered as a subset of the continuous space and the digital distance is then the restriction of the distance in the continuous space. Mainly from a computational viewpoint however, image processing has always been looking for local operators involving nearby pixels.

The constructive principle is to define the distance only for certain pairs of pixels, called *edges*. Each edge (a, b) is assigned a weight, its length $d(a, b)$. The set of weighted edges defines a *weighted graph* denoted by G. Then the distance is extended to any pair of pixels by taking the shortest path inside the graph. More precisely:

$$d(x, y) = \min \left\{ \sum_{i=1}^{n} d(x_{i-1}, x_i), \; (x_{i-1}, x_i) \in G, \; x_0 = x, \; x_n = y \right\}. \tag{3}$$

Fig. 4 depicts some usual graphs and the associated balls.

Three types of graphs have been investigated in mathematical morphology and have given rise to theoretical as well as practical applications:

- arbitrary graphs for the analysis of architecture of objects, like cells in histology. In this case the notion of underlying continuous distance is not used.

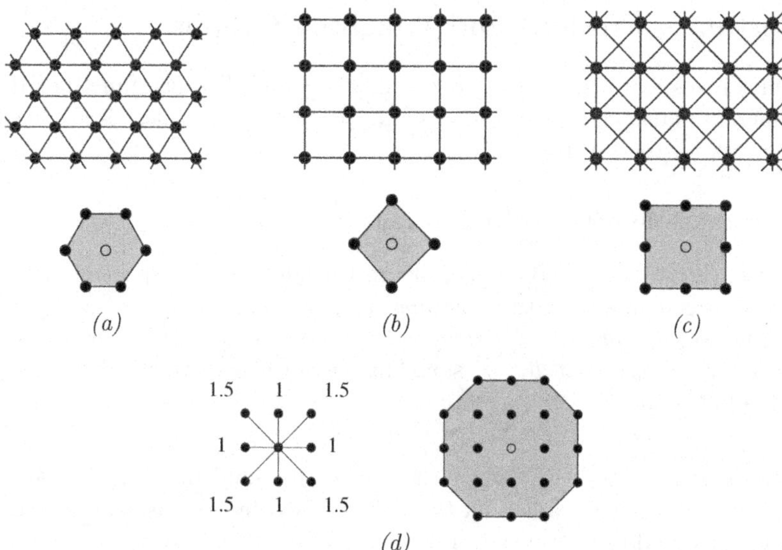

Fig. 4. Some examples of graphs and associated balls: (a) hexagonal lattice and unit hexagon, (b) square lattice in 4-connectivity and unit diamond, (c) square lattice in 8-connectivity and unit square, (d) weighted square lattice and octagon of size 2.5.

- graphs induced by some continuous distances,
- more recently, segmentation graphs, where vertices are regions and edges are dissimilarity measures (see the paper by F. Meyer in this book (page 315).

An important class of continuous distances inducing locally connected weighted graphs is the following. Take a convex polygon B which is symmetrical with respect to the origin. The continuous distance between x and y is defined as the size λ (homotethy factor) of the smallest ball λB centered at x and containing y. For example, on the square lattice, important cases include the square, the diamond and octagons. This type of digital distance derived from a continuous one plays a central role in the geodesic framework.

3 Digital Algorithms

Digital distances derived from weighted graphs have given rise to very efficient algorithms, some of them are described in this section. They take advantage of this local structure of the distance. They all require some restrictions on the weighted graph structure. For the first description of shortest path algorithm through these graphs, see for instance [16]. A complete review of distance algorithms may be found in the forthcoming book [23]. We describe the algorithms for the hexagonal distance, *i.e.* for images on an hexagonal lattice, each pixel having six neighbors at distance 1, as presented in fig. 5, and give some hints for the generalization to other graphs.

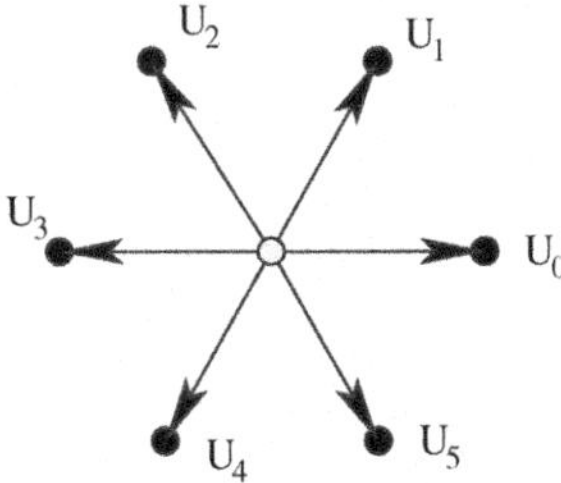

Fig. 5. Unit vectors in hexagonal grids.

3.1 Sequential Algorithms

The basic principle is here the use of only one image data. A scanning order is then defined, which has to be independent from the image content. Then each pixel is updated in turn, according to the predefined scanning order. This type of algorithm has been first proposed and extensively studied by Rosenfeld in [17].

For computing the distance function, the sequential algorithm requires only two scans of the image, one in raster order, one in antiraster order, yielding a computation time which is almost constant. The proof of the correctness of this algorithm is non-trivial and relies on complex properties of shortest path on the lattice. A complete description may be found in [3].

Sequential distance function

```
Input and output:
    I, binary image as input, gray-scale image as output
    /* The distance function is computed in I directly */
```
Scan I in raster order and let p be the current pixel {
    If $I(p) \neq 0$ then
        $I(p) \leftarrow \inf\{I(p + \mathbf{u}_1) + 1, I(p + \mathbf{u}_2) + 1, I(p + \mathbf{u}_3) + 1\};$
}
Scan I in anti-raster order and let p be the current pixel {
    If $I(p) \neq 0$ then
        $I(p) \leftarrow \inf\{I(p), I(p + \mathbf{u}_4) + 1, I(p + \mathbf{u}_5) + 1, I(p + \mathbf{u}_0) + 1\};$
}

By changing the size of the neighborhood and the values given to the neighbors, this algorithm is very easily extended to more circular distances. Fig. 6 gives the values and associated balls in some cases on the square lattice.

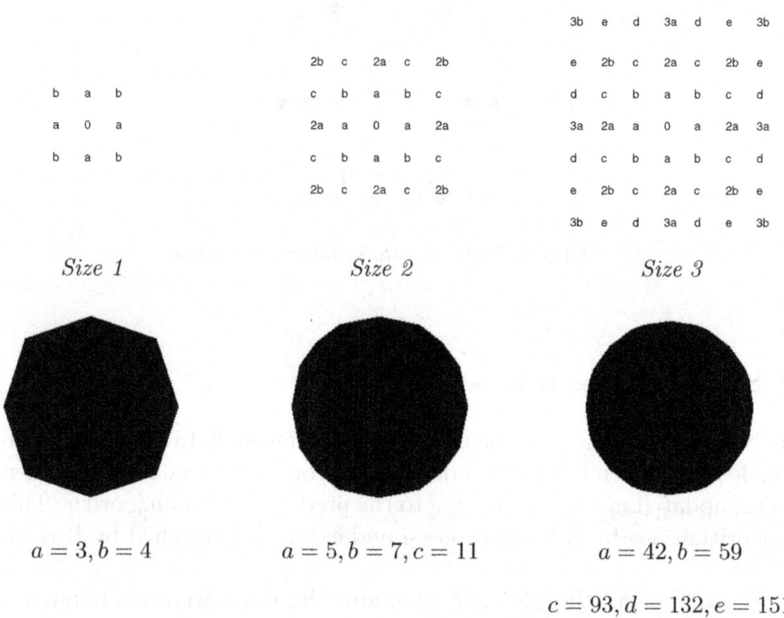

Fig. 6. More isotropic distances with sequential algorithms by increasing the size of the neighborhood and shape of a large disk in each case.

3.2 Fifo Algorithms

The basic principle is to use an auxiliary data structure, namely a first-in first-out queue in order to keep track of the order in which the pixels have to be updated (see [26]). In fact, the distances are computed in increasing values from the boundary of the set.

---

distance function using a queue of pixels

```
Input: I, binary image;
        /* The distance function is computed in I directly */
For every pixel p ∈ D_I, do {
    /* detection of the pixels to be initially put on the queue */
    If I(p) = 1 and ∃p' neighbor of p and I(p') = 0 {
        fifo_add(p);  I(p) ← 2;
```

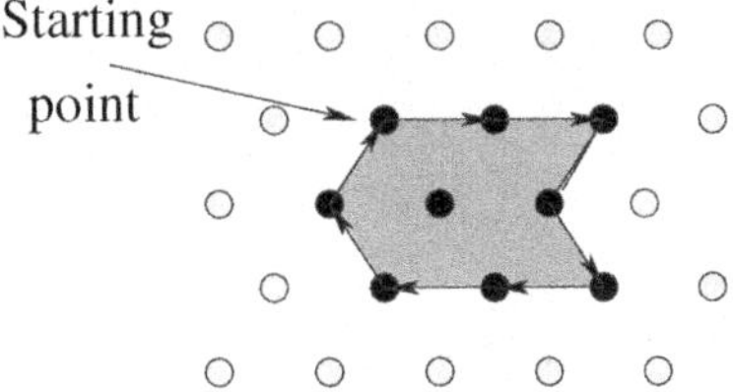

Fig. 7. Chain structure of the boundary. The sequence of directions from the starting point is here $(0, 0, 4, 5, 3, 3, 2, 1)$.

```
    }
}
While fifo_empty() = false {
    p ← fifo_first();
    For every p' neighbor of p {
        If  I(p') = 1 then {I(p') ← I(p) + 1;  fifo_add(p');}
    }
}
Subtract 1 to image I;
```

3.3 Loop and Chain Algorithms

This type of algorithm, based on geometrical considerations, is very similar to the fifo ones. The main difference lies in the data structure which is now one or several chains, which is a list of directions from one pixel to the next on the boundaries (see fig. 7). The directions are depicted in fig. 5. The distances are also computed in increasing order, but the computing of the new chains is computationally less complex [19]. The basic idea is that the boundary of the eroded set is contained in the parallel curve to the boundary of the set. The principle of the algorithm relies on two properties:

1. Parallel curves can be computed very simply on the list of directions. Fig. 8 presents the local computation of the parallel curve at distance one.
2. The unwanted parts of the parallel curve, *i.e.* the parts which do not belong to the boundary of the eroded set, are exactly composed of the pixels which already have been computed in previous steps and thus are removed by writing the parallel curve into the image.

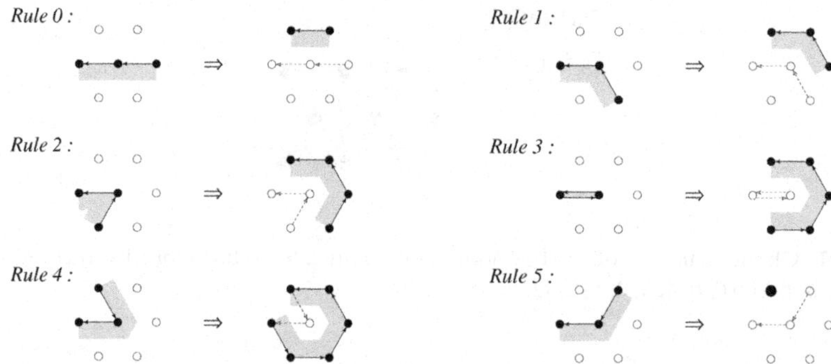

Fig. 8. Computation of the parallel curve. Every pair of successive directions is transformed according to the rules depicted here. The grey areas show on which side of the boundary the object lies.

---

distance function by chains and loops

```
Input: I, binary image;
                /* distance directly computed in I */
/* Track the contours of I and encode them as chains */
    Chains ← Track_All_Boundaries(I);
/* Compute the eroded chains */
    dist ← 2; /* variable containing the current distance */
    Repeat while there remain chains {
        Chains ← Chain_Erode(Chains);
        Chains ← AdjustChain(Chains, dist);
        dist ← dist + 1;
    }
```

The `AdjustChain` step simply consists in writing the chains on the image and in keeping the parts of the chains which are on pixels not having been visited before. See [23] for a precise statement and [19] for the proof of correctness.

3.4 An Approximate Euclidean Algorithm

We are able to compute the hexagonal distance so far. However, some applications require exact or nearly exact Euclidean distances between pixels. The sequential algorithm very easily allows the computation of almost exact Euclidean distances between pixels. It relies on the approximate idea that the region of pixels which are nearer from a given pixel (Voronoi region) is connected according to the lattice. This statement is true in the continuous space, but there are some counter examples on lattices. However, the computed distance is really Euclidean everywhere, except for a few pixels which are assigned a good approximation [4].

Euclidean distance vector in 4-connectivity

```
{ -  input  :   I, binary image (image (a)),
{ -  output :   J, vector image,
{                J(p) points to one nearest pixel outside X ;
/* Initializations */
    For every pixel p ∈ I do {
        If I(p) = 1 then J(p) ← o;
        else J(p) ← v∞;
    }
    Assign value v∞ to the frame of J;
/* See image (b) */
/* Top to bottom scan of the image */
    Scan the lines of I from top to bottom {
        Scan the current line from left to right {
            Let p the current pixel;
            J(p) ← v such that
                { v ∈ {J(p), J(p − ȷ) + ȷ, J(p − ı) + ı};
                { ||v|| = minimal norm of the above vectors;
/* Note that v is not necessarily unique... */
        }
        Scan the current line from right to left {
            Let p the current pixel;
            If ||J(p)||² > ||J(p + ı) − ı||² then J(p) ← J(p + ı) − ı;
        }
    }
/* See image (c) */
/* Bottom to top scan of the image */
    Scan lines of D_I from bottom to top {
        Scan the current line from right to left {
            Let p the current pixel;
            J(p) ← v such that
                { v ∈ {J(p), J(p + ȷ) − ȷ, J(p + ı) − ı};
                { ||v|| = minimal norm of the above vectors;
        }
        Scan the current line from left to right {
            Let p the current pixel;
            If ||J(p)||² > ||J(p − ı) + ı||² then J(p) ← J(p − ı) + ı;
        }
    }
```

```
/* See image (d) */
```

$\mathbf{v}_\infty$ is the vector with two infinite coordinates, $\imath$ is the unit horizontal vector pointing to the right and $\jmath$ is the unit vertical vector pointing to the bottom.

The different steps of the algorithm are depicted in fig. 9.

3.5 An Exact Algorithm

For computing the true Euclidean distance, the algorithms are more complex. Several have been proposed, based on loop and chains [26] or on the idea that the square of the Euclidean distance can be computed successively on rows and then on columns [13].

4 Some theoretical Uses of Distance function

Distance function has many other links with mathematical tools. We briefly present two of them:

-   the link with skeletons [8] and the design of a digital powerful and novel algorithms [15],
-   the link with Hausdorf distance on compact sets.

4.1 Distance Function and Skeletons

Loosely speaking, the skeleton is a set of (thin) lines describing an object. The main feature of this description is connectivity. If a set is connected, so is its skeleton. Any hole in a shape produces a closed loop in its skeleton. Mathematically speaking, the skeleton is defined as the locus of the maximal balls included in the set. Recall that a ball is maximal in X if it cannot be strictly contained in another ball also included in X. Let us denote by $Sq(X)$ the skeleton of an open set X. This assumption of openness is a technical but essential one.

On the other hand, the distance function d_X exhibits very strong regularity features. First, it is Lipschitz:

$$\|d_X(x) - d_X(y)\| \le d(x, y). \tag{4}$$

This inequation is an equality in the following sense: for all x such that $d_X(x) > 0$, there exists $y \ne x$ such that $d_X(y) > 0$ verifying

$$\|d_X(x) - d_X(y)\| = d(x, y). \tag{5}$$

The geometrical structure of such y is known: it is a union of line segments emanating from x.

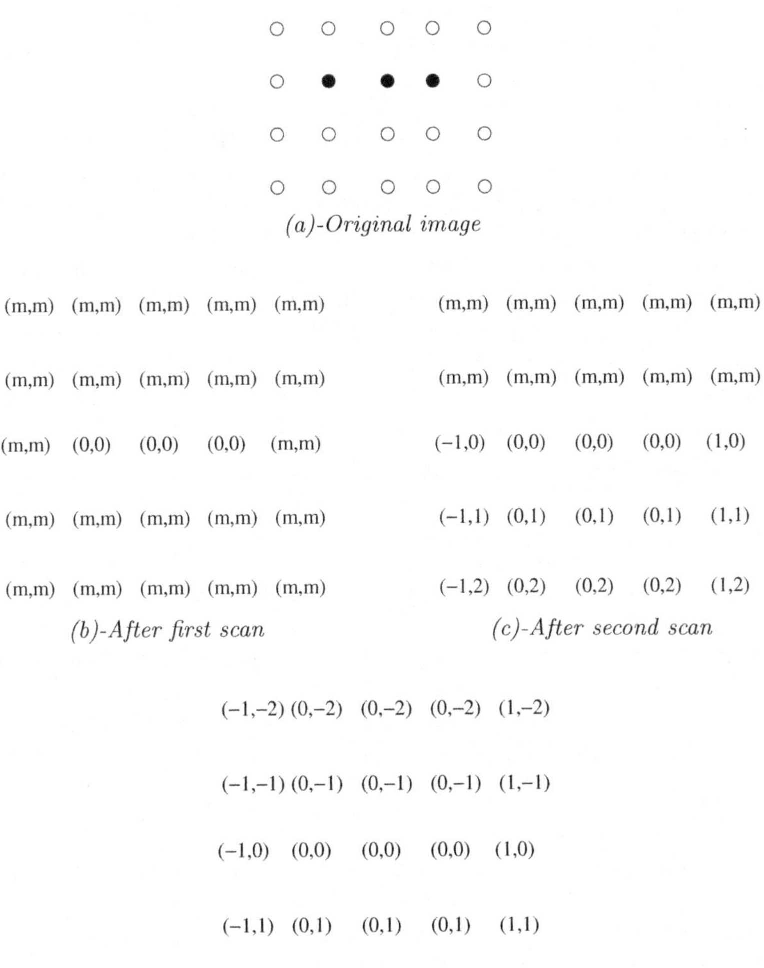

(a)-Original image

(b)-After first scan

(c)-After second scan

(d)-Final vector image

Fig. 9. Different steps in the Euclidean algorithm. m represents infinity. Positive directions for the coding of the vectors are left and above.

This last equation characterizes the distance function of an open set: d_X is the distance function on the set $\{x, d_X(x) > 0\}$.

More important is that outside the closure of the skeleton, the distance function is differentiable with unit gradient. In a reciprocal way, the closure of the skeleton is the closure of the points where the distance function is not differentiable. Loosely speaking, the skeleton corresponds to the crest points in the graph of the distance function or angular points in its level lines.

For a complete description and proofs of these non-trivial theorems, see [7, 8].

These very nice properties do not hold in a digital framework, because we do not have the notion of differentiability on a lattice. However, these ideas can be used and transformed on lattices, giving rise to the notion of *digital crest points* [15] which are the digital analog of non differentiable points. The main problem is that these crest points are generally not connected, so a very clever connection procedure, different from the perceptual graph, based on a detailed analysis of the digital structure of the distance function on the hexagonal lattice, has been designed. This algorithm needs three scans of the image and no auxiliary memory so that it is today one of the fastest skeleton algorithms.

4.2 Distance Function and Hausdorf Distance

Distances between sets is mainly of theoretical interest for the characterization of convergence of sequences of sets and for the definition of lower and upper semi-continuity for morphological operators. It may also be used in set optimization.

Many distances between compact sets may be defined, these distances being not equivalent (see for instance [18] for examples based on stereology). One of the commonly used distances is the Hausdorf distance $\varrho(X, Y)$ [6], defined as the maximum distance between a point in one set to the other set:

$$\varrho(X, Y) = \max\left(\max\left(d(x, Y),\ x \in X\right), \max\left(d(y, X),\ y \in Y\right)\right). \qquad (6)$$

The computation of this distance using the distance function is straightforward: compute the distance function to X^c and take the maximum over Y, then compute the distance function to Y^c and thake the maximum over X. The Hausdorf distance is the maximum of these two values.

5 The Geodesic Framework

Up to now, we have only investigated distances in the whole space or in the whole lattice. One very powerful tool developed in morphology has been geodesy [5]. The idea behind this notion is a generalization of the fact that the Euclidean distance between two points is the length of the line segment (shortest path) joining the two points. In the geodesic framework, the path is subject to some constraints: it has to lie inside a given set M, usually called the *mask*. When measuring the length of path inside M, the distance of points better reflects the shape of M, as depicted in fig. 10.

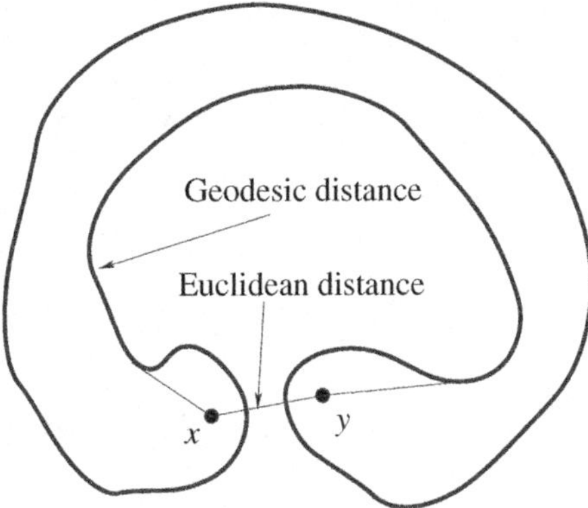

Fig. 10. In this elongated object, points x and y are at large geodesic distance, although they are close according to the Euclidean distance.

5.1 Digital-Continuous Antagonism

Mathematically speaking, given a mask M, a path is any continuous function from an interval $[a, b]$ into M. Then the *geodesic distance* $d^M(x, y)$ in M, is defined as:

$$d^M(x, y) = \min \left\{ \mathcal{L}(\gamma), \ \gamma \text{ path inside } M, \ \gamma(a) = x, \ \gamma(b) = y \right\}, \qquad (7)$$

where $\mathcal{L}(\gamma)$ stands for the length of the path. We do not enter into details in the continuous case, the interested reader may consult [22].

On a digital lattice, things are simpler, since in its definition, the distance derived from a weighted graph is the length of the shortest path inside the graph. This shortest path always exists, but may not be unique. In the geodesic framework with respect to a given mask M, the definition of the geodesic distance seems straightforward: we restrict the graph to the set of vertices belonging to M. For the hexagonal lattice equipped with 6-connectivity and the square lattice with the 4- or 8-connectivity, things work nicely and the three types of algorithms proposed at the beginning of the paper are easily extended. However, for more complicated graphs, the results are in opposition to the intuition. As depicted in fig. 11, a path may "jump" from one connected component to another.

How do we overcome this difficulty? On a theoretical basis, we propose the following solution [20] when the digital distance is the restriction of a distance in the continuous plane, which is always the case in practice: we should take advantage of the continuous case, where this jumping phenomenon does not appear. Two different steps are involved:

1. define a set $\mathcal{C}(M)$ in the continuous space associated with M,
2. compute the length of continuous paths entirely lying in $\mathcal{C}(M)$.

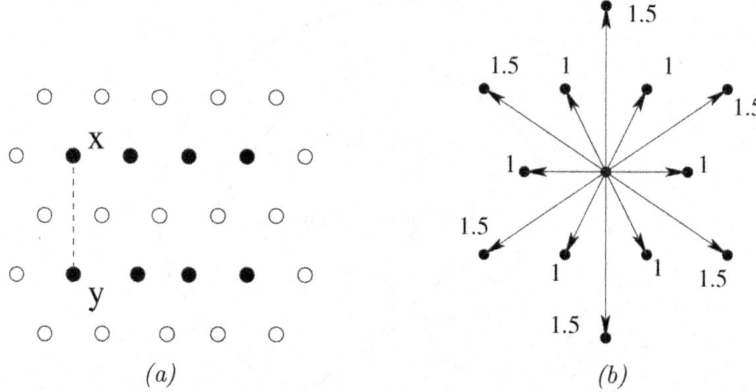

(a)   (b)

Fig. 11. (a): Points x and y are in two different connected components but their geodesic distance is not infinite according to the dodecagonal distance function defined on the square lattice in (b).

The first step is usually done by means of the hexagonal or square lattice, where $\mathcal{C}(M)$ is a polygon (fig. 12). The second consists in "drawing" the digital path used for the computing of the discrete distance as a polygonal line in the continuous plane. Only the polygonal lines entirely lying inside $\mathcal{C}(M)$ are considered in the minimum length.

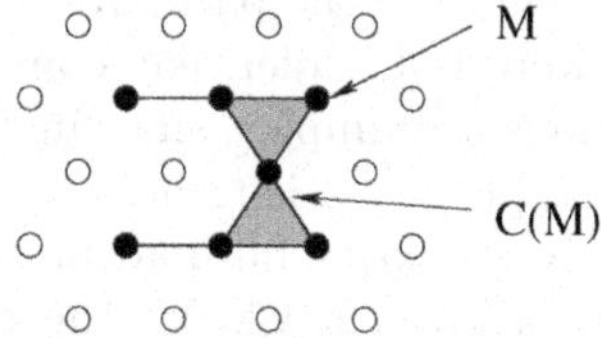

Fig. 12. Construction of a polygon in the continuous plane which respects the connectivity properties of the hexagonal lattice equipped with 6-connectivity.

The sequential algorithm extends to this setup and a connectivity computation of the neighborhood has to be done in each step.

5.2 An Example of Geodesic Transformation

As an example of the use of a geodesic transformation, we present now the *propagation function* and its use for *tip extraction*. Loosely speaking, a tip is a point at the very end of a particle, *i.e.* a point which is far away from the others.

Given a set X, the propagation function $T_X(x)$ is defined as the largest geodesic distance of x to any point in X according to the geodesic distance:

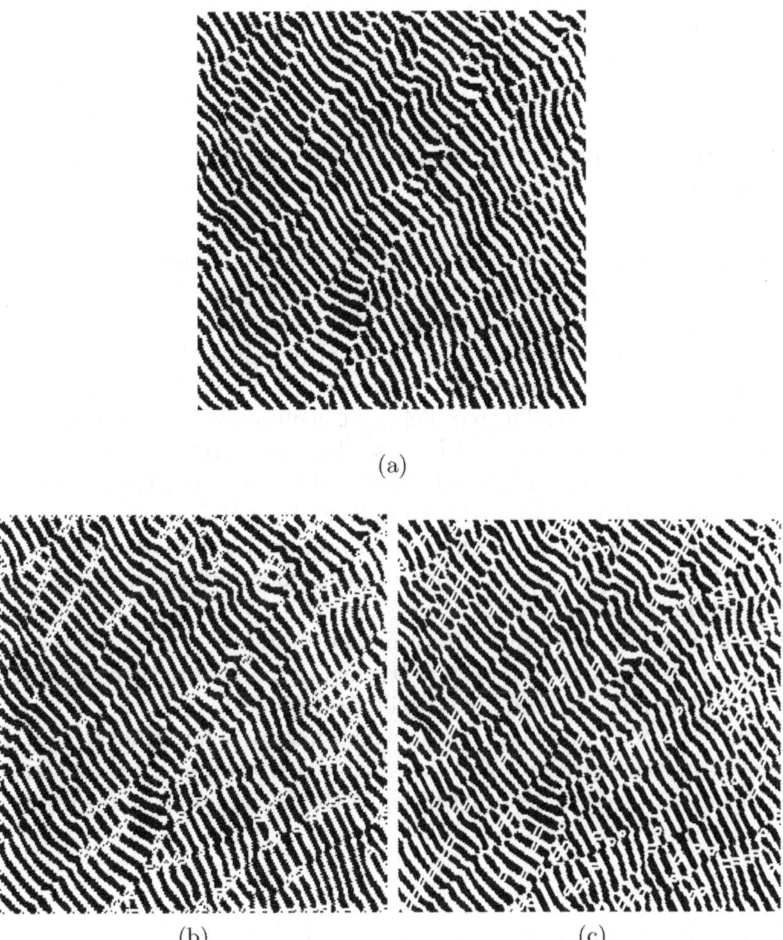

(a)

(b)   (c)

Fig. 13. Tip extraction as the regional maxima of the propagation function: (a) original image, (b) regional maxima of the propagation function, (c) regional minimum of the propagation function.

$$T_X(x) = \max\{d_X(x,y),\ y \in X\}. \tag{8}$$

This function gives rise to very interesting features:

- *geodesic length:* maximal value of $T_X(x)$, giving some idea of the "real length" of the object,
- *tips:* regional maxima of $T_X(x)$, points which are the farthest apart inside the set (see fig. 13),
- *geodesic center:* regional minimum of $T_X(x)$, (unique in each connected component) point inside X being in the "middle".

The principle of fast algorithm is that only a very few points $y \in X$ are useful in equation 8. For precise algorithms and other properties, see [20, 21].

6 Distance Function as Solution to an Evolutionary Equation

In this last section, we illustrate a very old remark on erosions and dilations with a disk and its outstanding developments with partial differentiable equations, known as the *level set* paradigm [25].

The remark is the following: the boundary of a dilated (or eroded) set by a disk of radius r is contained in the curve which is parallel to the boundary of the set and at distance r. The parallel curve is obtained by taking the point at distance r on the outer normal to the set boundary. In mathematical terms, care must be taken: the set has to be bounded by a differentiable curve, which is somewhat restrictive, since the boundary of a polygon is not everywhere differentiable. Some extensions have been proposed, the most complete using set-valued analysis (see [1, 9]). Note also that loop and chain algorithms are also based on the same idea.

Let us now take the viewpoint of front evolution and consider the size of the dilation as time. So, let $\gamma(s) = (x(s), y(s))$ be a closed smooth curve in the plane. The new curve at time t will be denoted by $\gamma_t(s) = (x_t(s), y_t(s))$. Define now a speed function, say F, which may depend on local and global properties of γ_t, expressing the speed at which the curve evolves along it normal. In other words, denoting by $n_t(s)$ the normal vector at $\gamma_t(s)$:

$$\frac{d}{dt}\left(\gamma_t(s)\right) = F(\gamma_t) \cdot n_t(s). \tag{9}$$

For a parallel curve $F = 1$. There are two main difficulties appearing:

1. If γ_t is not convex, γ_{t+h} will always exhibit some singular points (cusps) for sufficiently large h, where the curve is no more differentiable.
2. Fish tails appear, which are not part of the boundary of the dilated set.

For the first item, some mathematical solutions may be found and are not of interest here. For the second one, a very nice solution is proposed by *viscosity solutions*. They can be interpreted as the limit solution when adding some regularization term, a *viscous term* ε:

$$F(\gamma_t(s)) = 1 - \varepsilon \cdot \kappa\left(\gamma_t(s)\right), \tag{10}$$

where $\kappa(\gamma_t(s))$ is the curvature of γ_t at point $\gamma_t(s)$. As ε goes to 0, the usual dilated curve appears. For literature on viscosity solutions, see [2].

Now, other and useful F functions can be designed for defining new tranformations (see [25] as well in physics as in image processing).

From a computational viewpoint, efficient algorithm have been designed, like "fast marching" ones.

For coping with connectivity (merging of two connected components into one for instance), level set techniques have been introduced, defining the curves as the zero level of a given function. Then equation 10 is extended to the evolution of this function.

The paper "PDE for Morphological Operators" by F. Guichard, P. Maragos and J.-M. Morel in the present book (page 369 explains in details these concepts.

7 Conclusion

In this paper we have shown that distance functions have been one very fruitful notion for mathematical morphology. They include:

- theoretical aspects, like skeletons or set distances,
- algorithmical aspects with new classes of algorithms and new digital notions especially in the geodesic framework,
- new links with completely different areas like partial differential equations, which have their own scientific developments as level sets techniques.

References

1. J-P. Aubin and H. Frankowska. *Set-Valued Analysis*. Birkhauser, 1990. Systems and Control: Foundations and Applications.
2. M. Bardi, M.G. Crandall, L.C. Evans, H.M. Soner, and P.E Souganidis. *Viscosity Solutions and Applications*. Springer Verlag, 1995.
3. G. Borgefors. Distances transformations in digital images. *Computer Vision, Graphics and Image Processing*, 34:334–371, 1986.
4. P.E. Danielsson. Euclidean distance mapping. *Computer Graphics and Image Processing*, 14:227–248, 1980.
5. Ch. Lantuéjoul and F. Maisonneuve. Geodesic methods in quantitative image analysis. *Pattern Recognition*, 17(2):177–187, 1984.
6. G. Matheron. *Random Sets and Integral Geometry*. John Wiley and Sons, New York, 1975.
7. G. Matheron. Quelques propriétés topologiques du squelette. Technical Report 560, CGMM, École des Mines, April 1978.
8. G. Matheron. Examples of topological properties of skeletons. In J. Serra, editor, *Image Analysis and Mathematical Morphology, Volume 2: Theoretical Advances*. Academic Press, London, 1988.
9. J. Mattioli. Differential Relations of Morphological Operators. In *Mathematical Morphology and its applications to Signal Processing*, Barcelona, Spain, May, 12-14 1993.
10. J. Mattioli. *Problèmes inverses et relations différentielles en morphologie mathématique*. PhD thesis, Université Paris Dauphine, Paris, France, mai 1993.

11. J. Mattioli and M. Schmitt. Inverse problems for granulometries by erosion. *Journal of Mathematical Imaging and Vision*, 2:217–232, 1992.
12. J. Mattioli, M. Schmitt, É. Pernot, and F. Vallet. Shape discrimination based on mathematical morphology and neural networks. In *International Conference on Artificial Neural Networks (Helsinki)*, June 1991. p. 112-117.
13. A. Meijster, J.B.T.M. Roerdink, and W.H. Hesslink. A general algorithm fo r computing distance transforms in linear time. In *Mathematical Morphology and its Applications to Image and Signal Processing*, pages 331–340, Palo Alto, 2000.
14. F. Meyer. The perceptuel graph: A new algorithm. In *Proc. IEEE Conference on Acoustics, Speech and Signal Processing,*, Paris, 1982.
15. F. Meyer. Skeletons in digital spaces. In J. Serra, editor, *Image Analysis and Mathematical Morphology, Volume 2: Theoretical Advances*. Academic Press, London, 1988.
16. E.F. Moore. The shortest path through a maze. In *Proc. Int. Symposium on Theory of Switching*, volume 30, pages 285–292, 1991.
17. A. Rosenfeld and J.L. Pfaltz. Sequential operations in digital picture processing. *J. Assoc. Comp. Mach.*, 13(4):471–494, 1966.
18. L.A. Santalo. *Integral Geometry and Geometric Probability*. Addison Wesley, 1976.
19. M. Schmitt. Des algorithmes morphologiques à l'intelligence artificielle. Thèse Ecole des Mines de Paris, February 1989.
20. M. Schmitt. Geodesic arcs in non-euclidean metrics: Application to the propagation function. *Revue d'Intelligence Artificielle*, 3 (2):43–76, 1989.
21. M. Schmitt. Propagation function: constant time algorithms. *Journal of microscopy*, 178:272–281, June 1995.
22. M. Schmitt and J. Mattioli. *Morphologie Mathématique*. Logique - Mathématiques - Informatique. Masson, Décembre 1993.
23. M. Schmitt and L. Vincent. *Morphological image analysis: a practical and algorithmic handbook*. Cambridge University Press, To appear in 2005.
24. J. Serra. *Image Analysis and Mathematical Morphology*. Academic Press, London, 1982.
25. J.A. Sethian. *Level Sets Methods*. Cambridge University Press, 1996.
26. L. Vincent. Algorithmes morphologiques à base de files d'attente et de lacets : Extension aux graphes. Thèse École des Mines de Paris, May 1990.

Partial Differential Equations for Morphological Operators

Frederic Guichard[1], Petros Maragos[2], and Jean-Michel Morel[3]

[1] DO Labs
[2] School of ECE,National Technical University of Athens
[3] CMLA, Ecole Normale Superieure de Cachan,

1 Introduction

Two of G. Matheron's seminal contributions have been his development of size distributions (else called 'granulometries') and his kernel representation theory. The first deals with semigroups of multiscale openings and closings of binary images (shapes) by compact convex sets, a basic ingredient of which are the multiscale Minkowski dilations and erosions. The second deals with representing increasing and translation-invariant set operators as union of erosions by its kernel sets or as an intersection of dilations.

The semigroup structure of the basic multiscale morphological operators led to the development (by Alvarez et al. [2], Brockett & Maragos [9], and Boomgaard & Smeulders [60]) of Partial Differential Equations (PDEs) that can generate them on a continuum of scales. In parallel, the representation theory was extended by Maragos [36] to function operators as sup-inf of min-max filterings by elements of a kernel basis. These two seemingly unrelated research directions were later rejoined by Catte et al. [11] and by Guichard & Morel [22, 23] who used the basis representation of multiscale sup-inf operators to develop PDEs that can generate them based on variants of the mean curvature motion.

Many information extraction tasks in image processing and computer vision necessitate the analysis at multiple scales. Influenced by the work of Marr (and coworkers) [42], Koenderink [31] and Witkin [63], for more than a decade the multiscale analysis was based on Gaussian convolutions. The popularity of this approach was due to its linearity and its relationship to the linear isotropic heat diffusion PDE. The big disadvantage of the Gaussian scale-space approach is the fact that linear smoothers blur and shift important image features, e.g., edges. There is, however, a variety of *nonlinear* smoothing filters, including morphological open-closings (of the Minkowski type [43, 56] or of the reconstruction [53] and leveling type [47, 39]) and anisotropic nonlinear diffusion [51], which can smooth while preserving important image features and can provide a nonlinear scale-space.

Until the end of the 1990s, morphological image processing had been based traditionally on modelling images as sets or as points in a complete lattice of functions and viewing morphological image transformations as set or lattice operators. Further, the vast majority of implementations of multiscale morphological filtering had been discrete. In 1992, inspired by the modelling of the Gaussian scale-space via the linear heat diffusion PDE, three teams of researchers independently published nonlinear PDEs that model the continuous multiscale morphological scale-space. Specifically, Alvarez, Guichard, Lions and Morel [1] obtained PDEs for multiscale flat dilation and erosion, by compact convex structuring sets, as part of their general work on developing PDE-based models for multiscale image processing that satisfy certain axiomatic principles. Brockett and Maragos [8] developed PDEs that model multiscale morphological dilation, erosion, opening and closing by compact-support structuring elements that are either convex sets or concave functions and may have non-smooth boundaries or graphs, respectively. Their work was based on the semigroup structure of the multiscale dilation and erosion operators and the use of morphological sup/inf derivatives to deal with the development of shocks (i.e., discontinuities in the derivatives). In [59, Ch. 8], Boomgaard and Smeulders obtained PDEs for multiscale dilation and erosion by studying the propagation of the boundaries of 2D sets and the graphs of signals under multiscale dilation and erosion. Their work applies to convex structuring elements whose boundaries contain no linear segments, are smooth and possess a unique normal at each point. Refinements of the above three works for PDEs modelling multiscale morphology followed in [2, 3, 9, 38, 40, 60]. Extensions also followed in several directions including asymptotic analysis and iterated filtering by Guichard & Morel [22, 23], a unification of morphological PDEs using Legendre-Fenchel 'slope' transforms by Heijmans & Maragos [25], a common algebraic framework for linear and morphological scale-spaces by Heijmans & Boomgaard [26] and PDEs for morphological reconstruction operators with global constraints by Maragos and Meyer [47, 39].

To illustrate the basic idea behind morphological PDEs, we consider a 1D example, for which we define the multiscale flat dilation and erosion of a 1D signal $f(x)$ by the set $[-t, t]$ as the scale-space functions

$$\delta(x,t) = \sup_{|y| \leq t} f(x - y), \qquad \varepsilon(x,t) = \inf_{|y| \leq t} f(x + y).$$

The PDEs generating these multiscale flat dilations and erosions are [9]

$$\partial \delta / \partial t = |\partial \delta / \partial x|, \qquad \partial \varepsilon / \partial t = -|\partial \varepsilon / \partial x|,$$
$$\delta(x,0) = \varepsilon(x,0) = f(x). \tag{1}$$

In parallel to the development of the above ideas, there have been some advances in the field of differential geometry for evolving curves or surfaces using level set methods. Specifically, Osher & Sethian [50] have developed PDEs of the Hamilton-Jacobi type to model the propagation of curves, embedded as level curves (isoheight contours) of functions evolving in scale-space. The propagation was modelled using speeds along directions normal to the curve that contain a constant term and/or a term dependent on curvature. Furthermore, they developed robust numerical algorithms to solve these PDEs by using stable and shock-capturing schemes to solve similar, shock-producing, nonlinear wave PDEs that are related to hyperbolic conservation laws [32]. Kimia, Zucker & Tannenbaum [29] have applied and extended these curve evolution ideas to shape analysis in computer vision. Arehart, Vincent & Kimia [4] and Sapiro et al. [54] implemented continuous-scale morphological dilations and erosions using the numerical algorithms of curve evolution to solve the PDEs for multiscale dilation and erosion. There are several relationships between curve evolution and multiscale morphology, since the evolution with constant normal speed models multiscale set dilation, and the corresponding Hamilton-Jacobi PDEs contain the PDE of multiscale dilation/erosion by disks as a basic ingredient. Furthermore, the level sets used in curve evolution have previously been used extensively in mathematical morphology for extending set operations to functions [56], [41].

Multiscale dilations and erosions of binary images can also be obtained via distance transforms. Using Huygens' construction, the boundaries of multiscale dilations–erosions by disks can also be viewed as the wavefronts of a wave initiating from the original image boundary and propagating with constant normal speed in a homogeneous medium [7]. This idea can also be extended to heterogeneous media by using a weighted distance function, in which the weights are inversely proportional to the propagation speeds. In geometrical optics, these distance wavefronts are obtained from the isolevel contours of the solution of the Eikonal PDE. This ubiquitous PDE (or its solution as weighted distance) has been applied to solving various problems in image analysis and computer vision [27] such as shape-from-shading [52, 30], gridless halftoning, and image segmentation [61, 46, 49, 38, 40].

Modelling linear and morphological scale-space analysis via PDEs has several advantages, mathematical, physical, and computational. In particular, there exist several efficient numerical algorithms which implement morphology-related PDEs on a discrete grid [50, 58, 23]. Thus, one can have both the advantages of continuous modelling and discrete processing.

This chapter is organized as follows. In section 2, we review all first-order PDEs coming from the asymptotic form of classical multiscale dilations and erosions. In section 3, we state the most general results about PDEs associated with the rescaling of any local increasing operator. Section 4 treats the opposite viewpoint : instead of constructing the PDE by iterating local morphological operators, it starts with a *scale space* abstract set of axioms on multiscale image analysis. A scale space in this abstract setting is nothing but a scale indexed family of operators T_t, understood as operators smoothing more and more the image when the scale t increases. Under sound axioms, it can be proved that scale spaces are equivalent to the action of nonlinear or linear parabolic PDEs. A further classification of the PDEs is sketched, according to their invariance properties. Section 5 takes the last turn by focusing on the curve evolution interpretation of all that. Actually, all contrast invariant image scale spaces can be described as curve scale spaces applied to each level line of the image. This point of view has become popular under the name of "level set methods" and yields the nice geometric interpretation of contrast invariant scale spaces as "curvature flows". Needless to be said, this rich subject cannot be but sketched in one book chapter and actually deserves a long and mathematically clean presentation. Probably the presentations closest to our viewpoint here are F. Cao's book [10] and the book to appear [23].

2 PDEs for Multiscale Morphological Operators

The main tools of low-level morphological image processing are a broad class of nonlinear signal operators formed as parallel and/or serial interconnections of the two most elementary morphological signal operators, the Minkowski *dilation* $\oplus$ and the *erosion* $\ominus$:

$$(f \oplus g)(x) \triangleq \bigvee_{y \in \mathbb{E}} f(y) + g(x - y)$$

$$(f \ominus g)(x) \triangleq \bigwedge_{y \in \mathbb{E}} f(y) - g(y - x),$$

where $\bigvee$ and $\bigwedge$ denote supremum and infimum, and the signal domain can be continuous $\mathbb{E} = \mathbb{R}^d$ or discrete $\mathbb{E} = \mathbb{Z}^d$. The signal range is a subset of $\overline{\mathbb{R}} = \mathbb{R} \cup \{-\infty, +\infty\}$. Compositions of erosions and dilations yield two useful smoothing filters: the *opening* $f \mapsto (f \ominus g) \oplus g$ and *closing* $f \mapsto (f \oplus g) \ominus g$.

2.1 PDEs Generating Dilations and Erosions

Let $k : \mathbb{R}^2 \to \overline{\mathbb{R}}$ be a unit-scale upper-semicontinuous concave structuring function, to be used as the kernel for morphological dilations and erosions. Scaling both its values and its support by a scale parameter $t \geq 0$ yields a parameterized family of multiscale structuring functions

$$k_t(x, y) \triangleq \begin{cases} tk(x/t, y/t), & \text{for } t > 0 \ , \\ 0 \text{ at } (x, y) = (0,0) \text{ and } -\infty \text{ else}, & \text{for } t = 0, \end{cases} \tag{2}$$

which satisfies the semigroup property

$$k_s \oplus k_t = k_{s+t}. \tag{3}$$

Using k_t in place of g as the kernel in the basic morphological operations leads to defining the *multiscale* dilation and erosion of $f : \mathbb{R}^2 \to \mathbb{R}$ by k_t as the scale-space functions

$$\delta(x,y,t) \triangleq f \oplus k_t(x,y), \qquad \varepsilon(x,y,t) \triangleq f \ominus k_t(x,y), \tag{4}$$

where $\delta(x,y,0) = \varepsilon(x,y,0) = f(x,y)$.

In practice, a useful class of functions k consists of flat structuring functions

$$k(x,y) = \begin{cases} 0 & \text{for } (x,y) \in B, \\ -\infty & \text{for } (x,y) \notin B, \end{cases} \tag{5}$$

which are the $0/-\infty$ indicator functions of compact convex planar sets B. The general PDE[4] generating the multiscale flat dilations of f by a general compact convex symmetric B is [2, 9, 25]

$$\frac{\partial \delta}{\partial t} = \mathrm{sptf}_B(\delta_x, \delta_y), \tag{6}$$

where $\mathrm{sptf}_B(\cdot)$ is the support function of B:

$$\mathrm{sptf}_B(x,y) \triangleq \bigvee_{(a,b) \in B} ax + by. \tag{7}$$

Useful cases of structuring sets B are obtained by the unit balls $B_p = \{(x,y) : \|(x,y)\|_p \leq 1\}$ of the metrics induced by the L_p norms $\|\cdot\|_p$, for $p = 1, 2, \ldots, \infty$. The PDEs generating the multiscale flat dilations of f by B_p for three special cases of norms $\|\cdot\|_p$ are as follows:

$$B = \text{rhombus } (p = 1) \implies \delta_t = \max(|\delta_x|, |\delta_y|) = \|\nabla\delta\|_\infty, \tag{8}$$

$$B = \text{disk } (p = 2) \implies \delta_t = \sqrt{(\delta_x)^2 + (\delta_y)^2} = \|\nabla\delta\|_2, \tag{9}$$

$$B = \text{square } (p = \infty) \implies \delta_t = |\delta_x| + |\delta_y| = \|\nabla\delta\|_1, \tag{10}$$

with $\delta(x,y,0) = f(x,y)$. The corresponding PDEs generating mutliscale flat erosions are

$$B = \text{rhombus} \implies \varepsilon_t = -\|\nabla\varepsilon\|_\infty, \tag{11}$$

$$B = \text{disk} \implies \varepsilon_t = -\|\nabla\varepsilon\|_2, \tag{12}$$

$$B = \text{square} \implies \varepsilon_t = -\|\nabla\varepsilon\|_1, \tag{13}$$

with $\varepsilon(x,y,0) = f(x,y)$.

These simple but nonlinear PDEs are satisfied at points where the data are smooth, that is, the partial derivatives exist. However, even if the initial image or signal f is smooth, at finite scales $t > 0$ the above dilation or erosion evolution may create discontinuities in the derivatives, called *shocks*, which then continue propagating in scale-space. Thus, the multiscale dilations δ or erosions ε are *weak solutions* of the corresponding PDEs, in the sense put

---

[4] Notation often used for PDEs: $u_t = \partial u/\partial t$, $u_x = \partial u/\partial x$, $u_y = \partial u/\partial y$, $Du = \nabla u = (u_x, u_y)$, $\mathrm{div}(v,w)) = \nabla \cdot (v,w) = v_x + w_y$.

forth by Lax [32]. Ways to deal with these shocks include replacing standard derivatives with morphological derivatives [9] or replacing the PDEs with differential inclusions [44]. The most acknowledged viewpoint on this, however, is to use the concept of viscosity solutions. For first-order PDEs, a good exposition is given in Barles [5] or in the classic [14]. Probably the shortest, more pedagogic and up to date presentation of viscosity solutions is the recent one by Crandall [13].

Next, we provide two examples of PDEs generating multiscale dilations by *graylevel structuring functions*. First, if we use the compact-support spherical function

$$k(x,y) = \begin{cases} \sqrt{1 + x^2 + y^2} & \text{for } x^2 + y^2 \leq 1, \\ -\infty & \text{for } x^2 + y^2 > 1, \end{cases} \tag{14}$$

the dilation PDE becomes

$$\delta_t = \sqrt{1 + (\delta_x)^2 + (\delta_y)^2}. \tag{15}$$

As shown in [9], this can be proven by using the semigroup structure of dilations and the first-order Taylor's approximation for the difference between dilations at scales t and $t+dt$. Alternatively, it can be proven using slope transforms, as explained in the next section. As a second example of structuring function, if k is the infinite-support parabola

$$k(x,y) = -r(x^2 + y^2), \quad r > 0, \tag{16}$$

the dilation PDE becomes

$$\delta_t = [(\delta_x)^2 + (\delta_y)^2]/4r. \tag{17}$$

This can be proven using slope transforms.

2.2 Slope Transforms and Dilation PDEs

All of the above dilation (and erosion) PDEs can be unified using slope transforms. These transforms [37, 15] are simple variations of the Legendre-Fenchel transform. The word 'slope' was given only for insights because the eigenfunctions of a morphological dilation-erosion system are straight lines parameterized by their slope. Further, for morphological systems we can consider a new domain, called a 'slope domain', where morphological sup-inf convolutions in the time-space domain become addition of slope transforms in the slope domain.

Let the unit-scale kernel $k(x,y)$ be a general upper-semicontinuous concave function and consider its upper slope transform[5]

---

[5] In convex analysis, given a convex function $h(x)$ there uniquely corresponds another convex function $h^*(a) = \bigvee_x a \cdot x - h(x)$, called the *Legendre-Fenchel conjugate* of h. The lower slope transform of h, defined as $H^\wedge(a) = \bigwedge_x h(x) - a \cdot x$, is the dual of the upper slope transform. Obviously, the former is closely related to the conjugate function since $h^*(a) = -H^\wedge(a)$.

$$K^{\vee}(a,b) \triangleq \bigvee_{(x,y)\in\mathbb{R}^2} k(x,y) - (ax + by) \tag{18}$$

Then, as shown in [25, 44], the PDE generating multiscale signal dilations by k is

$$\partial\delta/\partial t = K^{\vee}(\delta_x, \delta_y) \tag{19}$$

Thus, the rate of change of δ in the scale (t) direction is equal to the upper slope transform of the structuring function evaluated at the spatial gradient of δ. Similarly, the PDE generating the multiscale erosion by k is

$$\partial\varepsilon/\partial t = -K^{\vee}(\varepsilon_x, \varepsilon_y). \tag{20}$$

For example, the PDE (6) modelling the general flat dilation by a compact convex set B is a special case of (19) since the support function (7) of B is the upper slope transform of the $0/-\infty$ indicator function of B. Likewise, the PDE (17) modelling multiscale dilations by parabolae results simply from (19) by noting that the upper slope transform of a concave parabola is a convex parabola.

All of the dilation and erosion PDEs examined are special cases of Hamilton-Jacobi equations, which are of paramount importance in physics. Such equations usually do not admit classic (i.e., everywhere differentiable) solutions. Viscosity solutions of Hamilton-Jacobi PDEs have been extensively studied by Crandall et al. [14]. The theory of viscosity solutions has been applied to morphological PDEs by Guichard & Morel [23]. Finally, Heijmans & Maragos [25] have shown via slope transforms that the multiscale dilation by a general upper-semicontinuous concave function is the viscosity solution of the Hamilton-Jacobi dilation PDE of Eq. (19).

2.3 PDEs Generating Openings and Closings

Let $u(x,y,t) = [f(x,y) \ominus tB] \oplus tB$ be the multiscale flat opening of an image f by the disk B. This standard opening can be generated at any scale $r > 0$ by running the following PDE [2]

$$u_t = -\max\left(\operatorname{sgn}(r-t), 0\right)\|\nabla u\|_2 + \max\left(\operatorname{sgn}(t-r), 0\right)\|\nabla u\|_2, \tag{21}$$

from time $t = 0$ until time $t = 2r$ with initial condition $u(x,y,0) = f(x,y)$, where $\operatorname{sgn}(\cdot)$ denotes the signum function. This PDE has time-dependent switching coefficients that make it act as an erosion PDE during $t \in [0,r]$ but as a dilation PDE during $t \in [r,2r]$. At the switching instant $t = r$ this PDE exhibits discontinuities. This can be dealt with by making appropriate changes to the time scale that make time 'slow down' when approaching the discontinuity at $t = r$, as suggested by Alvarez et al. [2]. Of course, the solution u of the above PDE is an opening only at time $t = r$, whereas the solutions at other times is not a opening. In a different work, Brockett & Maragos [9] have developed a partial differential-difference equation that models at all times the evolutions of multiscale openings of 1D images by flat intervals. This does not involve only local operations but also global features such as the support geometry of peaks of f at various scales.

The reconstruction openings have found many more applications than the standard openings in a large variety of problems. We next present a nonlinear PDE that can model and generate openings and closings by reconstruction. Consider a 2D reference signal $f(x, y)$ and a marker signal $g(x, y)$. If $g \leq f$ everywhere and we start iteratively growing g via incremental flat dilations with an infinitesimally small disk ΔtB but without ever growing the result above the graph of f, then in the limit we shall have produced the reconstruction opening of f (with respect to the marker g). The infinitesimal generator of this signal evolution $u(x, y, t)$ can be modelled via the following PDE, studied by by Maragos & Meyer [47, 39],

$$u_t(x, y, t) = \|\nabla u\| \mathrm{sgn}[f(x, y) - u(x, y, t)],$$
$$u(x, y, 0) = g(x, y), \tag{22}$$

where $\mathrm{sgn}(r)$ equals 1 if $r > 0$, -1 if $r < 0$ and 0 if $r = 0$. The mapping from the initial value $u_0(x, y) = u(x, y, 0)$ to the limit $u_\infty(x, y) = \lim_{t \to \infty} u(x, y, t)$ is the *reconstruction opening* filter. If we reverse the roles of f and g, in the limit we obtain the *reconstruction closing* of f with respect to the marker g. Now, if there is no specific order between f and g, the PDE has a sign-varying coefficient $\mathrm{sgn}(f - u)$ with spatiotemporal dependence, which acts as a global constraint that controls the instantaneous growth. The final result $u_\infty(x, y)$ is equal to the output from a more general class of morphological filters, called *levelings* [47], which have many useful scale-space properties and contain as special cases the reconstruction openings and closings. For stability of the solution of the leveling PDE, g has to be uniformly continuous in the viscosity sense.

3 Asymptotic of Increasing Operators

We consider a family $\mathcal{F}$ of functions from $\mathbb{E}$ into $\mathbb{R}$ representing a class of images. An operator S, from $\mathcal{F}$ into $\mathcal{F}$, is said **increasing** or **monotone** if $\forall f, g \in \mathcal{F}$, $(\forall \mathbf{x} \in \mathbb{E}, f(\mathbf{x}) \geq g(\mathbf{x})) \Longrightarrow (\forall \mathbf{x}, Sf(\mathbf{x}) \geq Sg(\mathbf{x}))$.

In all the following we will assume that S commutes with spatial translations of the image, in other words we assume that S is invariant by translation.

Note: It is a general property of the increasing and translation invariant operators to preserve the Lipschitz property of any Lipschitz function. Consequently, a possible choice for $\mathcal{F}$ can be made by considering the set of Lipschitz functions.

3.1 Increasing Operators

The following formulae, inspired from work of Matheron [43], Serra [56], and Maragos [36] gives us a general form for any increasing and translation invariant operator:

Let S be a increasing function operator defined of $\mathcal{F}$, invariant by translation and commuting with the addition of constants. There exists a family $\mathbb{F}_1(S)$ of functions from $\mathbb{E}$ into $\mathbb{R} \cup \{-\infty, +\infty\}$ such that for all functions f of $\mathcal{F}$, we have

$$Sf(\mathbf{x}) = \bigwedge_{g \in \mathbb{F}_1(S)} \bigvee_{\mathbf{y} \in \mathbb{E}} f(\mathbf{y}) - g(\mathbf{x} - \mathbf{y}).$$

Similarly, there exists another family of functions $\mathbb{F}_2(S)$ such that

$$Sf(\mathbf{x}) = \bigvee_{g \in \mathbb{F}_2(S)} \bigwedge_{\mathbf{y} \in \mathbb{E}} f(\mathbf{y}) - g(\mathbf{x} - \mathbf{y}).$$

The special cases where $\mathbb{F}$ are made of a single function g correspond to the classical Minkowski dilation and erosion that have already been presented in section 2.

Examples of classical increasing operators (or "filters") that cannot be represented with a $\mathbb{F}$ made of a single function are e.g. the **"median"** filter or the **"mean"** filter. In fact, it would be probably vain to try to classify all possible increasing filters. So, in this section, we wish to specify the general forms of the PDEs related to increasing filters.

3.2 Scaled and Local Increasing Operators

We consider a scaled increasing operator S_h, where the scale h is a positive real number. We say that S_h is a **local increasing operator** if for all u and v such that $u(\mathbf{y}) > v(\mathbf{y})$ for $\mathbf{y}$ in a neighborhood of $\mathbf{x}$ and $\mathbf{y} \neq \mathbf{x}$, then for h small enough we have

$$(S_h u)(\mathbf{x}) \geq (S_h v)(\mathbf{x})$$

Roughly speaking, a local increasing operator is a scale operator whose action is reduced when its scale decreases. Easy way to construct a local increasing operator S_h from an increasing operator S is to localize the action of the family of functions $\mathbb{F}$: e.g., one can set S_h as in [24]:

$$S_h(u)(\mathbf{x}) = \bigwedge_{g \in \mathbb{F}} \bigvee_{\mathbf{y} \in \mathbb{E}} (u(\mathbf{x} + \mathbf{y}) - h^\beta g(\mathbf{y}/h^\alpha)), \tag{23}$$

for some $\alpha, \beta \geq 0$. This construction, with adequate choices of α and β will transform e.g. the mean, median, erosion or dilation filters on a disk of radius 1, into their corresponding respective localized versions on a radius h disk. However, in general, this construction is not sufficient to get a local increasing operator from any increasing operator S.

We finally need some technical assumption stating that a very smooth image must evolve in a smooth way with the considered operator. Let us recall that we initially assume that the operator is translation-invariant, so that the analysis on its asymptotic could be done at $\mathbf{x}=0$ or any other point $\mathbf{x}$. So choosing any point $\mathbf{x}$, let $Q_{A,p,c}(\mathbf{y}) = \frac{1}{2}(A(\mathbf{y}-\mathbf{x}),\mathbf{y}-\mathbf{x}) + (p,\mathbf{y}-\mathbf{x}) + c$ be a quadratic form on $\mathbb{E}$. (If $\mathbb{E} = \mathbb{R}^N$ then A is a $N*N$ matrix $(A = D^2Q(\mathbf{x}))$, p a vector of $\mathbb{R}^N$ $(p = DQ(\mathbf{x}))$ and c a constant.)

We shall say that a local increasing operator is **regular** if there exists a function $F(A,p,c)$, continuous with respect to A, such that

$$\forall Q_{A,p,c}, \quad \frac{(S_hQ - Q)(\mathbf{x})}{h} \to F(A,p,c) \quad \text{when} \quad h \to 0.$$

In [2], Alvarez et al gave the general asymptotic shape of any local and increasing operator:

Fundamental Asymptotic Theorem: *Let S_h be a local regular increasing operator and F the real function associated with the regularity assumption. Then S_h satisfies*

$$((S_hu - u)/h)(\mathbf{x}) \to F(D^2u(\mathbf{x}), Du(\mathbf{x}), u(\mathbf{x})) \tag{24}$$

as h tends to 0^+ for every C^2 function u and every $\mathbf{x}$. In addition, F is nondecreasing with respect to its first argument : If $A \geq \tilde{A}$, for the ordering of symmetric matrices,

$$then, \quad F(A,p,c) \geq F(\tilde{A},p,c)). \tag{25}$$

This easy to prove theorem reduces the classification of all iterated local and increasing operators to the classification of all interesting functions F. In dimension 2, these real functions have six arguments. This number, however, can be drastically reduced when we impose obvious and rather necessary and useful invariance properties to the increasing operator.

This theorem also shows that the study of the asymptotic behavior of an increasing operator can be reduced to the study of its action on a parabolic function $(Q_{A,p,c})$.

4 The Scale-Space Framework

In this section, we consider an abstract framework, the "scale space", which at the end boils down, from the algorithmic viewpoint, to iterated filtering. Now, this framework will make it easier to classify and model the possible asymptotic behaviors of iterated increasing operators.

The scale space theory was founded (in a linear framework) by Witkin [63], Marr [42], and Koenderink [31]. Many developments have been proposed, see e.g. [33] for further references on that particular field.

We can see a "scale space" as a family of increasing operators $\{T_t\}_{t\geq 0}$, depending on a scale parameter t. Given an image $u_0(\mathbf{x})$, $(T_t u_0)(\mathbf{x}) = u(t, \mathbf{x})$ is the "image u_0 analyzed (in fact : smoothed) at scale t". For simplicity, $\mathcal{F}$ will be the set of Lipschitz functions on $\mathbb{E} = \mathbb{R}^N$.

We assume that the output at scale t can be computed from the output at a scale $t - h$ for very small h. This is natural, since a coarser analysis of the original picture is likely to be deduced from a finer one without any dependence upon the original picture. By that way the finest picture smoothing is the identity. T_t is obtained by composition of "transition filters", which we denote by $T_{t+h,t}$. For simplicity, we will assume here that $T_{t+h,t}$ will not depend on t, so that one can set $S_h = T_{t+h,t}$. (The general case can be found in [23]). We then say that the scale space $\{T_t\}_{t\geq 0}$ is **pyramidal** if there exists an operator S_h such that for all t one has:

$$T_{t+h} = S_h \circ T_t$$

Note that a much stronger version of the pyramidal structure is the semigroup property already presented in section 2.

Since the visual pyramid is assumed to yield more and more global information about the image and its features, it is clear that when the scale increases, no new feature should be created by the scale space : the image at scale t'>t must be simpler than the image at scale t. Furthermore, the transition operator S_h is assumed to act "locally", that is, to look at a small part of the processed image and in a monotone way. In other terms, S_h should be a regular and local increasing operator.

At last, we say that a scale-space $\{T_t\}_{t\geq 0}$ is **causal** if it is pyramidal and if its transition operator S_h is a translation invariant, regular and local increasing operator. To some extent, as increasing operators are the "basic" tools of morphology, causal scale-spaces can be seen as **Morphological Flows**. Operators seen in section 2.2 defined examples of causal scale-spaces or "morphological flows".

4.1 Causal Scale Space, Increasing Operators and PDEs

We consider a causal scale space $\{T_t\}_{t\geq 0}$ that commutes with addition of constants; i.e., for any constant C, we have $T_t(u+C) = T_t(u) + C$. We denote by F the asymptotic of the transition operator associated to S_h. We know from Eqn. (24) that F has the shape: $F(A, p, c)$. The commutation with addition of constants removes the dependence on c, which therefore yields for F a $F(A, p)$ shape.

The next theorems state the equivalence between causal scale-space and viscosity solutions of parabolic PDE. They require some technical assumptions on the shape of the function F that will be given later.

Theorem 1

Let T_t be a causal scale-space. Then for any Lipschitz function u_0: $u(t, .) = T_t(u)(.)$ is the viscosity solution of

$$\frac{\partial u}{\partial t} = F(D^2 u, Du) \tag{26}$$

with initial condition $u(0,.) = u_0$.

Theorem 2

The operator T_t that associates to a Lipschitz function u_0 the (unique) viscosity solution of the equation (26) at scale t is a increasing operator on Lipschitz functions and T_t defines a causal scale-space.

Proofs of these theorem has been given under some regularity conditions on function F. E.g. in [23], Guichard & Morel prove that the preceding theorems hold if F is assumed continuous for all $A, p \neq 0$ and such that there exists two continuous functions $G^+(A,p)$ and $G^-(A,p)$, with $G^+(0,0) = G^-(0,0) = 0$; $\forall A \geq 0$, $G^+(A,0) \geq 0$ and $G^-(-A,0) \leq 0$ and $\forall A, p$, $G^-(A,p) \leq F(A,p) \leq G^+(A,p)$. These conditions are in fact not so restrictive since they are satisfied by all equations mentioned in the present chapter.

4.2 Geometric and Contrast Invariant Scale Spaces

We shall now list a series of axioms which state some invariance for the scale space. We begin by considering a "contrast invariance" assumption, that the scale space should be independent from the (arbitrary) graylevel scale. We shall say that a scale space is **contrast invariant** if

$$g \circ T_t = T_t \circ g, \tag{27}$$

for any nondecreasing and continuous function g from $\mathbb{R}$ into $\mathbb{R}$. The contrast invariance is a particular formalization of the invariance of image analysis with respect to changes of illumination. This invariance has been stated in perception theory by Wertheimer [62], as early as 1923. In Mathematical Morphology, the contrast invariance is commented and used e.g. in Serra [56], or by Maragos et al [41]. Within the scale-space framework, Koenderink [31] insists on that invariance but did not proceed due to incompatibility with some imposed linearity property. We will see, in section 5, that in addition to this link with perception, "contrast invariance" generates an interesting link between function evolution and set or curve evolution.

Let R be an isometry of $\mathbb{R}^N$ and denote by Ru the function $Ru(\mathbf{x}) = u(R\mathbf{x})$. We shall say that a scale space T_t is **euclidean invariant** if for every isometry R of $\mathbb{R}^N$ into $\mathbb{R}^N$, $RT_t = T_t R$.

Finally, we state an axiom which implies the invariance of the scale space under any affine projection of a planar shape. Set for any such transform $Af(\mathbf{x}) = f(A\mathbf{x})$. We shall say that a scale space T_t is **affine invariant** if for any linear application A of $\mathbb{R}^N$ with $\det(A) = 1$, we have $AT_t = T_t A$.

If we impose the euclidean and contrast invariance, then $T_t u_0$ obeys a restricted form of the equation (26). A general study in dimension N can be found in [20]. We just recall from [2] the two dimensional case.

(i) Let T_t be a euclidean and contrast invariant causal scale space and u_0 be a Lipschitz function, then $u(t) = T_t(u_0)$ is the viscosity solution of

$$\frac{\partial u}{\partial t} = |Du|\beta(curv(u)), \tag{28}$$

where β is a continuous nondecreasing real function.
(ii) If the scale space is, in addition, affine invariant, then the only possible equation is, up to a rescaling,

$$\frac{\partial u}{\partial t} = |Du|(curv(u))^{1/3}. \tag{29}$$

where, for any C^2 function f and where $Df \neq 0$, $curv(f) = \kappa(f) = div(\frac{Df}{|Df|})$, is the curvature of the level line at the considered point.

Conversely, as proved in [23], the operator T_t that associates to a function u_0 the (unique) viscosity solution of the preceding equations at scale t is a euclidean and contrast invariant increasing operator on Lipschitz functions and the family T_t defines a euclidean and contrast invariant causal scale-space.

4.3   Iterations of Increasing Operators and PDEs

We have seen that the causal scale space framework ends up with some particular parabolic equations. However, this very formal definition of scale space might seem very restrictive to be of any interest. Question occurs on how to get a scale space from any scaled increasing operator ?

The following heuristic answers the question:

- choose a increasing operator S, e.g the mean, the median, the dilation, the erosion, etc...
- localize it: S_h, e.g by using equation (23),
- iterate it: Set $(T_n)_t = (S_h)^n$ with $hn = t$.

When $n \to \infty$ if the sequence $(T_n)_t$ converges to some operator T_t, then T_t is a causal scale-space. More precisely, consider u_0 a Lipschitz function and set $u_n(t) = (T_n)_t(u_0)$. If $u_n(t)$ converges when n tends to ∞, then $u(t) = lim_{n\to\infty}u_n(t)$ is the viscosity solution of equation (26) with F given by the asymptotic of S_h (equation (24)).

The shape of function F will necessary inherit from the invariance property of the increasing operator S. E.g. if S is contrast and euclidean invariant, then F is necessarily of the form $F(D^2u, Du) = |Du|\beta(curv(u))$, for some increasing function β.

Unfortunately convergence has not been proved for general forms of local and increasing operators S_h. Let us cite some basic examples: if S_h is the mean filter on a disk of radius h^2, then T_t will solve the heat equation

$$\frac{\partial u}{\partial t} = \Delta u$$

which confirms a well known result. If S_h is a median filter on a disk of radius h^2, then T_t will solve the mean curvature motion

$$\frac{\partial u}{\partial t} = |Du| \, curv(u) = |Du| \kappa$$

This last equation will be more deeply considered in the following section.

5 Curve Evolution and Morphological Flows

Consider at time $t = 0$ an initial simple, smooth, closed planar curve $\Gamma(0)$ that is propagated along its normal vector field at speed V for $t > 0$. Let this evolving curve (front) $\Gamma(t)$ be represented by its position vector $\mathbf{C}(p,t) = (x(p,t), y(p,t))$ and be parameterized by $p \in [0, J]$ so that it has its interior on the left in the direction of increasing p and $\mathbf{C}(0,t) = \mathbf{C}(J,t)$. The curvature along the curve is

$$\kappa = \kappa(p,t) \triangleq \frac{y_{pp} x_p - y_p x_{pp}}{(x_p^2 + y_p^2)^{3/2}}. \tag{30}$$

A general front propagation law (flow) is

$$\frac{\partial \mathbf{C}(p,t)}{\partial t} = V \mathbf{N}(p,t), \tag{31}$$

with initial condition $\Gamma(0) = \{\mathbf{C}(p,0) : p \in J\}$, where $\mathbf{N}(p,t)$ is the instantaneous unit *outward normal* vector at points on the evolving curve and $V = \mathbf{C}_t \cdot \mathbf{N}$ is the *normal speed*, with $\mathbf{C}_t = \partial \mathbf{C}/\partial t$. This speed may depend on local geometrical information such as the curvature κ, global image properties, or other factors independent of the curve. If $V = 1$ or $V = -1$, then $\Gamma(t)$ is the boundary of the dilation or erosion of the initial curve $\Gamma(0)$ by a disk of radius t.

An important speed model, which has been studied extensively by Osher and Sethian [50, 58] for general evolution of interfaces and by Kimia et al. [29] for shape analysis in computer vision, is

$$V = 1 - \epsilon \kappa, \qquad \epsilon \geq 0. \tag{32}$$

As analyzed by Sethian [58], when $V = 1$ the front's curvature will develop singularities, and the front will develop corners (i.e., the curve derivatives will develop shocks—discontinuities) at finite time if the initial curvature is anywhere negative. Two ways to continue the front beyond the corners are as follows: (1) If the front is viewed as a geometric curve, then each point is advanced along the normal by a distance t, and hence a "swallowtail" is formed beyond the corners by allowing the front to pass through itself. 2) If the front is viewed as the boundary separating two regions, an *entropy condition* is imposed to disallow the front to pass through itself. In other words, if the front is a propagating flame, then "once a particle is burnt it stays burnt" [58]. The same idea has also been used to model grassfire propagation leading to the medial axis of a shape [7]. It is equivalent to using Huygens' principle to construct the front as the set of points at distance t from the initial front. This can also be obtained from multiscale dilations of the initial front by disks of radii $t > 0$. Both the swallowtail and the entropy solutions are weak solutions. When $\epsilon > 0$, motion with curvature-dependent speed has a smoothing effect. Further, the limit of the solution for the $V = 1 - \epsilon\kappa$ case as $\epsilon \downarrow 0$ is the entropy solution for the $V = 1$ case [58].

To overcome the topological problem of splitting and merging and numerical problems with the Lagrangian formulation of Eq. (31), an Eulerian formulation was proposed by Osher and Sethian [50] in which the original curve $\Gamma(0)$ is first embedded in the surface of an arbitrary 2D Lipschitz continuous function $\phi_0(x, y)$ as its level set (contour line) at zero level. For example, we can select $\phi_0(x, y)$ to be equal to the signed distance function from the boundary of $\Gamma(0)$, positive (negative) in the exterior (interior) of $\Gamma(0)$. Then, the evolving planar curve is embedded as the zero-level set of an evolving space-time function $\phi(x, y, t)$:

$$\Gamma(t) = \{(x, y) : \phi(x, y, t) = 0\} \tag{33}$$
$$\Gamma(0) = \{(x, y) : \phi_0(x, y, 0) = \phi(x, y) = 0\}. \tag{34}$$

Geometrical properties of the evolving curve can be obtained from spatial derivatives of the level function. Thus, at any point on the front the curvature and outward normal of the level curves can be found from ϕ (assume $\phi < 0$ over curve interior):

$$\mathbf{N} = \frac{\nabla\phi}{\|\nabla\phi\|}, \quad \kappa = \mathrm{div}\left(\frac{\nabla\phi}{\|\nabla\phi\|}\right). \tag{35}$$

The curve evolution PDE of Eq. (31) induces a PDE generating its level function:

$$\partial\phi/\partial t = -V\|\nabla\phi\|,$$
$$\phi(x, y, 0) = \phi_0(x, y). \tag{36}$$

If $V = 1$, the above function evolution PDE is identical to the flat circular erosion PDE of Eq. (12) by equating scale with time. Thus, we can view this specific erosion PDE as a special case of the general function evolution PDE of Eq. (36) in which all level curves propagate in a homogeneous medium with unit normal speed. Propagation in a heterogeneous medium with a constant-sign $V = V(x, y)$ leads to the eikonal PDE.

5.1 Dilation Flows

In general, if B is an arbitrary compact, convex, symmetric planar set of unit scale and if we dilate the initial curve $\Gamma(0)$ with tB and set the new curve

$\Gamma(t)$ equal to the outward boundary of $\Gamma(0) \oplus tB$, then this action can also be generated by the following model [4, 54] of curve evolution

$$\frac{\partial \mathbf{C}}{\partial t} = \mathrm{sptf}_B(\mathbf{N})\mathbf{N} \tag{37}$$

Thus, the normal speed V, required to evolve curves by dilating them with B, is simply the support function of B evaluated at the curve's normal. Then, in this case the corresponding PDE (36) for evolving the level function becomes identical to the general PDE that generates multiscale flat erosions by B, which is given by (6) modulo a $(-)$ sign difference.

5.2 Curvature Flows

Another important case of curve evolution is when $V = -\kappa$; then,

$$\frac{\partial \mathbf{C}}{\partial t} = -\kappa \mathbf{N} = \frac{\partial^2 \mathbf{C}}{\partial s^2} \tag{38}$$

where s is the arc length. This propagation model is known as *Euclidean geometric heat* (or *shortening*) *flow*, as well as *mean curvature motion*. According to some classic results in differential geometry, smooth simple curves, evolving by means of (38), remain smooth and simple while undergoing the fastest possible shrinking of their perimeter [18], [19]. Furthermore, any non-convex curve converges first to a convex curve and from there it shrinks to a round point.

If the function $\phi(x, y, t)$ embeds a curve evolving by means of (38), as its level curve at a constant level, then it satisfies the evolution PDE

$$\partial \phi / \partial t = \mathrm{div}(\nabla \phi / ||\nabla \phi||)||\nabla \phi|| = \kappa ||\nabla \phi||$$

This smooths all level curves by propagation under their mean curvature. It has many interesting properties and has been extensively studied by many groups of researchers, including Osher & Sethian [50], Evans & Spruck [17], Chen, Giga & Goto [12] and Alvarez et al. [2].

Solutions of the Euclidean geometric heat flow (38) are invariant with respect to the group of Euclidean transformations (rotations and translations). Extending this invariance to affine transformations creates the *affine geometric heat flow* introduced by Sapiro and Tannenbaum [55]

$$\frac{\partial \mathbf{C}}{\partial t} = -\kappa^{1/3}\mathbf{N} = \frac{\partial^2 \mathbf{C}}{\partial \alpha^2} \tag{39}$$

where α is the affine arc length, i.e., a re-parameterization of the curve such that $\det[\mathbf{C}_\alpha \ \mathbf{C}_{\alpha\alpha}] = x_\alpha y_{\alpha\alpha} - x_{\alpha\alpha}y_\alpha = 1$. Any smooth simple non-convex curve evolving by the affine flow (39) converges to a convex one and from there to an elliptical point [55]. This PDE was also independently developed by Alvarez et al. [2] in the context of the affine morphological scale-space, already seen in section 4.2.

5.3 Morphological Representations of Curvature Flows

Matheron's famous representation theorem [43] states that any set operator Ψ on $\mathcal{P}(\mathbb{R}^d)$ that is translation-invariant (TI) and increasing can be represented as the union of erosions by all sets of its kernel $\mathrm{Ker}(\Psi) = \{X : \mathbf{0} \in \Psi(X)\}$ as well as an intersection of dilations by all sets of the kernel of the dual operator:

$$\Psi \text{ is TI and increasing} \implies \Psi(X) = \bigcup_{A \in \mathrm{Ker}(\Psi)} X \ominus A, \quad X \subseteq \mathbb{R}^d.$$

This representation theory was extended by Maragos [35, 36] to both function and set operators by using a basis for the kernel. As we have seen in section 3.1, according to the basis representation theory, every *TI, increasing*, and *upper-semicontinuous* (u.s.c.) operator can be represented as a supremum of morphological erosions by its basis functions. Specifically, let ψ be a signal operator acting on the set of extended-real-valued functions defined on $\mathbb{E} = \mathbb{R}^d$ or $\mathbb{Z}^d$. If $\mathrm{Ker}(\psi) = \{f : \psi(f)(\mathbf{0}) \geq 0\}$ defines the *kernel* of ψ, then its *basis* $\mathrm{Bas}(\psi)$ is defined as the collection of the minimal (w.r.t. $\leq$) kernel functions. Then [36]:

$$\psi \text{ is TI, increasing, and u.s.c.} \implies \psi(f) = \bigvee_{g \in \mathrm{Bas}(\psi)} f \ominus g$$

Dually, ψ can be represented as the infimum of dilations by functions in the basis of its dual operator $\psi^*(f) = -\psi(-f)$.

If the above function operator ψ is also flat (i.e., binary inputs yield binary outputs), with Ψ being its corresponding set operator, and commutes with thresholding, i.e.,

$$X_\lambda[\psi(f)] = \Psi[X_\lambda(f)], \quad \lambda \in \mathbb{R} \tag{40}$$

where $X_\lambda(f) = \{x \in \mathbb{R}^d : f(x) \geq \lambda\}$ are the *upper level sets* of f, then ψ is a supremum of flat erosions by the basis sets of its corresponding set operator Ψ [36]:

$$\psi(f) = \bigvee_{S \in \mathrm{Bas}(\Psi)} f \ominus S$$

where the basis $\mathrm{Bas}(\Psi)$ of the set operator Ψ is defined as the collection of the minimal elements (w.r.t. $\subseteq$) of its kernel $\mathrm{Ker}(\Psi)$.

Equation (40) implies that [57, p. 188] the operator ψ is 'contrast-invariant' or 'morphologically-invariant,' which means that [56, 1, 22]

$$\psi(g(f)) = g(\psi(f))$$

where $g : \mathbb{R} \to \mathbb{R}$ is any monotone bijective function, and $g(f)$ is the image of f under g. Such a function g is called an 'anamorphosis' in [56, 57], or a 'contrast-change' in [1, 22].

The above morphological basis representations have been applied to various classes of operators, including morphological, median, stack, and linear filters [35, 36, 41]. Moreover, one can define TI, increasing and contrast-invariant filters as supremum (or infimum) of flat erosions (or dilations) by sets belonging to some arbitrary basis $\mathbb{B}$. Catté, Dibos & Koepfler [11] selected as a basis the scaled version of a unit-scale isotropic basis (the set of all symmetric line segments of length 2)

$$\mathbb{B} \triangleq \{\{(x, y) : y = x \tan(\theta), |x| \leq |\cos(\theta)|\} : \theta \in [0, \pi)\} \tag{41}$$

and defined the following three types of multiscale flat operators $\mathcal{I}_t, \mathcal{S}_t, \mathcal{T}_t$:

$$\mathcal{I}_t(f) = \bigvee_{S \in \mathbb{B}} f \ominus \sqrt{2t}S \iff \partial \mathbf{C}/\partial t = -\max(\kappa, 0)\mathbf{N} \tag{42}$$

$$\mathcal{S}_t(f) = \bigwedge_{S \in \mathbb{B}} f \oplus \sqrt{2t}S \iff \partial \mathbf{C}/\partial t = \min(\kappa, 0)\mathbf{N} \tag{43}$$

$$\mathcal{T}_t(f) = [\mathcal{I}_{2t}(f) + \mathcal{S}_{2t}(f)]/2 \iff \partial \mathbf{C}/\partial t = -\kappa\mathbf{N} \tag{44}$$

If these operators operate on a level function embedding a curve $\mathbf{C}$ as one of its level lines, then this curve evolves according to the above following three flows [11]. Hence, the above multiscale operators, which are sup-of-erosions and inf-of-dilations by linear segments in all directions, are actually curvature flows. A generalization of this result was obtained, within the framework described in section 4, in Guichard and Morel [22], by assuming that $\mathbb{B}$ is any bounded and isotropic collection of planar sets. Furthermore, in slightly different settings it has been shown that, by iterating n times a median filter, based on a window of scale h, we asymptotically converge (when $h \to 0$, $n \to \infty$, with $nh = t$) to the curvature flow. The mathematical proof was given in [16], [6], following a conjecture of [45].

The above morphological representations deal with Euclidean curvature flow. Furthermore, by defining a unit-scale morphological basis $\mathbb{B}$ as a collection of convex symmetric sets invariant under the special linear group, it has been shown in [22] and in [20] that n iterations of morphological flat operators at scale h, which are sup-of-erosions, inf-of-dilations, or their alternate compositions, converge (when $h \to 0$, $n \to \infty$, with $nh = t$) to the affine curvature flow. An efficient implementation of the iterated affine invariant curve evolution has been proposed in [48]. It yields a fast implementation of the curve affine scale space and has proved its effectiveness in shape recognition [34]. An example of shape smoothing using this affine scale-space is shown in Fig. 1.

6 Conclusion

In this chapter we have presented some basic results from the theory of nonlinear geometric PDEs that can generate multiscale morphological operators.

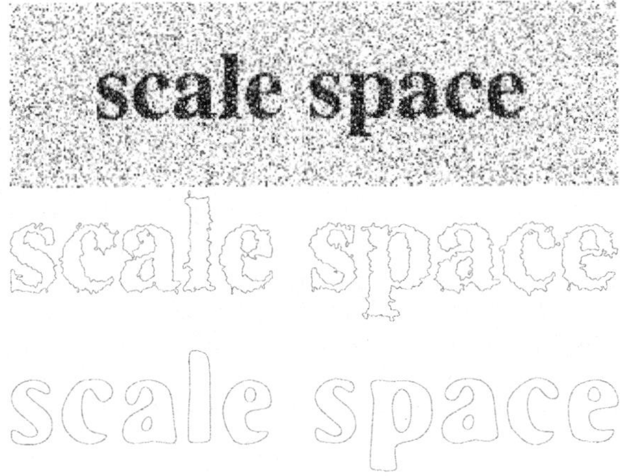

Fig. 1. Smoothing curves with the Affine Scale Space. Top: a text image corrupted by noise. Middle: thresholding the image reveals characters as irregular level lines. Bottom: the same level lines, smoothed with the affine scale space. The smoothing process produces curves almost independent of the noise, which is a requirement for robust pattern recognition. Algorithm used follows the affine erosion introduced in [48]. -Experiment courtesy of Lionel Moisan-

Further, we have outlined the relationships of these results with G. Matheron's development of size distributions and kernel representation theory.

Interpreting and modelling the basic morphological operators via PDEs opens up several new promising directions along which mathematical morphology can both assist and be assisted by other PDE-based theories and methodologies of image analysis and computer vision. Examples include scale-space analyses, variational methods of vision, level sets implementations of 2D/3D geometric flows, and their applications to major research problems such as image segmentation, object detection & tracking, and stereopsis.

References

1. L. Alvarez, F. Guichard, P.L. Lions, and J.M. Morel, "Axiomatisation et nouveaux opérateurs de la morphologie mathématique," *C. R. Acad. Sci. Paris*, pp. 265-268, t. 315, Série I, 1992.
2. L. Alvarez, F. Guichard, P.L. Lions, and J.M. Morel, "Axioms and Fundamental Equations of Image Processing," *Archiv. Rat. Mech.*, vol. 123 (3), pp. 199–257, 1993.
3. L. Alvarez and J.M. Morel, "Formalization and computational aspects of image analysis," *Acta Numerica*, pp. 1–59, 1994.
4. A. Arehart, L. Vincent and B. Kimia, "Mathematical Morphology: The Hamilton-Jacobi Connection," in *Proc. Int'l Conf. Comp. Vision*, pp. 215–219, 1993.

5. G. Barles, "Solutions de viscosité des équations de Hamilton-Jacobi. (Viscosity Solutions of Hamilton-Jacobi Equations)", *Mathématiques & Applications, Springer-Verlag*, vol. 194, 1994.

6. G. Barles and C. Georgelin, "A simple proof of convergence for an approximation scheme for computing motions by mean curvature", *SIAM Journal of numerical analysis*, vol. 32, pp. 454–500, 1995.

7. H. Blum, "A Transformation for Extracting New Descriptors of Shape," in Proceedings of the Symposium on *Models for the Perception of Speech and Visual Forms*, Boston, Nov. 1964, MIT Press, Cambridge, MA, 1967.

8. R. W. Brockett and P. Maragos, "Evolution Equations for Continuous-Scale Morphology," *Proc. IEEE Int'l Conf. Acoust., Speech, Signal Processing*, San Francisco, CA, March 1992.

9. R. Brockett and P. Maragos, "Evolution Equations for Continuous-Scale Morphological Filtering," *IEEE Trans. Signal Processing*, vol. 42, pp. 3377-3386, Dec. 1994.

10. F. Cao, "Geometric Curve Evolution and Image Processing", *Lecture Notes in Mathematics*, vol. 1805, Springer Verlag, February 2003.

11. F. Catté, F. Dibos and G. Koepfler, "A Morphological Scheme for Mean Curvature Motion and Applications to Anisotropic Diffusion and Motion of Level Sets," *SIAM J. of Num. Analysis*, vol. 32-6, pp. 1895-1909, décembre 1995.

12. Y.-G. Chen, Y. Giga and S. Goto", "Uniqueness and existence of viscosity solutions of generalized mean curvature flow equations.", *J. Differ Geom. 33, No.3 p 749-786*, 1991.

13. M. G. Crandall, "Viscosity Solutions : a primer", pp. 1–43, in M. Bardi, M.G. Crandall, L.C. Evans, H.M. Soner and P.E. Souganidis, "Viscosity Solutions and Applications", Montecatini Terme, 1995, edited by I.C. Dolcetta and P.L. Lions, *Lecture Notes in Mathematics* vol. 1660, Springer.

14. M. G. Crandall, H. Ishii and P.-L. Lions, "User's Guide to Viscosity Solutions of Second Order Partial Differential Equations," *Bull. Amer. Math. Soc.*, vol. 27, pp. 1–66, July 1992.

15. L. Dorst and R. van den Boomgaard, "Morphological Signal Processing and the Slope Transform," *Signal Processing*, vol. 38, pp. 79–98, July 1994.

16. L. C. Evans, "Convergence of an algorithm for mean curvature motion", *Indiana University Mathematics Journal*, vol. 42, pp. 553–557, 1993.

17. L. C. Evans and J. Spruck, "Motion of Level Sets by Mean Curvature. I," *J. Diff. Geom.*, vol. 33, pp. 635–681, 1991.

18. M. Grayson , "The heat equation shrinks embedded plane curves to round points", *J. Differential Geometry* , vol. 26, pp. 285–314, 1987.

19. M. Gage, and R. S. Hamilton, "The heat equation shrinking convex plane curves", *J. Differ. Geom.*, vol. 23, pp. 69–96, 1986.

20. F. Guichard "Axiomatization of images and movies scale-space", PhD Thesis, University Paris Dauphine, 1994.

21. F. Guichard and J.-M. Morel, "Partial differential equations and image iterative filtering", in *The State of the Art in Numerical Analysis,* Duff. I.S. (ed.), 1996

22. F. Guichard and J.-M. Morel, "Geometric Partial Differential Equations and Iterative Filtering," in *Mathematical Morphology and Its Applications to Image and Signal Processing* (H. Heijmans and J. Roerdink, Eds.), Kluwer Acad. Publ., 1998, pp. 127–138.

23. F. Guichard and J.-M. Morel, *Image Analysis and P.D.E.s*, book to be published.

24. F. Guichard and J.-M. Morel, "A Note on Two Classical Enhancement Filters and Their Associated PDE's", *International Journal of Computer Vision*, vol. 52(2), pp. 153–160, May 2003.

25. H.J.A.M. Heijmans and P. Maragos, "Lattice Calculus and the Morphological Slope Transform," *Signal Processing*, vol. 59, pp. 17–42, 1997.

26. H.J.A.M. Heijmans and R. van den Boomgaard, "Algebraic Framework for Linear and Morphological Scale-Spaces", *J. Vis. Commun. and Image Repres.*, vol.13, pp.269-301, 2002.

27. B.K.P. Horn, *Robot Vision*, MIT Press, Cambridge, MA, 1986.

28. H. Ishii, "A generalization of the Bence, Merriman and Osher algorithm for motion by mean curvature.", *Proceedings of the international conference on curvature flows and related topics held in Levico, Italy'"*, GAKUTO Int. Ser., Math. Sci. Appl., vol. 5, pp. 111–127, 1994.

29. B. Kimia, A. Tannenbaum, and S. Zucker, "Shapes, Shocks and Deformations I: The components of Two-dimensional Shape and the Reaction-Diffusion Space", *Int'l J. Comp. Vision*, vol. 15(3), pp. 189-224, 1995.

30. R. Kimmel, N. Kiryati, and A. M. Bruckstein, "Sub-Pixel Distance Maps and Weighted Distance Transforms," *J. Math. Imaging and Vision*, vol. 6, pp. 223-233, 1996.

31. J. J. Koenderink, "The Structure of Images," *Biol. Cybern.*, vol. 50, pp. 363-370, 1984.

32. P. D. Lax, "Hyperbolic Systems of Conservation Laws and the Mathematical Theory of Shock Waves," SIAM Press, Philadelphia, 1973.

33. T. Lindeberg, *Scale-Space Theory in Computer Vision*, Kluwer Academic Publishers, 1994.

34. J-L. Lisani, P. Monasse, L. Moisan, and J-M. Morel. "On the theory of planar shape". *SIAM J. on Multiscale Modeling and Simulation*, vol. 1(1):1-24, 2003.

35. P. Maragos, *A Unified Theory of Translation-Invariant Systems with Applications to Morphological Analysis and Coding of Images*, Ph.D. Thesis, Georgia Inst. Technology, Atlanta, June 1985.

36. P. Maragos, "A Representation Theory for Morphological Image and Signal Processing," *IEEE Trans. Pattern Analysis and Machine Intelligence*, vol. 11, pp. 586–599, June 1989.

37. P. Maragos, "Morphological Systems: Slope Transforms and Max–Min Difference and Differential Equations," *Signal Processing*, vol. 38, pp. 57–77, July 1994.

38. P. Maragos, "Differential Morphology and Image Processing", *IEEE Trans. Image Processing*, vol.5, pp.922-937, June 1996.

39. P. Maragos, "Algebraic and PDE Approaches for Lattice Scale-Spaces with Global Constraints", *Int'l J. Comp. Vision*, vol.52 (2/3), pp.121-137, May 2003.

40. P. Maragos and M. A. Butt, "Curve Evolution, Differential Morphology and Distance Transforms as Applied to Multiscale and Eikonal Problems", *Fundamentae Informatica*, vol.41, pp.91-129, Jan. 2000.

41. P. Maragos and R. W. Schafer, "Morphological Filters – Part II: Their Relations to Median, Order-Statistic, and Stack Filters," *IEEE Trans. Acoust., Speech, and Signal Processing*, vol. 35, pp. 1170–1184, Aug. 1987. "Corrections," *IEEE Trans. ASSP*, vol. 37, no. 4, p. 597, Apr. 1989.

42. D. Marr, *Vision*, Freeman, San Francisco, 1982.

43. G. Matheron, *Random Sets and Integral Geometry*, Wiley, New York, 1975.

44. J. Mattioli, "Differential Relations of Morphological Operators," *Proc. Int'l Workshop on Math. Morphology and its Application to Signal Processing*, J. Serra and P. Salembier, Eds., Univ. Polit. Catalunya, Barcelona, Spain, May 1993.

45. B. Merriman, J. Bence, and S. Osher, "Diffusion generated motion by mean curvature", American Mathematical Society", *Computational Crystal Growers Workshop*, pp. 73–83, 1992.

46. F. Meyer, "Topographic Distance and Watershed Lines," *Signal Processing*, vol. 38, pp. 113-125, July 1994.

47. F. Meyer and P. Maragos, "Nonlinear Scale-Space Representation with Morphological Levelings", *J. Visual Communic. and Image Representation*, vol.11, pp.245-265, 2000.

48. L. Moisan, "Affine Plane Curve Evolution : a Fully Consistent Scheme", *IEEE Transactions On Image Processing*, vol. 7, 3, pp. 411–420, March 1998.

49. L. Najman and M. Schmitt, "Watershed of a Continuous Function," *Signal Processing*, vol. 38, pp. 99-112, July 1994.

50. S. Osher and J. Sethian, "Fronts Propagating with Curvature-Dependent Speed: Algorithms Based on Hamilton-Jacobi Formulations," *J. Comput. Physics*, vol. 79, pp. 12–49, 1988.

51. P. Perona and J. Malik, "Scale-Space and Edge Detection Using Anisotropic Diffusion," *IEEE Trans. Pattern Anal. Mach. Intellig.*, vol. 12, pp. 629–639, July 1990.

52. E. Rouy and A. Tourin, "A Viscosity Solutions Approach to Shape from Shading," *SIAM J. Numer. Anal.*, vol. 29 (3), pp. 867-884, June 1992.

53. P. Salembier and J. Serra, "Flat Zones Filtering, Connected Operators, and Filters by Reconstruction", *IEEE Trans. Image Processing*, vol. 4, pp.1153-1160, Aug. 1995.

54. G. Sapiro, R. Kimmel, D. Shaked, B. Kimia, and A. Bruckstein, "Implementing Continuous-scale Morphology via Curve Evolution," *Pattern Recognition*, vol. 26, pp. 1363–1372, 1993.

55. G. Sapiro and A. Tannenbaum, "Affine Invariant Scale-Space," *Int'l J. Comp. Vision*, vol. 11, pp. 25–44, 1993.

56. J. Serra, *Image Analysis and Mathematical Morphology*, Acad. Press, NY, 1982.

57. J. Serra, editor, *Image Analysis and Mathematical Morphology. Vol. 2*, Acad. Press, NY, 1988.

58. J. A. Sethian, *Level Set Methods and Fast Marching Methods*, Cambridge Univ. Press, 1999.

59. R. van den Boomgaard, *Mathematical Morphology: Extensions towards Computer Vision*, Ph.D. Thesis, Univ. of Amsterdam, The Netherlands, 1992.

60. R. van den Boomgaard and A. Smeulders, "The Morphological Structure of Images: The Differential Equations of Morphological Scale-Space," *IEEE Trans. Pattern Anal. Mach. Intellig.*, vol.16, pp.1101-1113, Nov. 1994.

61. P. Verbeek and B. Verwer, "Shading from shape, the eikonal equation solved by grey-weighted distance transform," *Pattern Recogn. Lett.*, vol.11, pp.618-690, 1990.

62. M. Wertheimer, "Untersuchungen zur Lehre der Gestalt, II", *Psychologische Forschung*, vol4. p 301-350, 1923.

63. A. P. Witkin, "Scale-Space Filtering," *Proc. Int'l Joint Conf. Artif. Intellig.*, Karlsruhe, Germany, 1983.

Index

Lecture Notes in Statistics

For information about Volumes 1 to 128,
please contact Springer-Verlag

129: Wolfgang Härdle, Gerard Kerkyacharian, Dominique Picard, and Alexander Tsybakov, Wavelets, Approximation, and Statistical Applications. xvi, 265 pp., 1998.

130: Bo-Cheng Wei, Exponential Family Nonlinear Models. ix, 240 pp., 1998.

131: Joel L. Horowitz, Semiparametric Methods in Econometrics. ix, 204 pp., 1998.

132: Douglas Nychka, Walter W. Piegorsch, and Lawrence H. Cox (Editors), Case Studies in Environmental Statistics. viii, 200 pp., 1998.

133: Dipak Dey, Peter Müller, and Debajyoti Sinha (Editors), Practical Nonparametric and Semiparametric Bayesian Statistics. xv, 408 pp., 1998.

134: Yu. A. Kutoyants, Statistical Inference For Spatial Poisson Processes. vii, 284 pp., 1998.

135: Christian P. Robert, Discretization and MCMC Convergence Assessment. x, 192 pp., 1998.

136: Gregory C. Reinsel, Raja P. Velu, Multivariate Reduced-Rank Regression. xiii, 272 pp., 1998.

137: V. Seshadri, The Inverse Gaussian Distribution: Statistical Theory and Applications. xii, 360 pp., 1998.

138: Peter Hellekalek and Gerhard Larcher (Editors), Random and Quasi-Random Point Sets. xi, 352 pp., 1998.

139: Roger B. Nelsen, An Introduction to Copulas. xi, 232 pp., 1999.

140: Constantine Gatsonis, Robert E. Kass, Bradley Carlin, Alicia Carriquiry, Andrew Gelman, Isabella Verdinelli, and Mike West (Editors), Case Studies in Bayesian Statistics, Volume IV. xvi, 456 pp., 1999.

141: Peter Müller and Brani Vidakovic (Editors), Bayesian Inference in Wavelet Based Models. xiii, 394 pp., 1999.

142: György Terdik, Bilinear Stochastic Models and Related Problems of Nonlinear Time Series Analysis: A Frequency Domain Approach. xi, 258 pp., 1999.

143: Russell Barton, Graphical Methods for the Design of Experiments. x, 208 pp., 1999.

144: L. Mark Berliner, Douglas Nychka, and Timothy Hoar (Editors), Case Studies in Statistics and the Atmospheric Sciences. x, 208 pp., 2000.

145: James H. Matis and Thomas R. Kiffe, Stochastic Population Models. viii, 220 pp., 2000.

146: Wim Schoutens, Stochastic Processes and Orthogonal Polynomials. xiv, 163 pp., 2000.

147: Jürgen Franke, Wolfgang Härdle, and Gerhard Stahl, Measuring Risk in Complex Stochastic Systems. xvi, 272 pp., 2000.

148: S.E. Ahmed and Nancy Reid, Empirical Bayes and Likelihood Inference. x, 200 pp., 2000.

149: D. Bosq, Linear Processes in Function Spaces: Theory and Applications. xv, 296 pp., 2000.

150: Tadeusz Caliński and Sanpei Kageyama, Block Designs: A Randomization Approach, Volume I: Analysis. ix, 313 pp., 2000.

151: Håkan Andersson and Tom Britton, Stochastic Epidemic Models and Their Statistical Analysis. ix, 152 pp., 2000.

152: David Ríos Insua and Fabrizio Ruggeri, Robust Bayesian Analysis. xiii, 435 pp., 2000.

153: Parimal Mukhopadhyay, Topics in Survey Sampling. x, 303 pp., 2000.

154: Regina Kaiser and Agustín Maravall, Measuring Business Cycles in Economic Time Series. vi, 190 pp., 2000.

155: Leon Willenborg and Ton de Waal, Elements of Statistical Disclosure Control. xvii, 289 pp., 2000.

156: Gordon Willmot and X. Sheldon Lin, Lundberg Approximations for Compound Distributions with Insurance Applications. xi, 272 pp., 2000.

157: Anne Boomsma, Marijtje A.J. van Duijn, and Tom A.B. Snijders (Editors), Essays on Item Response Theory. xv, 448 pp., 2000.

158: Dominique Ladiray and Benoît Quenneville, Seasonal Adjustment with the X-11 Method. xxii, 220 pp., 2001.

159: Marc Moore (Editor), Spatial Statistics: Methodological Aspects and Some Applications. xvi, 282 pp., 2001.

160: Tomasz Rychlik, Projecting Statistical Functionals. viii, 184 pp., 2001.

161: Maarten Jansen, Noise Reduction by Wavelet Thresholding. xxii, 224 pp., 2001.

162: Constantine Gatsonis, Bradley Carlin, Alicia Carriquiry, Andrew Gelman, Robert E. Kass Isabella Verdinelli, and Mike West (Editors), Case Studies in Bayesian Statistics, Volume V. xiv, 448 pp., 2001.

163: Erkki P. Liski, Nripes K. Mandal, Kirti R. Shah, and Bikas K. Sinha, Topics in Optimal Design. xii, 164 pp., 2002.

164: Peter Goos, The Optimal Design of Blocked and Split-Plot Experiments. xiv, 244 pp., 2002.

165: Karl Mosler, Multivariate Dispersion, Central Regions and Depth: The Lift Zonoid Approach. xii, 280 pp., 2002.

166: Hira L. Koul, Weighted Empirical Processes in Dynamic Nonlinear Models, Second Edition. xiii, 425 pp., 2002.

167: Constantine Gatsonis, Alicia Carriquiry, Andrew Gelman, David Higdon, Robert E. Kass, Donna Pauler, and Isabella Verdinelli (Editors), Case Studies in Bayesian Statistics, Volume VI. xiv, 376 pp., 2002.

168: Susanne Rässler, Statistical Matching: A Frequentist Theory, Practical Applications and Alternative Bayesian Approaches. xviii, 238 pp., 2002.

169: Yu. I. Ingster and Irina A. Suslina, Nonparametric Goodness-of-Fit Testing Under Gaussian Models. xiv, 453 pp., 2003.

170: Tadeusz Calinski and Sanpei Kageyama, Block Designs: A Randomization Approach, Volume II: Design. xii, 351 pp., 2003.

171: D.D. Denison, M.H. Hansen, C.C. Holmes, B. Mallick, B. Yu (Editors), Nonlinear Estimation and Classification. x, 474 pp., 2002.

172: Sneh Gulati, William J. Padgett, Parametric and Nonparametric Inference from Record-Breaking Data. ix, 112 pp., 2002.

173: Jesper Møller (Editor), Spatial Statistics and Computational Methods. xi, 214 pp., 2002.

174: Yasuko Chikuse, Statistics on Special Manifolds. xi, 418 pp., 2002.

175: Jürgen Gross, Linear Regression. xiv, 394 pp., 2003.

176: Zehua Chen, Zhidong Bai, Bimal K. Sinha, Ranked Set Sampling: Theory and Applications. xii, 224 pp., 2003

177: Caitlin Buck and Andrew Millard (Editors), Tools for Constructing Chronologies: Crossing Disciplinary Boundaries, xvi, 263 pp., 2004

178: Gauri Sankar Datta and Rahul Mukerjee , Probability Matching Priors: Higher Order Asymptotics, x, 144 pp., 2004

179: D.Y. Lin and P.J. Heagerty (Editors), Proceedings of the Second Seattle Symposium in Biostatistics: Analysis of Correlated Data, vii, 336 pp., 2004

180: Yanhong Wu, Inference for Change-Point and Post-Change Means After a CUSUM Test, xiv, 176 pp., 2004

181: Daniel Straumann, Estimation in Conditionally Heteroscedastic Time Series Models , x, 250 pp., 2004

182: Lixing Zhu, Nonparametric Monte Carlo Tests and Their Applications, xi., 192 pp., 2005

183: Michel Bilodeau, Fernand Meyer, and Michel Schmitt (Editors), Space, Structure and Randomness, xiv., 416 pp., 2005

Made in the USA
Monee, IL
07 July 2026

56546072R00233